AF760233

BACTERIOLOGY AND MICROBIOLOGY

Editor

Hossam Nejem

scitus
academics

Bacteriology and Microbiology

Edited by **Hossam Nejem**

This Book is Printed on Acid-Free Paper. ♾

This book contains information obtained from highly regarded resources. Copyright for individual articles remains with the authors as indicated. All chapters are distributed under the terms of the Creative Commons Attribution License, which permits unrestricted use, distribution, and reproduction in any medium, provided the original author and source are credited.

ISBN: 9781681179230

Library of Congress Control Number: 2017948952

© 2018 by
SCITUS Academics LLC,
616, Corporate Way, Suite 2, 4766,
Valley Cottage, NY 10989

www.scitusacademics.com

Notice

Reasonable efforts have been made to publish reliable data and views articulated in the chapters are those of the individual contributors, and not necessarily those of the editors or publishers. Editors or publishers are not responsible for the accuracy of the information in the published chapters or consequences of their use. The publisher believes no responsibility for any damage or grievance to the persons or property arising out of the use of any materials, instructions, methods or thoughts in the book. The editors and the publisher have attempted to trace the copyright holders of all material reproduced in this publication and apologize to copyright holders if permission has not been obtained. If any copyright holder has not been acknowledged, please write to us so we may rectify.

Printed in United States of America.

Preface

Microorganisms are extremely important in our everyday lives. Some are responsible for a significant proportion of the diseases affecting not only humans, but also plants and animals, while others are vitally important in the maintenance and modification of our environment. Still others play an essential role in industry, where their unique properties have been harnessed in the production of food, beverages and antibiotics. Scientists also have learned how to exploit microorganisms in the field of molecular biology, which makes an enormous impact both industrially and medically. Microbiology also encompasses immunology, the study of the body's ability to mount defenses against infectious microbes. Bacteria are single-celled microorganisms that lack a nuclear membrane, are metabolically active and divide by binary fission. Medically they are a major cause of disease. Superficially, bacteria appear to be relatively simple forms of life; in fact, they are sophisticated and highly adaptable. These organisms exist widely in both parasitic and free-living forms. Because they are ubiquitous and have a remarkable capacity to adapt to changing environments by selection of spontaneous mutants, the importance of bacteria in every field of medicine cannot be overstated.

'Bacteriology and Microbiology' is concerned with microbial (bacterial, fungal, viral) diseases of domesticated vertebrate animals (livestock, companion animals, and poultry,) that supply food, other useful products or companionship. In addition, Microbial diseases of wild animals living in captivity, or as members of the feral fauna will also be considered if the infections are of interest because of their interrelation with humans (zoonoses) and/or domestic animals. It describes representative bacteria that are pathogenic and can cause infectious diseases to animals. Studies of antimicrobial resistance are also included, provided that the results represent a substantial advance in knowledge.

Table of Contents

CHAPTER 1

Clinical mastitis in ewes; bacteriology, epidemiology and clinical features

Tormod Mørk*[1], Steinar Waage[2], Tore Tollersrud[1], Bjørg Kvitle[1] and Ståle Sviland[1]

[1]*Department of Animal Health, National Veterinary Institute, PO Box 8156 Dep, N-0033 Oslo, Norway and* [2]*Department of Production*

[2]*Animal Clinical Sciences, Norwegian School of Veterinary Science, PO Box 8146 Dep, N-0033 Oslo, Norway*

ABSTRACT

Background

Clinical mastitis is an important disease in sheep. The objective of this work was to identify causal bacteria and study certain epidemiological and clinical features of clinical mastitis in ewes kept for meat and wool production.

Methods

The study included 509 ewes with clinical mastitis from 353 flocks located in 14 of the 19 counties in Norway. Clinical examination and collection of udder secretions were carried out by veterinarians. Pulsed-field gel electrophoresis (PFGE) was performed on 92 *Staphylococcus aureus* isolates from 64 ewes.

Results and conclusion

S. aureus was recovered from 65.3% of 547 clinically affected mammary glands, coagulase-negative staphylococci from 2.9%, enterobacteria, mainly *Escherichia coli*, from 7.3%, *Streptococcus* spp. from 4.6%, *Mannheimia haemolytica* from 1.8% and various other bacteria from 4.9%, while no bacteria were cultured from 13.2% of the samples. Forty percent of the ewes with unilateral clinical *S. aureus* mastitis also had a subclinical *S. aureus* infection in the other mammary gland. Twenty-four of 28 (86%) pairs of *S. aureus* isolates obtained from clinically and subclinically affected mammary glands of the same ewe were indistinguishable by PFGE. The number of identical pairs was significantly greater than expected, based on the distribution of different *S. aureus* types within the flocks. One-third of the cases occurred during the first week after lambing, while a second peak was observed in the third week of lactation. Gangrene was present in 8.8% of the clinically affected glands; *S. aureus* was recovered from 72.9%, *Clostridium perfringens* from 6.3% and *E. coli* from 6.3% of the secretions from such glands. This study shows that *S. aureus* predominates as a cause of clinical ovine mastitis in Norway, also in very severe cases. Results also indicate that *S. aureus* is frequently spread between udder halves of infected ewes.

1. BACKGROUND

Mastitis is an important disease in sheep. Clinical cases are often severe; systemic signs are present and the condition is obviously painful. Clinically affected glands frequently suffer partial or complete damage and do not resume normal function. Reduced milk yield leads to decreased growth of the lambs [1, 2, 3]. Additional losses associated with clinical mastitis are costs of treatment and culling of ewes due to permanent udder damage [3, 4, 5, 6, 7]. In very severe cases, gangrene may develop in the mammary gland and the ewe may die. Thus, mastitis has a major impact on both economy and animal welfare in sheep production.

Although a wide range of microorganisms may cause ovine mastitis, most cases are reported to be due to staphylococci [8]. Several reports indicate that coagulase-negative staphylococci (CNS) are the most common cause of subclinical mastitis in dairy ewes [9, 10, 11, 12, 13, 14], while both CNS and *Staphylococcus aureus* are frequent causes in meat sheep [5, 15, 16]. With regard to organisms associated with clinical mastitis, there are fewer reports published. *S. aureus* has been reported to be the most common causal organism in both meat [5, 15, 17, 18, 19] and dairy ewes [13, 20, 21].

Mannheimia haemolytica [5, 18, 19, 22], *Escherichia coli* [13, 18, 19] and various streptococci [15, 18, 19] are other important causative organisms.

Differences in climate, production forms and management practices may give rise to differences both in the epidemiology, bacteriology and clinical manifestations of mastitis. In Norway, sheep are kept exclusively for meat and wool production. They are housed during the winter and early spring, including the lambing season.

The objective of this study was to identify bacteria associated with clinical ovine mastitis in Norway. In addition, certain epidemiological and clinical features of the disease were studied.

2. METHODS

2.1 Animals and clinical data

Udder secretions were collected and clinical data recorded from 509 ewes with clinical mastitis. The ewes belonged to 353 flocks located in 14 counties in Norway (Figure 1). The geographical distribution of the cases is shown in Table 1. Clinical mastitis was present in one gland in 471 ewes and in both glands in 38 ewes. The study was carried out in 2002, 2003 and 2004. Only cases that occurred between 1 week prepartum and 8 weeks postpartum were included. In Norway, lambing generally takes place in April and May.

2.2 Data and sample collection

Thirty-two veterinary practitioners contributed to the study. When called to a case of clinical mastitis, the veterinarian was to examine the ewe, collect udder secretions and record information regarding the identity, age, date of parturition, the number of lambs and the clinical condition of the ewe on a standardized form. Clinical data included the rectal temperature, an assessment of the severity of systemic signs (graded as none, weak, moderate or severe) and local clinical signs of the affected gland, including whether or not gangrene was present (i.e., cold and blue udder and teat skin).

Prior to treatment, samples were collected aseptically from the clinically affected glands in 10-ml sterile plastic vials by the veterinary practitioner according to the International Dairy Federation's standards [23]. Additionally, samples were taken from the clinically unaffected gland of 252 of the ewes with unilateral clinical mastitis. The samples were sent by mail to the laboratory as soon as possible after sampling, or frozen and stored at -20°C until submission.

Table 1: Distribution by region and county of 547 milk samples obtained from ovine mammary glands with clinical mastitis, and of the 509a ewes and 353 flocks from which the samplesoriginated.

Region	County	No. of flocks	No. of ewes	No. of glands
East	Akershus	13	27	31
	Hedmark	68	121	128
	Oppland	60	85	86
	Buskerud	4	5	6
South	Aust-Agder	34	56	57
	Vest-Agder	4	6	7
	Rogaland	40	48	53
West	Hordaland	22	26	28
	Sogn og Fjordane	25	25	29
	Møre og Romsdal	32	39	40
North	Sør-Trøndelag	33	46	50
	Nord-Trøndelag	6	10	13
	Nordland	3	3	4
	Troms	9	12	15

[a] Four hundred and seventy-one ewes with unilateral and 38 with bilateral intramammary infection.

If a ewe experienced more than one episode of mastitis during the observation period, only the first episode was included in the study.

2.3 Microbiological methods

The samples were examined at the National Veterinary Institute or at the TINE Mastitis Laboratory in Molde, Norway, and bacteria were identified according to the recommendations of the International Dairy Federation [23] with additions. The National Veterinary Institute and the TINE Mastitis Laboratory are both quality assured in accordance with NS-EN ISO/IEC 17025. Briefly, the secretions were brought to room temperature, assessed visually and characterized by appearance before they were mechanically shaken and 10 µl plated on Bacto Blood Agar Base No 2 (Difco Laboratories, Detroit, MI, USA) containing 5% washed bovine erythrocytes and incubated for 48 hours in a 5% CO_2 atmosphere at 37°C. Cultures were read at 24 and 48 hours. If growth was not detected after incubation for 24 hours, the original sample was incubated for 4 hours at 37°C and 50 µl aliquots plated and incubated for 24 hours under aerobic (5% CO_2 atmosphere) and anaerobic conditions.

Figure 1 Map of Norway showing the location of the sheep flocks from which cases of clinical mastitis were obtained. Thin lines show county boundaries and thick lines region boundaries. the sheep flocks from which cases of clinical mastitis were obtained

Bacterial species were identified tentatively by their gross colony morphology and by Gram staining, and further confirmatory tests were used as necessary. All suspected staphylococcal colonies were tested using the tube coagulase test (Becton Dickinson Microbiology Systems, Bedford, MA, USA). Coagulase-positive staphylococci were streaked on peptone agar (p-agar) (Difco, Sparks, MD) supplemented with 7 mg/l of acriflavin (Sigma-Aldrich Chemie, Steinheim, Germany) [24], and incubated at 37°C for 24 hours. Bacterial growth in the full length of the streak on p-agar was considered confirmative of *S. aureus*. Isolates identified as *S. aureus* were stored at -70°C in Bacto Heart Infusion Broth (Difco) with 15% glycerol. *E. coli* was identified by the lactose and indole tests, and other enterobacteria were identified to the species or genus level by using a microtube identification system (API 20 E®; bioMérieux S.A., Marcy-l'Etoile, France). *Streptococcus uberis*, *Streptococcus dysgalactiae*, *Streptococcus agalactiae*, *Streptococcus* spp. and *Enterococcus* spp. were distinguished by the CAMP reaction, the aesculin and inulin tests and by culture on the bromthymolblue lactose-sucrose agar. Bacteria within the family *Pasteurellaceae* were identified to the species level by the CAMP reaction, the indole, mannitol, sorbitol, trehalose, dulcitol, oxidase and beta-galactosidase tests and the haemolysis patterns. *Clostridium perfringens* was differentiated from other *Clostridium* spp. by colony morphology, immobility and the presence of a zone of partial haemolysis and a zone of complete haemolysis. *Arcanobacterium pyogenes* was identified by Gram staining and the presence of pinpoint colonies surrounded by a narrow zone of clear haemolysis at 48 hours. None of the samples were from ewes with arthritis, conjunctivitis or pneumonia; therefore, the mammary secretions were not checked for the presence of mycoplasms.

Total DNA was prepared and pulsed-field gel electrophoresis (PFGE) performed as described previously [25] on 92 *S. aureus* isolates from 21 flocks in which at least one ewe had bilateral *S. aureus* intramammary infection (IMI) and where at least two ewes experienced clinical *S. aureus* mastitis. The band patterns were compared visually. Isolates with indistinguishable patterns were considered identical PFGE types while those with at least one band difference were considered to be different types.

2.4 Statistical methods

The chi-square test was used to compare the frequencies of cases within different time intervals in relation to parturition and the relative proportions of clinical *S. aureus* cases and gangrenous mastitis cases in ewes of different parity and with different number of lambs.

The distribution of pairs of *S. aureus* PFGE types within flocks (equal vs. unequal) in ewes with bilateral IMI was compared with the corresponding distribution that would be expected if all *S. aureus* isolates found within each flock were paired randomly. All isolates from the flocks that supplied two or more cases of clinical *S. aureus* mastitis, of which at least one ewe had bilateral IMI, were included, and Fisher's exact test was used to test the probability of identical distributions of the observed and expected pairs.

$P < 0.05$ was considered statistically significant.

3. RESULTS

3.1 Bacteriological and epidemiological findings

The distribution of bacteria cultured from secretions from the glands with clinical mastitis is shown in Table 2. *S. aureus* was the predominant pathogen and was found in 65.3% of the samples from affected glands. In the samples from the southern, eastern, western and northern regions, *S. aureus* was found in 76.0%, 59.0%, 63.9% and 69.5%, respectively.

Information about the date of parturition was received from 318 of the 471 cases of unilateral clinical mastitis. The distribution of the observed clinical mastitis cases in relation to the time of parturition is shown in Figure 2. The relative proportion of cases was greatest during the first week after lambing. Sixty-four (20.1%) of the 318 ewes for which the times of lambing and treatment were recorded, were treated for clinical mastitis during the first two days after parturition. The proportion of cases was significantly greater in the first ($P < 0.005$) and the third ($P < 0.05$) week postpartum as compared with the second week postpartum.

The proportion of the clinical *S. aureus* cases among all the clinical cases did not differ significantly between weeks of lactation, between ewes of different parities or between ewes with different number of lambs (data not shown).

In 180 ewes with unilateral clinical *S. aureus* mastitis, from which samples from both mammary glands were examined, *S. aureus* was found in secretions from 72 (40.0%) of the glands without clinical signs. In 72 ewes with unilateral clinical mastitis not caused by *S. aureus*, a subclinical *S. aureus* infection was found to be present in 10 (13.9%) of the glands without clinical signs.

Table 2 Bacteriological findings

Bacteriological finding	n	%
Staphylococcus aureus	357	65.3
Coagulase-negative staphylococci	16	2.9
Streptococcus uberis	9	1.6
Streptococcus dysgalactiae[a]	8	1.5
Streptococcus spp.[b]	8	1.5
Enterococcus spp.	4	0.7
Escherichia coli	35	6.4
Klebsiella pneumoniae	2	0.4
Enterobacter spp.	3	0.5
Mannheimia haemolytica	10	1.8
Arcanobacterium pyogenes	4	0.7
Clostridium perfringens	7	1.3
Pasteurella spp.[c]	4	0.7
No growth	72	13.2
Contaminated samples[d]	8	1.5

[a] Subsp. *dysgalactiae*.

[b] Other than *Str. dysgalactiae* and *Str. uberis*.

[c] *Pasteurella mairii* (2 samples), *Pasteurella multocida* (2 samples).

[d] If blood agar plates contained more than two different types of colonies.

3.2 S. aureus PFGE types in ewes with bilateral IMI

A total of 22 different PFGE types were found among 92 *S. aureus* isolates from 21 farms. Comparisons of *S. aureus* pairs revealed that 24 (86.0%) of 28 pairs had indistinguishable band patterns (Table 3). Given the PFGE types of all *S. aureus* isolates within each flock, expected flock-specific distributions of PFGE type pairs based on all possible random pairwise combinations of isolates were arranged. The number of pairs with identical types observed in the flocks was significantly greater than that expected, when assuming a random distribution of isolates ($P < 0.0001$).

3.3 Clinical signs

Of the 471 cases of unilateral clinical mastitis systemic signs were recorded in 325 (Table 4) and the rectal temperature was measured in 342. Moderate or severe systemic signs were present in 159 (48.9%) ewes and 193 (56.4%) ewes had a rectal temperature above 40.0°C. Thirty-seven (11.8%) out of 313 ewes showed no systemic signs and had a rectal temperature below 40.1°C.

Gangrene was present in 48 (8.8%) of 547 clinically affected udder halves and *S. aureus* was found in 35 (72.9%), *C. perfringens* in 3 (6.3%), *E. coli* in 3 (6.3%) and *A. pyogenes* in 1 (2.1%). Bacteria were not found in 4 (8.3%) samples from such cases and two samples were contaminated. The degree of systemic influence was recorded in 33 ewes with gangrenous mastitis. Thirty (90.9%) of these cases exhibited moderate or severe signs. The proportion of the gangrenous mastitis cases among all the clinical cases did not differ significantly between weeks of lactation, between ewes of different parities or between ewes with different number of lambs (data not shown).

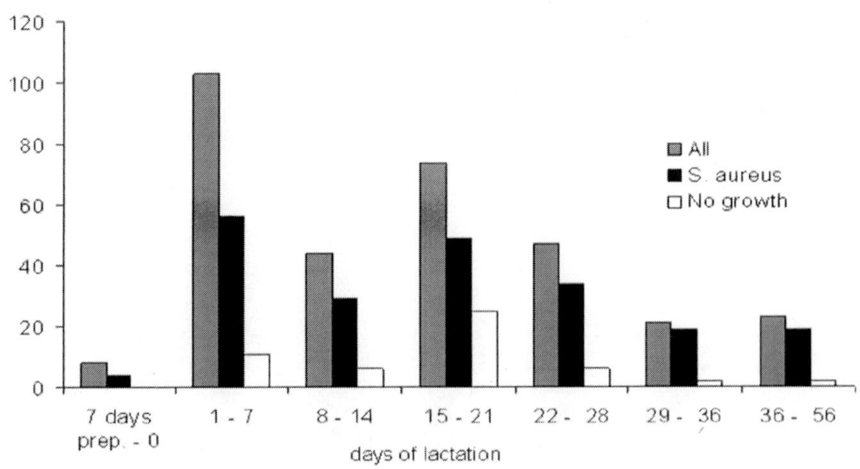

Figure2 Distribution of 318 cases of clinical mastitis in relation to weeks of lactation.

Only flocks in which at least one ewe experienced bilateral S. aureus IMI and two ewes experienced clinical mastitis are included. The observed pairs of PFGE types from the ewes with bilateral S. aureus IMI are shown. Also shown is the expected distribution of pairs with equal and unequal PFGE type combinations when assuming random pairwise distribution of the observed isolates within each flock.

Table 3: Distribution within flocks of 22 different PFGE types of 92 S. aureus isolates from ewes with intramammary infection (IMI).

				Expected pairs (n)	
Flock	Ewes with IMI (n)	*S. auerus* types (n) present in flock	Bilateral PFGE type combinations	Equal	Unequal
F1	6	K, P, Q (4), R	QQ	6	15
F2	9	I, J, L (7), M	LL	21	24
F3	2	R (3)	RR	3	0
F4	3	R (4)	RR	6	0
F5	2	H (2), O (2)	HH, OO	2	4
F6	2	D (2), H	DD	1	2
F7	2	H, I, U	IU	0	3
F8	2	H (3), X	HH, HX	3	3
F9	2	D, H (3)	DH, HH	3	3
F10	2	I, M, V	IV	0	3
F11	6	H (5), I, W	HH	10	11
F12	3	H, I (3)	II	3	3
F13	3	H (4), I (2)	HH, HH, II	7	8
F14	3	F, J (3)	JJ	3	3
F15	2	H, S (2)	SS	1	2
F16	2	D (3)	DD	3	0
F17	4	H (2), I (2), N, G (2)	HH, II, GG	3	18
F18	2	C (2), E	CC	1	2
F19	2	C (2), G	CC	1	2
F20	3	I (2), R, T	II	1	5
F21	2	A (2), B	AA	1	2

Table 4 Clinical signs

Pathogen	1 No. (%)	2 No. (%)	3 No. (%)	4 No. (%)	Total
All	51 (15.7)	115 (35.4)	119 (36.6)	40 (12.3)	325
S. aureus	30 (13.5)	77 (34.5)	89 (39.9)	27 (12.1)	223
Enterobacteria	2 (7.7)	9 (34.6)	11 (42.3)	4 (15.4)	26
No growth	11 (34.4)	12 (37.5)	4 (12.5)	5 (15.6)	32

[a] 1 = no systemic signs, 2 = weak systemic signs, 3 = moderate systemic signs, 4 = severe systemic signs.

4. DISCUSSION

A random and representative selection of sheep flocks was not deemed feasible for this type of study. However, in order to have a reasonably representative geographical spread of the mastitis cases ewes from 14 of the 19 counties in Norway were included. Thus, the present flocks were housed and pastured under various conditions, and the variations in climatic conditions, flock size and management routines of Norwegian sheep production were reasonably well represented. During the lambing season and subsequent weeks the ewes are paid close attention. Later, most flocks are moved to pastures in the forests or mountains. For this reason, the study was restricted to cases that occurred between 1 week prepartum and 8 weeks postpartum in order to obtain clinical cases of recent origin.

S. aureus was found in 65% of the samples from clinically affected glands. The dominance of *S. aureus* as a cause of clinical ovine mastitis has also been shown in regional studies in Norway, where *S. aureus* was isolated from udder secretions of between 64 and 87% of the ewes [26, 27]. In the present study, the largest proportion of *S. aureus* was found in the southern region (76%). A similar proportion of such cases (75%) was previously observed in a study including cases from one of the municipalities of this region [28]. Studies of clinical mastitis in meat sheep in other countries have found varying, though mostly relatively great, proportions of *S. aureus* infections (20–60%) [5, 17, 18, 19, 20, 22]. In a study of dairy sheep in Jordan, Lafi et al. [13] found that 22% of the clinical mastitis cases were caused by *S. aureus*.

The main *S. aureus* reservoirs in sheep are suggested to be infected mammary glands and teat lesions [29]. However, *S. aureus* can also be cultured from intact teat skin and other body sites [10, 30, 31]. In dairy flocks, transfer during milking is considered an important mechanism for the spread of this organism [29]. In flocks of meat sheep, transmission of *S. aureus* between ewes could be a result of the herdsman transmitting *S. aureus* between ewes during manual udder control, or the udder being exposed to bedding material contaminated from infected ewes [6, 7]. Some lambs occasionally suck other ewes than their dam, which might be a mechanism for the spread of *S. aureus* [32]. In Norway, routine examination of teats and udders are performed after weaning, and ewes with palpable abnormalities or which have experienced clinical mastitis are usually slaughtered before the breeding season. This contributes to decreasing the reservoir of *S. aureus*, but it obviously does not eliminate it.

Studies in other countries have reported prevalences of subclinical *S. aureus* IMI to be between 1 and 6% [12, 13, 15, 33]. This indicates that subclinically infected glands are an important reservoir of *S. aureus* that can only be detected through bacteriological examination. In this study, 14% of the ewes with unilateral clinical mastitis caused by other pathogens than *S. aureus* had a subclinical *S. aureus* infection in the other mammary gland, while 40% of the ewes with clinical *S. aureus* infection in one gland had a subclinical *S. aureus* infection in the other. PFGE typing showed that 86% of the pairs of isolates from ewes with bilateral *S. aureus* IMI were indistinguishable. This percentage was much higher than what would be expected if the isolates from each of the flocks were paired at random, thus demonstrating a considerably greater tendency for spread of *S. aureus* between the udder halves of a ewe than between ewes within a flock.

The very low percentage of *M. haemolytica* in the clinically affected glands in this study (1.8%) contrasts results of clinical mastitis surveys in meat sheep in the UK and Ireland, where the proportions of cases caused by this organism were found to be approximately 50% and 21%, respectively [5, 18]. Enterobacteria, mainly *E. coli*, were obtained from 7.3% of the clinically affected udder halves, which is similar to the proportions found in other studies on meat sheep [5, 18, 22]. However, the number of clinical cases caused by Gram negative bacteria and *A. pyogenes* may be underestimated because the samples were frozen before bacteriological analysis [34, 35].

It is noteworthy that nearly 85% of the ewes exhibited systemic signs and that gangrene was present in as much as 9% of the clinically affected udder halves, clearly showing that ovine clinical mastitis must be considered a serious animal welfare problem. However, mild cases of clinical mastitis are most likely underrepresented in this study. According to Norwegian legislation, antibiotic treatment of animals must be initiated by a veterinarian and, for economical reasons, farmers might be reluctant to call for a veterinary surgeon to treat mild clinical cases.

Most mastitis cases occurred close to lambing. One-third of the ewes developed clinical mastitis during the first week after lambing, and a second peak, although somewhat smaller, was observed in the third week postpartum. This is in accordance with data from the Norwegian Sheep Recording System [36]. Likewise, studies in the UK and Ireland found that cases of acute clinical mastitis occurred most frequently during the first week of lactation, while a second peak occurred between the third and fourth week [5] or the fourth and seventh week [18] after lambing. Clinical cases around parturition might be newly acquired IMI or aggravations of existing subclinical infections [37]. The proportion of very severe cases, in which gangrene had developed, was not greater among cases occurring close to lambing as compared with those occurring later. The second peak could be explained by increased milk demand from the lambs and the eruption of incisors, which increases the risk of teat lesions. It has been reported that teat lesions frequently are present in ewes with clinical mastitis three to four weeks after lambing [38].

5. CONCLUSION

This study shows that *S. aureus* is the most common cause of clinical mastitis in sheep in Norway and that this organism is frequently spread between glands in infected ewes. Further studies identifying predisposing factors, including reservoirs, transmission routes and factors facilitating *S. aureus* infection of the ovine mammary gland, are needed in order to improve strategies to reduce the occurrence of mastitis in sheep.

COMPETING INTERESTS

The author(s) declare that they have no competing interests.

ACKNOWLEDGEMENTS

Of the veterinary surgeons who kindly contributed to the study, the authors would particularly like to thank Gunvald Mosdøl, Olav Koltveit, Grethe Steihaug, Dag Aasland, Jan I. Holøymoen, Anne Aaby, Anders Avdem, Berit C. Brændvang, Aud Solberg, Odd S. Skregelid and Knut I. Dragset.

REFERENCES

1. Fthenakis GC, Jones JET: The effect of experimentally induced subclinical mastitis on milk yield of ewes and on the growth of lambs. Br Vet J. 1990, 146: 43-49.
2. Johnston WS, Maclachlan GK, Murray IS: A survey of sheep losses and their causes on commercial farms in the north of Scotland. Vet Rec. 1980, 106: 238-240.
3. Larsgard AG, Vaabenoe A: Genetic and environmental causes of variation in mastitis in sheep. Small Rumin Res. 1993, 12: 339-347. 10.1016/0921-4488(93)90069-T.
4. Gibson IR, Hendy PG: Mastitis in dry ewes. Vet Rec. 1976, 98: 511-512.

5. Jones JET: Mastitis in sheep. Breeding for Disease Resistance in Farm Animals. Edited by: Owen JB, Axfor RFE. 1991, Bangor: Tucson, AZ, CAB International, 412-423
6. Kirk JH, Glenn JS: Mastitis in ewes. Compend Cont Educ Pract Vet. 1996, 18: 582-591.
7. Watson DJ, Buswell JF: Modern aspects of sheep mastitis. Br Vet J. 1984, 140: 529-534.
8. Bergonier D, Berthelot X: New advances in epizootiology and control of ewe mastitis. Livest Prod Sci. 2003, 79: 1-16. 10.1016/S0301-6226(02)00145-8.
9. Ariznabarreta A, Gonzalo C, San Primitivo F: Microbiological quality and somatic cell count of ewe milk with special reference to staphylococci. J Dairy Sci. 2002, 85: 1370-1375.
10. Burriel AR: Dynamics of intramammary infection in the sheep caused by coagulase-negative staphylococci and its influence on udder tissue and milk composition. Vet Rec. 1997, 140: 419-423.
11. Fthenakis GC: Prevalence and aetiology of subclinical mastitis in ewes of Southern Greece. Small Rumin Res. 1994, 13: 293-300. 10.1016/0921-4488(94)90078-7.
12. Gonzalo C, Ariznabarreta A, Carriedo JA, San Primitivo F: Mammary pathogens and their relationship to somatic cell count and milk yield losses in dairy ewes. J Dairy Sci. 2002, 85: 1460-1467.
13. Lafi SQ, Al-Majali AM, Rousan MD, Alawneh JM: Epidemiological studies of clinical and subclinical ovine mastitis in Awassi sheep in northern Jordan. Prev Vet Med. 1998, 33: 171-181. 10.1016/S0167-5877(97)00048-2.
14. Pengov A: The role of coagulase-negative *Staphylococcus* spp. and associated somatic cell counts in the ovine mammary gland. J Dairy Sci. 2001, 84: 572-574.
15. Watson DL, Franklin NA, Davies HI, Kettlewell P, Frost AJ: Survey ofintramammary infections in ewes on the New England Tableland of New South Wales. Aust Vet J. 1990, 67: 6-8.
16. Hariharan H, Donachie W, Macaldowie C, Keefe G: Bacteriology and somatic cell counts in milk samples from ewes on a Scottish farm. Can J Vet Res. 2004, 68: 188-192.
17. Kryzanowski J, Wawron W, Malinowski V, Gluszak J, Orlik S: Bacterial flora isolated from the secretion of mastitic udders of ewes and their sensitivity to antibiotics. Med Vet. 1983, 39: 462-464.
18. Onnasch H, Healy AM, Brophy PO, Kinsella A, Doherty ML: A study of mastitis in sheep. Res Vet Sci. 2002, 72: 42-10.1016/S0034-5288(02)90120-7.
19. Quinlivan TD: Survey observations on ovine mastitis in New Zealand stud Romney flocks. 2. The bacteriology of ovine mastitis. N Z Vet J. 1968, 16: 153-160.
20. Al-Samarrae SAG, Sharma VK, Yousif AA: Mastitis in sheep in Iraq. Vet Rec. 1985,116: 323
21. Kirk JH, Glenn JS, Maas JP: Mastitis in a flock of milking sheep. Small Rumin Res. 1996, 22: 187-191. 10.1016/S0921-4488(96)00881-4.
22. Bocklisch H, Wetzstein D: Clinical, diagnostic laboratory and therapeutic studies of mastitis in a large sheep breeding flock (article in German). Tierärztl Prax. 1994, 22 (6): 524-528.
23. International Dairy Federation: Laboratory methods for use in mastitis work. 1981, Brussels: IDF, Document 132
24. Capurro A, Concha C, Nilsson L, Östensson K: Identification of coagulase-positive staphylococci isolated from bovine milk. Acta Vet Scand. 1999, 40: 315-321.
25. Mørk T, Tollersrud T, Kvitle B, Jørgensen HJ, Waage S: Genetic diversity of *Staphylococcus aureus* isolated from ovine intramammary infections in Norway. Vet Microbiol. 2005, 106: 265-273. 10.1016/j.vetmic.2004.12.020.
26. Kvitrud A, Lysne I: Examinations of mastitis in sheep (article in Norwegian). Nord Vet Med. 1959, 11: 129-140.
27. Sæter EA, Eieland E: Ovine mastitis in mountain districts in Norway (article in Norwegian). Nord Vet Med. 1961, 13: 32-44.
28. Indrebø A, Mosdøl G: Mastitis in the ewe. Microbiological findings and sensitivity to antibiotics (article in Norwegian). Nor Vet Tidsskr. 1991, 103: 107-113.\
29. Vautor E, Abadie G, Guibert JM, Huard C, Pepin M: Genotyping of *Staphylococcus aureus* isolated from various sites on farms with dairy sheep using pulsed-field gel electrophoresis. Vet Microbiol. 2003, 96: 69-79. 10.1016/S0378-1135(03)00207-4.

30. Scott MJ, Jones JET: The carriage of *Pasteurella haemolytica* in sheep and its transfer between ewes and lambs in relation to mastitis. J Comp Pathol. 1998, 118: 359-363. 10.1016/S0021-9975(07)80011-9.
31. Vautor E, Abadie G, Guibert JM, Chevalier N, Pepin M: Nasal carriage of *Staphylococcus aureus* in dairy sheep. Vet Microbiol. 2005, 106: 235-239. 10.1016/j.vetmic.2004.11.019.
32. Bergonier D, de Cremoux R, Rupp R, Lagriffoul G, Berthelot X: Mastitis of dairy small ruminants. Vet Res. 2003, 34: 689-716. 10.1051/vetres:2003030.
33. Deutz A, Pernthaner A, Schlerka G, Baumgartner W: Investigations on cell count of milk and distribution of bacterial mastitis in Lower Austrian sheep and goat flocks (article in German). Wien Tierärztl Mschr. 1990, 77: 70-77.
34. Sanchez A, Contreras A, Jimenez J, Luengo C, Corrales JC, Fernandez C: Effect of freezing goat milk samples on recovery of intramammary bacterial pathogens. Vet Microbiol. 2003, 94: 71-77. 10.1016/S0378-1135(03)00066-X.
35. Schukken YH, Grommers FJ, Smit JA, Vandegeer D, Brand A: Effect of freezing on bacteriologic culturing of mastitis milk samples. J Dairy Sci. 1989, 72: 1900-1906.
36. Vatn S: Epidemiology of mastitis in Norwegian sheep. Proceedings of the 6th International Sheep Veterinary Congress:17–21 June 2005; Hersonissos. Edited by: Fthenakis GC, McKaller FA. 2005, 6th International Sheep Veterinary Congress, 318-319.
37. Watkins GH, Burriel AR, Jones JET: A field investigation of subclinical mastitis in sheep in southern England. Br Vet J. 1991, 147: 413-420. Indrebø A: Mastitis and teat injuries in the ewe in relation to age, partus and number of lambs (article in Norwegian). Nor Vet Tidsskr. 1991, 103: 197-204.

CHAPTER 2

A Study into Digital Dermatitis Transmission and Bacterial Associated Pathological Changes Involved in the Disease

N. Capion[1*], M. Boye[2], C. Ekstrøm[3], K. Dupont[2], T. K. Jensen[2]

[1]*Department of Large Animal Sciences, Faculty of Health and Medical Sciences, University of Copenhagen, Copenhagen, Denmark*

[2]*National Veterinary Institute, Technical University of Denmark, Copenhagen, Denmark*

[3]*Department of Biostatistics, University of Southern Denmark, Odense, Denmark*

ABSTRACT

Digital dermatitis is a highly prevalent painful lesion affecting the feet in dairy cattle. Even though the pathogenesis has been subject of investigation since 1974, there is still a lack of knowledge about the spread of the disease among cows within a herd as well as between herds. The purpose of this study was to monitor transmission of digital dermatitis under experimental conditions between naïve heifers and affected animals, to monitor the changes in clinical appearance, microbial colonisation of the skin as lesions progressed and to apply a q-PCR for the detection of Treponema spp. in faecal samples. Eight heifers with clinical normal digital skin were housed with 5 heifers with severe digital dermatitis lesion for 8 weeks on a solid concrete floor with an accumulating layer of slurry. Digital skin was examined daily and lesions were clinically scored. Skin biopsies were taken from the healthy heifers at introduction and weekly from all lesions for histopathological evaluation and fluorescence in situ hybridization. None of the healthy heifers developed digital dermatitis and in 4 out of 5 infected heifers the lesions healed during the study. All samples from healthy skin were negative for Treponema spp. and one sample were positive for Dichelobacter nodosus. Colonization of healthy skin could not be identified in this study. There was no significant relation between clinical scoring of the lesions and histopathological score and the presence of Treponema spp. There were however a significant relation between the prevalence of Treponema spp. in the skin and severity of changes in epidermis and dermis. By qPCR all the healthy heifers were found to excrete Treponema spp. in their faeces.

Keywords: Digital Dermatitis; Treponema spp.; Histopathology; Infection Dynamics

1. INTRODUCTION

Digital dermatitis (DD) is the most common cause of lameness in Danish dairy herds. A study performed in 2003 showed that 89% of Danish Holstein dairies have cows infected with DD and the mean herd level prevalence was 21% [1].

The pathogenesis of DD has been intensely investigated during the last decades. DD appears to be polymicrobial, since a wide variety of bacteria have been isolated from lesions [2,3]. Recently laboratory techniques, has made the identification of bacteria such as spirochetes possible within the epidermis and dermis using cultureindependent methods. More than 16 different Treponema phylotypes have been identified in DD lesions. The most commonly found bacteria in association with DD are treponemes belonging to three different clusters T. phagedenis-like [4-7], T. refringens-like [8,9] and T. denticola-like [10]. Even though the microbiological background for DD has been intensely investigated; there is still uncertainty about the spread of the disease. The ability of the causative agents to survive in the environment is unknown and therefore the knowledge of contamination within a herd is yet to be elucidated. DD has been called contagious and infectious [11], however, experimental transmission is difficult [12]. Only two times has an experimental spread of DD been reported as successful. Both studies had to macerate the skin and have anaerobic conditions unlike conditions seen with natural infection [13,14].

It is obvious from clinical studies and reports of outbreaks that DD do spread among cows in a herd; however the routes of disease spreading have not been elucidated. The development of the lesion and colonization of the pathogens in the skin prior to clinical DD have not been investigated, which means that the initial causative agents have not been identified. Studies have examined healthy skin in the perimeter of lesions and have not been able to show the presence of Treponema spp. [8]. Several studies have described finding Treponema spp. in the GI tract of bovine [15-18] and these treponemes are probably shed in feces, however the treponemes found associated with digital dermatitis have not been isolated from GI samples or feces [15]. The reservoir for DD Treponemes remains uncertain.

Naïve young stock/young heifers have previously been described to be especially susceptible to DD [19-21] and moisture and detrimental effect of urine and faeces are known risk factors for DD [21,22].

The purpose of this study was 1) to monitor transmission of digital dermatitis under experimental conditions between naïve heifers and affected animals; 2) to evaluate the relation between clinical macroscopic appearance (lesion stage) of DD and histopathological changes and microbial colonization of the skin and 3) to apply a qPCR for the detection of Treponema spp. in faecal samples.

2. MATERIALS AND METHODS

2.1. Animals

Two maiden Holstein heifers 13 months of age from a DD-free tie stall herd (herd 1) and six maiden Holstein heifers 15 months of age from a DD-free free stall herd (herd 2) were housed together in a stall with solid concrete floor that measured 14 m by 8 m. Twelve hours a day the heifers has access to an additional area covered with deep straw bedding measuring 6 m by 8 m. During a period of four weeks the eight heifers build up a layer of approximately 40 cm of slurry on the concrete floor. After the four week introduction-period one first lactation Holstein cow (herd 3) with clinical lameness score 4/5 [23] and DD on both hind legs, from an endemic infected free stall herd, were introduced to the eight heifers. Following another four week period four maiden Holstein heifers (herd 4) from an endemic infected free stall herd were introduced to the group. All heifers had severe DD in the hind legs and a lameness score of 3/5 or 4/5 [23].

2.2. Examination

Before introduction to the experimental housing facility the heifers were examined in a trimming chute. Photographs were taken of all four limbs, the claws were trimmed and all lesions were recorded. Punch biopsies (6 mm, Kruuse, Denmark) were taken in the digital skin proximal of the heel horn on all hind legs following local anaesthesia of the skin, with 10 ml of Lidocain (20 mg/ml). No treatment or bandage was applied following biopsy sampling.

After introduction to the group all animals were examined on a daily basis. The examination included evaluation of general appearance, temperature, heart rate, respiratory rate, rumen contractions, faeces consistency, hydration, and they received a lameness score and were palpated for soreness of the digital skin. Once a week the heifers were examined in the trimming chute. The examination in the trimming chute included inspection and palpation of the digital skin. Faeces were removed with a gloved hand simultaneously with the palpation, leaving the skin visible but not clean, skin/lesions were scored for lesion stage and biopsies were sampled from all lesions.

2.3. Classification of DD Lesion Stage

DD was recorded using a standardised scoring system comprising five stages (M0-M4) [24]. If different stages of DD were present, the lesion was scored according to the most predominant stage of DD. M0: Normal skin, M1: Early stage of DD with a circumscribed granulomatous area, 0 to 4 cm in diameter, which lies at the epithelial surface or up to 2 mm underneath it. M1 is generally not painful on palpation. M2: Classical ulcerative stage of DD, with granulomatous tissue. M2 is often painful on palpation. M3: Classical ulceration of DD in the process of healing covered by a scab. M3 is generally not painful on palpation. M4: The chronic stage of DD are dyskeratotic or hyperkeratotic and can present themselves with a proliferative aspect. M4 is generally not painful on palpation.

2.4. Biopsies

Biopsies were fixed in 10% neutral buffed formalin, dehydrated and embedded in paraffin blocks. Serial tissue sections (4 μm) were cut and mounted on Super Frost + slides (Menzel-Gläser, Germany) for haematoxylin and eosin (H&E) stain and fluorescence in situ hybridization (FISH).

2.5. Histopathological Evaluation

All samples were evaluated histopathologically in H&E stained sections. Epidermal changes were scored 0 to 4 (0 = normal epidermis, 1 = hyperkeratosis and epidermal hyperplasia, 2 = focal, 3 = moderate, and 4 = severe acanthotic epidermis with or without parakeratotic papillomatous proliferation and ballooning degeneration of keratinocytes, exocytosis and/or erosion of dermal papilla).

Additionally, the biopsies were graded according to the inflammatory response in dermis as mild or absent (score 1), moderate (score 2), or (score 3) severe.

2.6. Fish

The biopsies were hybridized using oligonucleotide probes specific for Domain bacterium, Dichelobacter nodosus; Treponema phagedenis as well as Treponema spp.; (a Treponema group probe) as previously reported [8].

The hybridized sections were all read and scored by one of the senior authors. The total bacterial (probe for Domain bacterium) colonization of the epidermis was scored from 0 to 3 (0 = no invasive bacteria, 1 = low number of invasive bacteria, 2 = moderate number of bacteria, and 3 = high number of invasive bacteria). Similarly, the prevalence Treponema spp. and T. phagedenis-like organisms were scored from 0 to 3, whereas the prevalence of D. nodosus was scored 0 = no hybridization, or 1 = positive hybridization.

2.7. Collection of Fecal Samples and DNA Extraction

Forty eight samples of faeces were collected from the 8 heifers from herd 1 and 2 during 6 consecutive days (one sample per animal per day). The samples were frozen immediately and stored at –20°C.

DNA was extracted from the samples using the QIAamp DNA Stool Mini Kit (QIAGEN, Hilden, Germany): 180 - 220 mg of frozen faeces was transferred to ASL lysis buffer and DNA was extracted following manufacturer's protocol. The purified DNA was quantified (the yield per sample was approximately 6 μg with no variation between the samples) and diluted in nuclease-free water (Applied Biosystems, Foster City, CA, USA) to the concentration of 6.5 ng × μL^{-1}.

DNA for standard serial dilutions was obtained from a Treponema phagedenis isolate (isolate VI) provided by the Dept. of Biomedical Sciences and Veterinary Public Health, Swedish University of Agricultural Sciences, Uppsala, Sweden.

DNA was extracted using Easy-DNA Kit, protocol #3 (Invitrogen, Taastrup, Denmark).The quality of the purified DNA was verified both photometrically using a (NanoDrop 2000) spectrophotometer (Thermo Scientific, Wilmington. DE, USA) and by performing a PCR with general and Treponema spp. specific 16S RNA gene primers.

2.8. qPCR

The serial dilutions with well-defined number of DNA copies were created by cloning Treponema phagedenis 16S RNA gene fragment into pCR4Blunt-TOPO vector (Zero Blunt® TOPO® PCR Cloning Kit for Sequencing, Invitrogen)

PCR was performed in 4 × 50 µL reaction mixture containing: 0.5 µM of each primer, dNTP (each nucleotide 400 µM), $MgSO_4$ (2.0 mM), pfu DNA polymerase (0.05 U × μL^{-1}; Promega Corporation, Madison, WI, USA), template DNA from Treponema phagedenis isolate (40 ng per 50 µL of mix).

The cloning was performed according to the Invitrogen protocol for Zero Blunt® TOPO® PCR Cloning Kit. Five selected transformants were cultured overnight, harvested and the plasmids were purified using Plasmid Mini Kit (Qiagen, Cat. No.12123). The purified plasmids were quantified (Nanodrop) and digested with Pst I (Promega, cat. no. R4114) following Promega's protocol. The plasmid digest was then purified with MinElute PCR Purification. The insert sequence was controlled by sequencing purified plasmid using the Treponema primers. Sequencing was performed on a 3130 Genetic Analyzer (Applied Biosystems, USA) using BigDye Terminator v3.1 Cycle Sequencing Kit (Applied Biosystems, cat. no. 4337455) according to the manufacturer's protocol. The samples were run in triplicates.

qPCR was performed using a Rotor-Gene 6000 (Corbett Research, Sydney, Australia).

For qPCR on DNA extracted from faeces qPCR was run as single reactions for standard dilution and as triplicates for the samples. The conditions of PCR were the same as described above.

2.9. Statistical Analyses

To determine whether clinical classification of lesion stage, changes in epidermis and, inflammatory response in dermis were influenced by the prevalence of Treponema spp., prevalence of T. phagedenis-like, and prevalence of D. nodosus a multinomial (ordered) regression model was employed. The model was fitted using generalized estimating equations (GEE) with a working independence covariance structure based on the specific lesion within cow in order to accommodate any effect of repeated observations on a single lesion. Misspecification of the correct correlation structure was not critical with respect to the fixed effect estimates and would result in a loss of efficiency. Clusters defined by leg within cow were considered to see if this influenced the standard error of the estimated parameters.

Due to the size of the dataset and the low number of animals with DD a forward selection scheme was used to identify the associated predictors and the biological effect of the associated predictors was subsequently quantified.

2.10. Generalized Estimation Equation

For the generalized estimating equation (GEE) procedure, a multinomial error distribution and a cumulative logistic link function was employed. Initially, a working independence covariance structure with clusters based on individual legs was used. However, clusters based on individual lesions or on the broader grouping defined by cows were also considered. Explanatory variables using a forward selection approach was included and subsequently the full model was reduced using model reduction.

3. RESULTS

None of the 8 naïve heifers developed digital skin lesions during the 12 week observation period. The two heifers from tie stall (herd 1) developed hyperkeratosis of the digital skin above the heel bulbs. There were one case of lameness due to distortion and swelling of the fetlock joint on a foreleg. Otherwise the naïve heifers remained healthy. However, one biopsy was positive for Dichelobacter nodosus in the final biopsy sample collected from one of the healthy heifers at the end of the 12th week. The clinical evaluation of the skin showed hyperkeratosis and this was confirmed in the histology sample upon analysis.

Table 1 shows the status of DD lesions during the 12 weeks of study. The first lactation cow (herd 3) introduced after four weeks had DD on both hind legs. On left hind leg there was a lesion on an interdigital hyperplasia and a lesion between the heel bulbs. On the right hind leg there was a lesion between and above the heel bulbs. Both lesions between and above the heel bulbs healed completely, during the eight weeks in the experimental setting, where the legs were constantly covered in manure. However, the

DD on the interdigital skin remained throughout the study. Three of the four heifers from herd 4, had lesions that healed completely during the study period.

Table 1. Total number of digital dermatitis lesions on the individual heifers from introduction to the study population to the end of the study period.

Herd	Cow	Number of DD lesions											
		Week											
		1	**2**	**3**	**4**	**5**	**6**	**7**	**8**	**9**	**10**	**11**	**12**
1	1	0	0	0	0	0	0	0	0	0	0	0	0*
1	2	0	0	0	0	0	0	0	0	0	0	0	0
2	3	0	0	0	0	0	0	0	0	0	0	0	0
2	4	0	0	0	0	0	0	0	0	0	0	0	0
2	5	0	0	0	0	0	0	0	0	0	0	0	0
2	6	0	0	0	0	0	0	0	0	0	0	0	0
2	7	0	0	0	0	0	0	0	0	0	0	0	0
2	8	0	0	0	0	0	0	0	0	0	0	0	0
3	9					3	3	3	2	1	1	1	1
4	10									3	3	3	2
4	11									3	3	3	2
4	12									1	1	1	1
4	13									2	2	2	1

*This sample had a positive hybridisation of *D. nodosus*.

3.1. Treponema spp.

Because of few observations for some categories, lesion stages M1 and M2, lesion stages M3 and M4, and change in epidermis score 0 and 1 were combined when analysing the prevalence of Treponema spp. The score for inflammation in dermis and for changes in epidermis were included in the full model as well as examination date. Classification of lesion stage was not significant ($p = 0.18$).

The final model included either changes in epidermis score ($p = 0.02$) or inflammation in dermis score ($p = 0.03$) but not both. When individual legs were used as cluster the result were the same ($p = 0.04$ for changes in epidermis and $p = 0.02$ for inflammation in dermis).

Table 2 shows that a moderate inflammatory response in dermis (score 2) had an odds ratio resulting in a lower prevalence number of Treponema spp. of 0.56 relative to an extensive to diffuse inflammatory response in dermis (score 3) (i.e. odds ratio of 1.78 for a higher prevalence of Treponema spp.). The odds ratio of absent or mild inflammatory response in dermis (score 1) (relative to 3) was 7.63 for a lower prevalence of Treponema spp.

The results from the final model for the prevalence of Treponema spp. including changes in epidermis (Table 3) show that changes in epidermis score 2 had an odds ratio resulting in a lower prevalence of Treponema spp. of 62.85 relative to more severe changes in epidermis (score 3). The odds ratio of category 1 (relative to 3) was 355.28 for a lower prevalence of Treponema spp. These extremely large odds ratios were due to the small size of the dataset relative to the number of categories.

Table 2. The estimates for the final model for the prevalence of Treponema spp. including inflammation in dermis score 1, 2 and 3.

Parameter	Estimate	Standard Error	95% Confidence Limits		Z	Pr > |Z|
Intercept 1	−2.11	1.028	−4.11	−0.12	−2.07	0.04
Intercept 2	−1.83	0.97	−3.72	0.06	−1.90	0.06
Intercept 3	−0.919	1.15	−3.17	1.35	−0.79	0.43
Score 1	2.03	0.75	0.57	3.49	2.72	0.01
Score 2	−0.58	1.12	−2.78	1.63	−0.51	0.61
Score 3	0.00	0.00	0.00	0.00		

Table 3. The estimates for the final model for the prevalence of Treponema spp. including changes in epidermis score 1, 2 and 3.

Parameter	Estimate	Standard Error	95% Confidence Limits		Z	Pr > |Z|
Intercept 1	−4.54	0.56	−5.63	−3.45	−8.17	<0.01
Intercept 2	−4.06	0.81	−5.65	−2.48	−5.03	<0.01
Intercept 3	−2.16	0.71	−3.55	−0.77	−3.04	<0.01
Score 1	5.87	1.12	3.68	8.07	5.24	<0.01
Score 2	4.14	1.23	1.73	6.55	3.36	0.01
Score 3	0.00	0.00	0.00	0.00		

3.2. Treponema phagedenis-Like Organisms

Bnges in epidermis categories 0 and 1 were combined when analysing T. phagedenis.

The main effect of inflammation in dermis score and changes in epidermis score were both significant (p-values of 0.02 and 0.03) while macroscopically classification of DD was not ($p = 0.07$). However, the size of the data did not allow for an inclusion of multiple explanatory variables, and even individual legs as clusters were considered there was no significance.

The estimates for the classification of lesion stage main effects are shown below (Table 4) although the overall effect was not significant.

3.3. Dichelobacter nodosus

There were only two possible categories for D. nodosus so a logistic regression model was used with a cluster that is defined either by leg or lesion.

In order to be able to estimate the parameters the changes in epidermis scores 0 and 1, as well as lesion stages M3 and M4 were grouped. The single observation belonging to the healthy individual with M0 was disregarded. Inflammation in dermis score could not be used directly in a logistic regression setting as there was only observed variation for score 1. Consequently, D. nodosus and inflammation in dermis score were compared using Fisher's exact test (which on the other hand disregarded the fact that there was repeated measurements on the individuals). The size of the data did not allow for inclusion of multiple explanatory variables, which left only the main effects of the remaining two explanatory variables of interest.

Neither changes in epidermis score ($p = 0.89$) nor classification of lesion stage were significant ($p = 0.09$). However, looking at changes in epidermis and examination data combined then examination date was borderline significant ($p = 0.04$).

Only one of the analyses showed any systematic effect of time although measurements taken on the same cluster were potentially positively correlated. The statistical effect, however, was very small. It should be stressed that the number of observations was quite small for this study and given the number of categories for the response and explanatory variables so in most cases it had not been possible to make analyses that accommodate all the available predictors. It is also worth emphasizing that many of the results were just borderline significant (below the 5% level), and that while they all point in the same direction (which suggests that the statistical significance might have been greater if the dataset had been larger), also keeping in mind that multiple test/analyses have been made.

Table 4. The main effects of DD lesion stage (M2, M3 and M4) on the prevalence of T. phagedenis.

Parameter	Estimate	Standard Error	95% Confidence Limits		Z	Pr > |Z|
Intercept 1	−2.62	0.55	−3.70	−1.54	−4.77	<0.01
Intercept 2	−2.36	0.48	−3.29	−1.43	−4.96	<0.01
Intercept 3	−0.83	0.69	−2.18	0.53	−1.19	0.23
M2	0.17	1.31	−2.41	2.74	0.13	0.90
M3	2.95	0.41	2.14	3.76	7.13	<0.01
M4	0.00	0.00	0.00	0.00		

3.4. Treponema spp. qPCR on Faeces

The standard dilutions Ct values were consistent with expected 10 folds concentration differences between consecutive standards and ranged from 7.18 for the 10^7 standard dilution to 26.92 for the 10^1 standard dilution. The Ct values for the samples lay between 14.73 and 18.05.

The t_m (melting temperature) for standard dilutions (Treponema phagedenis) was 86.5°C - 86.8°C. The t_ms for samples ranged from 84.5°C to 86.7°C. The wider melting temperature range of the samples was most probably caused by the fact that this was a mixture of Treponema spp. therefore one would expect some variation in the amplified region of the 16S RNA gene.

The number of Treponema spp. copies in the samples was estimated by absolute quantification using standard serial dilutions. Treponema species were found in all samples with the number varying from 4000 to 38,000 copies. Although Treponema spp. was present in all samples there was no significant variation in the number of copies detected.

4. DISCUSSION

Based on the result of this study, the experimental transfer of DD between heifers did not occur. The area the heifers were kept in was rather crowded, with approximately 8.6 m^2 per animal. New animals were introduced to the group every 4 week, causing disturbance and probably introducing new pathogens. One might speculate that the environment the heifers were kept in probably did not particularly favour treponemes, since none of the naïve animals was infected. Another thought could be that the skin was exposed to urea instead of ammonium gas and thereby was in better condition to limit the DD lesions and the proliferation of treponemes within the epidermis. The possibility of the naïve heifers being unsusceptible to DD is also present. There has been some evidence that a rather large variation in susceptibility between Holstein heifers exists [25].

The PCR showed large amounts of treponemes in the faeces, hence in the manure covering the floor. Even though previous studies have failed to identify DD-treponemes in the bovine GI-canal [15] and the present study failed to classify the exact Treponema spp. in the faecal samples it still poses the question whether the manure is in fact the reservoir for DD. The finding that DD did not spread in this group of animals, in spite of several risk factors supports the theory that DD indeed has a complicated multifactorial etiology.

However, it is quite interesting that despite the rather unhygienic conditions these heifers were exposed to for 12 weeks none of them developed signs of DD or even interdigital dermatitis or heel horn erosion. And epidermal colonisation of bacteria could only be found in one biopsy from hyperkeratotic skin.

With regards to the gradual healing of lesions in the group of infected cows, the frequent biopsy sampling could have induced an inflammatory response that acted synergistically with the cow and sped up the healing process. This completely adverse direction this part of the study has taken only shows that there are still many unknowns about this particular disease.

The most common way of diagnosing DD in practice is by clinically identifying the lesion and more advanced scoring the lesion according to a scoring system. The results show that the relationship between clinical appearance of the lesion, the lesion stage and the degree of histopathological changes or the degree of treponemal colonisation of the tissue is very poor. This is in contrast to the previous findings of this research group. In 2007 this group found that the larger and the more variable the colonisation the more severe were the lesion stage and the clinical evaluation based on data collected in a crosssectional study [8]. One may speculate that the DD lesions in the present study are atypical lesions due to the effect of the environment that actually improves or heals the lesions. In the present study we saw a discontinuation of a typical DD lesion instead of a development of new lesions. Furthermore data were based on repeated measurements on the same animals, providing more information from each lesion.

However, when analysing at the histopathological changes in epidermis and dermis there is a correlation between more treponemes and more severe changes in dermis and epidermis. The implications of these new findings could mean that the colonisation of the skin is not very diverse in this group of heifers.

5. CONCLUSIONS

None of the healthy heifers exposed to DD infected animals developed DD, evaluated both on a clinical or histopathological level. In only one of the 5 infected heifers the lesions remained unchanged during the study, the rest of the lesions healed.

In this study the clinical appearance of lesions (lesion stage) is not significantly associated with the prevalence of Treponema spp., the histopathological changes in epidermis and the level of inflammation in dermis. One of variables histopathological changes in epidermis or the level of inflammation in dermis (but not both) is significantly associated with the prevalence of Treponema spp. The effects of histopathological changes in epidermis and the level of inflammation in dermis on the prevalence of Treponema spp. have more or less the same positive directions: the likelihood of the prevalence of Treponema spp. increases with increasing level of histopathological changes in epidermis and severity of the inflammatory response.

It should be stressed that the number of observations is quite small for this study and that many of the results are just borderline significant.

All faecal samples collected rectally from the healthy heifers were positive for Treponema spp. on real-time PCR.

ACKNOWLEDGEMENTS

The study was funded by the Danish Research Council for Technology and Production. We thank the farmers for letting us borrow the cows.

REFERENCES

1. N. Capion, M. Boye, C. Ekstrøm, K. Dupont, T. K. Jensen The study was funded by the Danish Research Council for Technology and Production. We thank the farmers for letting us borrow the cows.
2. N. Capion, S. M. Thamsborg and C. Enevoldsen, "Prevalence of Foot Lesions in Danish Holstein Cows," Veterinary Record, Vol. 163, No. 3, 2008, pp. 80-85. doi:10. 1136/vr.163.3.80
3. N. J. Evans, J. M. Brown, I. Demirkan, R. D. Murray, W. D. Vink, R. Blowey, C. A. Hart and S. D. Carter, "Three Unique Groups of Spirochetes Isolated from Digital Dermatitis Lesions in UK Cattle," Veterinary Microbiology, Vol. 130, No. 1-2, 2008, pp. 141-150. doi:10.1016/j.vetmic.2007.12.019
4. C. M. Schroeder, K. W. Parlor, T. L. Marsh, N. K. Ames, A. K. Goeman and R. D. Walker, "Characterization of the Predominant Anaerobic Bacterium Recovered from Digital Dermatitis Lesions in Three Michigan Dairy Cows," Anaerobe, Vol. 9, No. 3, 2003, pp. 151-155. doi:10.1016/S1075-9964(03)00084-2
5. D. J. Trott, M. R. Moeller, R. L. Zuerner, J. P. Goff, W. R. Waters, D. P. Alt, R. L. Walker and M. J. Wannemuehler, "Characterization of Treponema Phagedenis-Like Spirochetes Isolated from Papillomatous Digital Dermatitis Lesions in Dairy Cattle," Journal of Clinical Microbiology, Vol. 41, No. 6, 2003, pp. 2522-2529. doi:10.1128/JCM.41.6.2522-2529.2003
6. M. Nordhoff, A. Moter, K. Schrank and L. H. Wieler, "High Prevalence of Treponemes in Bovine Digital Dermatitis-A Molecular Epidemiology," Veterinary Microbiology, Vol. 131, No. 3-4, 2008, pp. 293-300. doi:10.1016/j.vetmic.2008.04.019
7. F J. N. Evans, J. M. Brown, I. Demirkan, P. Singh, B. Getty, D. Timofte, D. Vink, R. D. Murray, R. W. Blowey, R. Birtles, C. A. Hart and S. D. Carter, "Association of Unique, Isolated Treponemes with Bovine Digital Dermatitis Lesions," Journal of Clinical Microbiology, Vol. 47, No. 3, 2009, pp. 689-696. doi:10.1128/JCM.01914-08
8. T. Yano, R. Yamagami, K. Misumi, C. Kubota, K. K. Moe, T. Hayashi, K. Yoshitani, O. Ohtake and N. Misawa, "Genetic Heterogeneity among Strains of Treponema Phagedenis-Like Spirochetes Isolated from Dairy Cattle with Papillomatous Digital Dermatitis in Japan," Journal of Clincal Microbiology, Vol. 47, No. 3, 2009, pp. 727- 733. doi:10.1128/JCM.01574-08
9. K. Klitgaard, M. Boye, N. Capion and T. K. Jensen, "Evidence of Multiple Treponema Phylotypes Involved in Bovine Digital Dermatitis as Shown by 16S rRNA Gene Analysis and Fluorescence In Situ Hybridization," Journal of Clinical Microbiology, Vol. 46, No. 9, 2008, pp. 3012-3020. doi:10.1128/JCM.00670-08
10. M. Rasmussen, N. Capion, K. Klitgaard, T. Rogdo, T. Fjeldass, M. Boye and T. K. Jensen, "Bovine Digital Dermatitis: Possible Pathogenic Consortium Consisting of Dichelobacter nodosus and Multiple Treponema Species," Veterinary Mircobiology, Vol. 160, No. 1-2, 2012, pp. 151-161. doi:10.1016/j.vetmic.2012.05.018
11. B. K. Choi, H. Nattermann, S. Grund, W. Haider and U. B. Gobel, "Spirochetes from Digital Dermatitis Lesions in Cattle Are Closely Related to Treponemes Associated with Human Periodontitis," International Journal of Systematic Bacteriology, Vol. 47, No. 1, 1997, pp. 175-181. doi:10.1099/00207713-47-1-175
12. R. Laven, "Control of Digital Dermatitis in Cattle," In Practice, Vol. 23, No. 6, 2001, pp. 336-341. doi:10.1136/inpract.23.6.336
13. H. F. Bassett, M. Monaghan, P. Lenhan, M. L. Doherty and M. E. Carter, "Bovine Digital Dermatitis," Veterinary Record, Vol. 126, 1990, pp. 164-165.
14. D. H. Read and R. L. Walker, "Papillomatous Digital Dermatitis (Footwarts) in California Dairy Cattle: Clinical and Gross Pathologic Findings," Journal of Veterinary Diagnostic Investigation, Vol. 10, No. 1, 1998, pp. 67-76. doi:10.1177/104063879801000112
15. Gomez, N. Cook, N. D. Bernardoni, J. Rieman, A. F. Dusick, R. Hartshorn, M. T. Socha, D. H. Read and D. Döpfer, "An Experimental Infection Model to Induce Digital Dermatitis

Infection in Cattle," Journal of Dairy Science, Vol. 95, No. 4, 2012, pp. 1821-1830. doi:10.3168/jds.2011-4754
16. J. N. Evans, J. M. Brown, R. D. Murray, B. Getty, R. Birtles, C. A. Hart and S. D. Carter, "Characterization of Novel Bovine Gastrointestinal Tract Treponema Isolates and Comparison with Bovine Digital Dermatitis Treponemes," Applied and Environmental Microbiology, Vol. 77, No. 1, 2011, pp. 138-147. doi:10.1128/AEM.00993-10
17. T. Shibahara, T. Ohya, R. Ishii, Y. Ogihara, T. Maeda, Y. Ishikawa and K. Kadota, "Concurrent Spirochaetal Infections of the Feet and Colon of Cattle in Japan," Australian Veterinary Journal, Vol. 80, No. 8, 2002, pp. 497-502. doi:10.1111/j.1751-0813.2002.tb12474.x
18. T. B. Stanton and E. Canale-Parola, "Treponema Bryantii sp. nov., a Rumen Spirochete That Interacts with Cellulolytic Bacteria," Archives of Microbiology, Vol. 127, No. 2, 1980, pp. 145-156. doi:10.1007/BF00428018
19. B. J. Paster and E. Canale-Parola, "Treponema Saccharophilum sp. nov., a Large Pectinolytic Spirochete from the Bovine Rumen," Applied Environmental Microbiology, Vol. 50, No. 2, 1985, pp. 212-219.
20. D. J. Argaez-Rodriguez, D. W. Hird, J. H. deAnda, D. H. Read and A. Rodriguez-Lainz, "Papillomatous Digital Dermatitis on a Commercial Dairy Farm in Mexicali, Mexico: Incidence and Effect on Reproduction and Milk Production," Preventive Veterinary Medicine, Vol. 32, No. 3-4, 1997, pp. 275-286. doi:10.1016/S0167-5877(97)00031-7
21. M. Holzhauer, C. Hardenberg, C. J. M. Bartels and K. Frankena, "Herdand Cow-Level Prevalence of Digital Dermatitis in the Netherlands and Associated Risk Factors," Journal of Dairy Science, Vol. 89, No. 2, 2006, pp. 580-588. doi:10.3168/jds.S0022-0302(06)72121-X
22. Rodriguez-Lainz, P. Melendez-Retamal, D. W. Hird, D. H. Read and R. L. Walker, "Farmand Host-Level Risk Factors for Papillomatous Digital Dermatitis in Chilean Dairy Cattle," Preventive Veterinary Medicine, Vol. 42, No. 2, 1999, pp. 87-97. doi:10.1016/S0167-5877(99)00067-7
23. Rodriguez-Lainz, D. W. Hird, R. L. Walker and D. H. Read, "Papillomatous Digital Dermatitis in 458 Dairies," Journal of the American Veterinary Medical Association, Vol. 209, No. 8, 1996, pp. 1464-1467.
24. D. J. Sprecher, D. E. Hostetler and J. B. Kaneene, "A Lameness Scoring System that uses Posture and Gait to Predict Dairy Cattle Reproductive Performance," Theriogenology, Vol. 47, No. 6, 1997, pp. 1179-1187. doi:10.1016/S0093-691X(97)00098-8
25. D. Döpfer, A. Koopmans, F. A. Meijer, I. Szakall, Y. H. W. Schukken, W. Klee, R. B. Bosma, J. L. Cornelisse, A. J. A. M. van Asten and A. A. H. M. ter Huurne, "Histological and Bacteriological Evaluation of Digital Dermatitis in Cattle, with Special Reference to Spirochaetes and Campylobacter Faecalis," Veterinary Record, Vol. 140, No. 24, 1997, pp. 620-623. doi:10.1136/vr.140.24.620
26. N. Capion, M. Boye, C. T. Ekstrøm and T. K. Jensen, "Infection Dynamics of Digital Dermatiits in First-Lactation Holstein Cows in an Infected Herd," Journal of Dairy Science, Vol. 95, No. 11, 2012, pp. 1-8. doi:10.3168/jds.2012-5335

CHAPTER 3

Uterine Microbiology and Antimicrobial Susceptibility in Isolated Bacteria from Mares with Fertility Problems

A Albihn[1], V Båverud[2] and U Magnusson[3]

[1]*Department of Disease Control and Biosecurity*

[2]*Department of Bacteriology, National Veterinary Institute,*

[3]*Department of Obstetrics and Gynaecology, Centre for Reproductive Biology, Uppsala, Sweden.*

ABSTRACT

Albihn A, Båverud V, Magnusson U: Uterine microbiology and antimicrobial susceptibility in isolated bacteria from mares with fertility problems. Acta vet. scand. 2003, 44, 121-129.Uterine microbiology and antimicrobial susceptibility was investigated in 239 mares with fertility problems in a prospective study in Sweden. Uterine swab samples were collected with double guarded swabs and transported overnight before being cultured. The Minimum Inhibitory Concentrations (MIC) was determined for a panel of antimicrobials. From 152 of the 239 mares at least one bacterial species was isolated, most frequently *E. coli* (104 isolates), *β*-haemolytic streptococci (31) and fungi (16). *β*-haemolytic streptococci were more frequently ($p < 0.01$) associated with clinical endometritis than with repeat breeding. The opposite was true for E. coli ($p < 0.01$). Among β-haemolytic streptococcal isolates some resistance was noted for 4 of 11 tested antibiotics, however, all isolates were susceptible to the widely used penicillin G. Among *E. coli* isolates enrofloxacin was the only of the 10 tested antibiotics for which no resistance was noted. Resistance was most commonly noted to cephalothin (39% of the isolates), streptomycin (22%), trimethoprim/sulphamethoxazole (15%) and ampicillin (11%). In conclusion, we show that both *E. coli* and *β*-haemolytic streptococci are frequently associated with fertility problems in mares and that antimicrobial resistance is a common feature of *E. coli* but also recognised for *β*-haemolytic streptococcal uterine isolates.

Keywords: equine subfertility, endometritis, uterine, swab samples, bacteriology, resistance to antibiotics, E. coli, β-haemolytic, streptococci Streptococcus, zooepidemicus

1. INTRODUCTION

Uterine infections have long been recognised as one of the major causes of reduced fertility in the mare (*Asbury* 1986). These infections are most often caused by opportunistic micro-organisms and a variety of species have been isolated (*Shin et al.* 1979, *Ricketts et al.* 1993).

The uterine infections often cause endometritis. Antibiotics are one component often used in the treatment of endometritis (*Perkin* 1999). For clinicians there is a need of rapid microbiological diagnosis so that adequate treatment of the infection can be performed while the mare is still in oestrus (*Ricketts & Mackintosh* 1987, *Ricketts* 1989). Therefore some mares are treated with antibiotics without a preceding microbiological investigation, whereas sometimes bacteriological cultivation is performed by the clinicians themselves. If the treatment is performed without a microbiological diagnosis, the choice of antibiotic is often based on data from earlier studies, e.g. *Shin et al.* (1979) and *Ricketts et al.* (1993). However, the bacterial species isolated, as well as their susceptibility to antibiotics, may vary over time as well as from one population of horses to another (*Sternberg* 1999). The variation may be attributable to differences in antibiotic treatment policies, studfarm management, breed and clinical history of the sampled mares as well as microbiological culture routines. The present survey of uterine microbiology and antimicrobial susceptibility in mares selected for having fertility problems was conducted in Sweden where artificial insemination is commonly used. Also the policies for clinical use of antibiotics are regarded to be strict in Sweden (*Franklin* 1999,

SVARM 2001). The aim of the present study was to determine the most common bacterial species in uterine samples from Swedish mares with fertility problems and the antimicrobial susceptibility of isolated bacteria. Such data should serve as a basis for updated recommendations on how to treat uterine infections in the mare.

2. MATERIALS AND METHODS

2.1 Sampling

Clinicians representing different types of stud farms, geographically located all over Sweden, were invited to send uterine samples from mares in oestrus to the National Veterinary Institute (SVA) for culture free of charge. During the spring and summer 1996 and 1997 swabs from 239 mares were submitted from 36 different clinicians.

2.2 Mares

For all 239 mares included in the study, at least one fertility problem was noted. From the preceding season 89 barren mares and 33 abortions/resorbtions were recorded. During the current season 121 repeat breeding mares were recorded. Repeat breeding was defined as starting a new oestrus cycle after artificial insemination (AI) or being bred by a stallion once or repeatedly during oestrus in at least one oestrus cycle (with a normal or changed length of the luteal phase). Clinical signs of endometritis during the current season were noted in 89 mares. The designation clinical signs of endometritis in this study included at least one of the following criteria: vulvular discharge or fluid in the uterine lumen during the luteal phase, the latter diagnosed with ultrasonography, inflammatory cells on a cytological smear sample or significant bacterial growth. Reproductive problems during the current season may be combined with barreness or resorbtion/abortion during the preceding season. Repeated breeding was often combined with some of the other reproductive disorders. The average age of the mares was 12.2 years. For 32 mares the age was not recorded, 32 mares were from 3–7 years, 85 from 8 to 13 years and 90 from 14 to 24 years old. The dominating breeds of the included mares were Swedish Warmblood (80), Standardbred Trotter (75), North-Swedish Trotter (39) and Thoroughbred (25). Twenty mares were either of other breeds or their breed was not recorded.

Since many clinicians were involved in the present study, it was important to establish a simple, straightforward sampling protocol in order to achieve good quality and reliable data. Hence, we limited the study to bacteriological sampling. For these samples, the external genitalia were carefully washed with soap and water and thereafter dried with paper. In order to minimise contamination of the sample by bacteria from the vagina and perineum the sampling was performed using a gloved hand in the vagina and double-guarded, occluded swabs enabling sampling from the uterus solely (Equi-Vet, Kruuse, Marslev, Denmark). The uterine culture swabs were transported in Amies' modified media with charcoal (SVA, Uppsala, Sweden) (*Amies* 1967, *Engvall* 1985) at ambient temperature and cultured within 24 h. This medium has been widely used in Sweden as an all-purpose transport medium for equine gynaecological swabs.

2.3 Bacteriology

All samples were cultured on 5% horse blood agar (SVA) and lactose bromocresol purple agar (SVA). The samples were inoculated on the agar plates and diluted by an inoculation loop obtaining 3 levels of dilution on the agar plate. Each bacteriological culture was inspected and bacterial growth was registered after 24 and 48 h incubation at 37°C.

Growth of Pseudomonas (P.) aeruginosa, Klebsiella (K.) pneumoniae, haemolytic Escherichia (E.) coli and β-haemolytic streptococci was always considered to be of significance (*Shin et al.* 1979, *Ricketts et al.* 1993, *Langoni et al.* 1997). Other bacterial isolates were typed and considered as significant if growth was in pure culture or dominating on the agar plate. From the same sample, 2 bacterial species might be isolated and typed. Bacterial growth was evaluated on horse blood agar plates according to the following guidelines: abundant, >100 CFU (colony forming units)/plate; moderate, 21–100 CFU/plate; sparse, 10–20 CFU/plate; insignificant, <10 CFU/plate. Conventional methods for isolation and identification of microorganisms were used (*Quinn et al.* 1994).

2.4 Mycology

Samples from 233 of the 239 mares were also cultured for fungi on sabouraud dextrose agar 2% (Difco laboratories, Detroit, M) with chloramphenicol (0.5 µg/ml, Fluka Chemi, Buchs, Switzerland) and incubated at 30°C for 5 days. In the identification of fungus yeast was not identified to species except for Candida albicans according to *McGinnis* (1980).

2.5 Antimicrobial susceptibility testing

A microtiter plate system (VETMIC™, SVA) was used for the antimicrobial susceptibility tests. The test was done according to the manufacturer's instruction. In brief, each well was inoculated with 50 µl of Mueller Hinton broth (Merck, KgaA, Darmstadt, Germany) to which 10^3 to 10^4 CFU of the bacteria to be tested were added. The wells were sealed with transparent adhesive tape and incubated at 35–37°C for 16–18 h. The lowest concentration of an antibiotic completely inhibiting bacterial growth was registered as Minimum Inhibitory Concentrations (MIC). Results were categorised by using the breakpoints for resistant, intermediate and sensitive recommended by the NCCLS for bacteria isolated from animals (1999). Currently, no recommendations are available from the NCCLS for spiramycin, streptomycin, fusidic acid, nitrofurantoin and enrofloxacin. Therefore, for these antimicrobials the values recommended by the manufacturer were used. Quality control strains included *E. coli* ATCC 25922, *P. aeruginosa* ATCC 27853, *Enterococcus faecalis* ATCC 29212 and *Staphylococcus aureus* ATCC 29213. The MICs of the quality control strains were always within recommended ranges (NCCLS, 1997).

2.6 Statistical analysis

The analysis of frequencies of the various fertility problems connected with microbiological diagnosis was made by chi-square analysis within the frequency procedure in SAS [25].

3. RESULTS

3.1 Bacteriology

From 152 positive samples out of 239 sampled mares one or 2 significant bacterial species were isolated and identified. Ninety-two (38%) of the positive samples yielded growth of one single species in pure or almost pure culture. Thirty-one (13%) mares yielded 2 species and 29 (12%) mares yielded 1 species dominating on the agar plate together with sparse non-specific mixed culture. In samples from the mares without significant growth 57 (24%) yielded no growth at all and 30 (13%) growth of non-specific mixed culture.

The bacterial species mostly isolated was *E. coli*, yielding 104 isolates, thereof 64 isolates in pure or almost pure culture (Table 1).

Table 1 Species isolated in microbiological cultivation from uterine swabs from 152 mares. Two bacterial species might be isolated and typed from the same mare.

Microorganisms	Number	%
Actinobacillus spp/Pasteurella spp	1	<1
Corynebacterium spp	2	1
Enterobacter aerogenes	4	3
Enterobacter agglomerans	3	2
Enterobacter spp	1	<1
Enterococcus spp	2	1
Escherichia coli, non-haemolytic	99	64
Escherichia coli, haemolytic	5	3
Klebsiella pneumoniae	1	<1
Gramnegative coccus	12	8
Gramnegative rods, inactive	11	7
Pseudomonas aeruginosa	1	<1
Pseudomonas spp	5	3
Sphingomonas paucimobilis[1]	1	<1
Staphylococcus spp, coagulase neg	3	2
Streptococcus spp, α-haemolytic	1	<1
Streptococcus, β-haemolytic	31	20
Streptococcus equi subsp. *equi*	1	
Streptococcus dysgalactiae subsp. *equisimilis*	4	
Streptococcus equi subsp. *zooepidemicus*	21	
Streptococcus, β-haemolytic[3]	5	
Fungi[2]	16	
Yeast, not *Candida albicans*	13	
Yeast[3]	2	
Mould	1	

[1]previous name *Pseudomonas paucimobilis*
[2]233 out of 239 mares were cultured for fungi
[3]not further typed

When the bacterial growth was quantified, 72 *E. coli* isolates yielded abundant, 21 moderate and 11 sparse growth. From two mares, two different isolates of *E. coli* were isolated. Only 5 of the 104 *E. coli* isolates were haemolytic *E. coli,* 3 of these in pure culture. When the bacterial growth was quantified, 3 of these haemolytic *E. coli* yielded abundant, 1 moderate and 1 sparse growth.

The second most frequently isolated species was β-haemolytic streptococci yielding 31 isolates, thereof 12 grew in pure culture (Table 1). When the bacterial growth was quantified, 12 Streptococcus isolates yielded abundant, 12 moderate and 7 sparse growth.

3.2 Relation between fertility problems and microbiological diagnosis

From repeat breeding mares, as well as from mares with clinical symptoms of endometritis, *E. coli* was the most frequently isolated species. *β*-haemolytic streptococci were more frequently ($p < 0.01$) associated with clinical endometritis than with repeat breeding (68 versus 23% of *β*-haemolytic streptococcal isolates). The opposite was true for *E. coli* ($p < 0.01$) (38 versus 53% of *E. coli* isolates).

3.3 Mycology

From 15 mares yeast was isolated (13 of these further typed as not being Candida albicans) and from 1 mare mould was isolated. These were all in mixed culture with bacteria.

3.4 Antimicrobial susceptibility

Among the 104 E. coli isolates, resistance was most common to cephalothin, streptomycin, trimethoprim/sulphamethoxazole (TMP) and ampicillin (Table 2). Several isolates were resistant to more than one antimicrobial. Enrofloxacin was the only one of the 10 tested antimicrobial agents for which no resistance was noted. The 5 haemolytic E. coli isolates were all classified as susceptible to TMP and gentamicin (Table 2).

Among the 31 *β*-haemolytic streptococcal isolates resistance was most common to gentamicin, neomycin, oxytetracycline, and to TMP (Table 3). All isolates were classified as susceptible to the *β*-lactam antibiotics, penicillin G and ampicillin, and also to cephalothin, erythromycin, spiramycin and chloramphenicol.

Table 2 Distribution (no. of isolates) of MIC values for 104 Escherichia coli isolates of the 10 antibiotics tested. Vertical lines show the break points between sensitive (S), intermediate (I) and resistant (R). The S-isolates are to the left and the R-isolates to the right of the lines.

Antimicrobial tested	MIC (mg/L)										Range tested[1]	S-I-R % isolates
	≤0,12	0,25	0,5	1	2	4	8	16	32	>32		
Ampicillin[2]					20^{H1}	50^{H2}	19	3	11^{H2}		2-16	86 3 11
Cephalothin						5	14	44^{H1}	31^{H4}	10	4-32	18 43 39
Chlor-amphenicol					4	22	71^{H5}	6	1		2-16	94 6 1
Enrofloxacin		101^{H5}		1	2						0,25-2	97 3 0
Gentamicin				48^{H2}	NT	52^{H3}		3	1		1-16	96 0 4
Neomycin[2]					58^{H2}	NT	38^{H3}	1	2	4	2-32	93 3 4
Nitro-furantoin[3]						3	17	76^{H5}	7	1	4-32	99 - 1
Oxy-tetracycline				6	NT	79^{H4}	12^{H1}	2	5		1-16	81 12 7
Streptomycin					1	NT	51^{H1}	23^{H3}	6	23^{H1}	2-32	51 27 22
Trimethoprim-sulpha-methoxazole[3,4]	79	NT	6^{H5}	3	NT	NT	1	15			0,12-8	85 - 15

[1] When the MIC value was above the range tested, the value for the next titration step (the value just above the range) was used.
[2] One strain not tested.
[3] The vertical line shows the break point between S and R, no I sensitivity is given.
[4] The MIC value for trimethoprim tested in combination with sulfamethoxazol (1:20) is given.
[H1-5] The number of haemolytic *E. coli* isolates.
NT = not tested, the titration step is not included in the VetMIC™ system.

4. DISCUSSION

Mares included in this study were selected for reproductive disorders and only 32% of them yielded no significant growth of microorganisms. This figure should be compared with studies where mares have not been selected for reproductive problems. In these studies 70% (*Redaelli & Codazza* 1977), 68% (*Shin et al.*1979) and 61% (*Ricketts et al.* 1993) of the mares yielded no significant growth. Other possible reasons for the difference in outcome between studies could be that different breeds have been studied and/or there was different breeding management between the studied populations. Also the sampling technique and microbiological culture routines may influence the results.

Table 3 Distribution (no. of isolates) of MIC values for 31 β-haemolytic streptococcal isolates of the 11 antibiotics tested. Vertical lines show the break points between sensitive (S), intermediate (I) and resistant (R) isolates. The S are to the left and the R to the right of the lines.

Antimicrobial tested	MIC (mg/L)											Range tested[1]	S-I-R % isolates
	≤0,06	0,12	0,25	0,5	1	2	4	8	16	32	>32		
Ampicillin		30	1	NT	NT							0,12-16	100 0 0
Cephalothin							31					4-16	100 0 0
Chlor-amphenicol						15	15	1				2-16	100 - 0
Clindamycin[2]					28	NT	3					1-4	90 0 10
Erythromycin[3]				31								0,5-4	100 0 0
Gentamicin						NT	6	9	10	6		1-16	19 29 52
Neomycin						1	NT	3	5	10	12	2-32	13 48 39
Oxytetra-cycline					9	NT	15	6		1		1-16	29 48 23
Penicillin G	31		NT	NT		NT	NT					0,06-8	100 0 0
Spiramycin[4]							31					4-32	100 - 0
Trimethoprim - sulphameth-oxazole[4,5]		17	5	6		NT	NT	1	2			0.12-8	90 3 7

[1] When the MIC value was above the range tested, the value for the next titration step (the value just above the range) was used.
[2] All isolates equal to or less than 1 were regarded as S since this was the lowest concentration tested.
[3] All isolates equal to or less than 0.5 were regarded as S since this was the lowest concentration tested.
[4]The vertical line shows the break point between S and Rt. no I-sensitivity is given.
[5]The MIC value for trimethoprim tested in combination with sulfamethoxazol (1:20) is given.
NT = not tested, the titration step is not included in the VetMIC™ system.

Our data suggest that *E. coli* is the microorganism most frequently associated with fertility problems in the mare and that *β*-haemolytic streptococci are the second most frequent. Further, *E. coli* seems to be more associated with repeat breeding without clinical symptoms than with clinical symptoms of endometritis. The opposite relation seems to apply to *β*-haemolytic streptococci. This finding may be of interest for clinical considerations.

The overall dominance of *E. coli* relative to β-haemolytic streptococci in uterine swab samples was consistent in both years of sampling. The dominance of *E. coli* is in contrast to studies from other countries, where *β*-haemolytic streptococci have been the bacteria most commonly isolated. We cannot give a causal explanation to this dominance of *E. coli* in the present study. Other studies have mostly been performed in normal populations of mares (*Shin et al.* 1979, *Ricketts et al.* 1993), which may be one explanation of this difference. But also a study in barren mares *(Langoni et al.1997)*shows this dominance of β-haemolytic streptococci.

Further, it has earlier been suggested that non-haemolytic *E. coli* is a non-pathogen in the equine uterus (*Barrelet* 1995). In our study, 99 of the 104 *E. coli* isolates were non-haemolytic *E. coli*. Given the clinical history of these 99 mares, and the fact that most of the non-haemolytic isolates yielded abundant growth in pure or almost pure culture, we suggest that also non-haemolytic *E. coli* in the equine uterus may cause fertility problems.

In the present study, fungi were the third most frequent microbiological finding. Yeasts or mould were always isolated together with bacteria. Unfortunately, it was not possible to tell whether fungi or bacteria caused the primary infection. Fungal infections of the non-pregnant equine uterus were earlier said to be uncommon (*Redaelli & Codazza* 1977), but more recently fungal infections have been believed to be more frequent, possibly due to the widespread use of antibiotics and the increasingly intensive management and manipulation of reproduction in mares (*Blue* 1987, *Le Blanc*1997).

Staphylococcus aureus is reported to be a rather frequently isolated species from the equine uterus in a normal population of mares (*Rickettset al.* 1993, *Shin et al.* 1979). In our study this species was not isolated at all, only coagulase-negative Staphylococcus spp. were isolated from 3 mares. Notable in the present study is also that well-known uterine pathogens such as P. aeruginosa and K. pneumoniae were only isolated from one mare each.

The bacterial species isolated may be influenced by the stud farm management and the breeding regime used. In Sweden as well as in this study, the two dominating breeds are the Swedish Warmblood and the Standardbred Trotter. These breeds are mainly bred by AI, 71% and 88% for the Swedish Warmblood (Hästavel, 1998) and the Standardbred Trotters (STC, 1999), respectively.

Also the sampling technique influences the culture results. In the present study, most of the isolated *E. coli* and β-haemolytic streptococci yielded moderate or abundant growth of the isolated bacteria, indicating that these isolates represented an infection in the uterus rather than a vulvovestibular contaminant (*Hinrichs etal.* 1988, *Waelchli et al.* 1992). In this study, samples were transported overnight and when a time-span of this kind exists between collections and cultures the choice of transport medium is likely to influence the culture results (*Shin et al.* 1979, *Ricketts et al.* 1993).

A general global rise in antibiotic resistance has been linked to an increased use of antibiotics (*Fox* 1997, *Swartz* 1997). In stud farm practice antibiotics have long been used both prophylactically before breeding, as a treatment of endometritis (*Kenney et al.* 1975, *Shin et al.* 1979) as well as in semen extenders (*Burns et al.* 1975, *Kenney et al.* 1975). In the present study, resistance to several commonly used antimicrobials was recorded. Notably as much as 15% of the *E. coli* isolates were resistant to TMP and 4% of the isolates to gentamicin. With respect to gentamicin, *McCue et al.* (1991) report a sensitivity in only 86% of *E. coli* isolates from equine endometrial swabs collected in the US. Our corresponding figure is 96%. This difference might be due to differences in how often the drug is used. The distribution of MIC-values for resistant *E. coli* isolates is rather consistent when comparing isolates from different species and organs (*SVARM* 2001). All β-haemolytic streptococcal isolates were uniformly sensitive to 6 of 11 tested antibiotics, which is in accordance with *Shin et al.* (1979). As expected, all streptococcal isolates were sensitive to β-lactam antibiotics.

5. CONCLUSION

The key findings from this study of mares with a history of fertility problems were firstly: *E. coli* was the overall most frequently isolated bacterial species, while uterine pathogens such as P. aeruginosa and K. pneumonia were rare. Secondly: β-haemolytic streptococci were more frequently associated with clinical endometritis than with repeat breeding, whereas the opposite applied to *E. coli*. Thirdly: the noticed resistance to antibiotic suggests that a proper microbiological diagnosis and antimicrobial susceptibility testing are required for successful antimicrobial therapy.

NOTES

Reprints may be obtained from: A. Albihn, Department of Disease Control and Biosecurity, National Veterinary Institute, SE-751 89 Uppsala, Sweden. E-mail: Ann.Albihn@sva.se, tel: +46 18 67 40 00, fax: +46 18 67 44 45.

ACKNOWLEDGEMENTS

The authors are grateful to Eva Tysén and Roland Mattson for excellent technical assistance and to Drs. Anders Engvall, Kerstin Darenius and Anders Gunnarsson for valuable comments on the manuscript.

REFERENCES

1. Amies CR: A modified formula for the preparation of Stuart's transport medium. Can J Publ Health. 1967, 58: 296-300.
2. Asbury AC: Endometritis in the mare. Current Therapy in Theriogenology. Edited by: Morrow DA. 1986, WB Sounders, Philadelphia, USA, 718-722.
3. Barrelet A: Laboratory aids to routine gynaecological management. Proc Equine Stud Medicine and AI Course, British Equine Vet Assoc. 1995, New-market, UK, 52-56.
4. Blue MG: Mycotic endometritis in mares. Review and clinical observations. N Z vet J. 1987, 35: 181-183.
5. Burns SJ, Simpson RB, Snell JR: Control of microflora in stallion semen with a semen extender. J Reprod Fertil Suppl. 1975, 23: 139-142.
6. Engvall A: Survival of contagious equine metritis organisms (CEMO) in different transport media as influenced by storage time, temperature and contaminating flora. Zbl Vet Med B. 1985, 32: 454-459.
7. FASS VET: Swedish list of approved veterinary drugs, Drug information service of pharmaceutical companies in Sweden, Kungsbacka, Sweden. (Läkemedel för veterinärmedicinskt bruk.). 2001
8. Fox J: Antibiotic resistance on the rise globally. Am Soc Microbiol News. 1997, 63: 665-
9. Franklin A: Current status of antibiotic resistance in animal production. Acta vet scand. 1999, 92 (Suppl): 23-28.
10. Hinrichs K, Cummings MR, Sertich PL, Kenney RM: Clinical significance of aerobic bacterial flora of the uterus, vagina, vestibule and clitorial fossa of clinically normal mares. JAVMA. 1998, 193: 72-75.
11. Horse breeding: 1998, Swedish Horse Breeders Association, Uppsala, Sweden, (Hästavel.), 252-
12. Kenney RM, Bergman RV, Cooper WL, Morse GW: Minimal contamination techniques for breeding mares: techniques and preliminary findings. Proc Am Assoc Equine Pract. 1975, 327-336.
13. Langoni H, Alvarenga MA, Papa FO, Sakamoto C, Baldini S, Listoni FJP: Aerobic, microaerobic and anaerobic bacteria in equine endometritis. Pferdeheilkunde. 1997, 13: 558-
14. LeBlanc MM: The equine endometrium and the pathophysiology of endometritis. Proc Reprod Pathol. 1997, 78-84.
15. McCue PM, Hughes JP, Jang SS, Biberstein EL: Antimicrobial susceptibility patterns for equine endometrial isolates. California Veterinarian. 1991, 45: 23-26.
16. McGinnis MR: Yeast Identification. Laboratory Handbook of Medical Mycology. 1980, Academic Press, Inc Ltd, London. GB, 337-411.
17. NCCLS: Methods for dilution antimicrobial susceptibility test for bacteria that grow aerobically: approved standard, document M7-A4, Table 3. 1997, National Committee of Clinical Laboratory Standards, Villanova, Pennsylvania, USA
18. NCCLS: Performance standards for antimicrobial disk and dilution susceptibility test for bacteria isolated from animals. Approved standard. M31-A. 1999, National Committee for Clinical Laboratory Standards, Wayne, USA
19. Perkins NR: Equine reproductive pharmacology. Vet Clin North Am Equine Pract. 1999, 15: 687-704.
20. Quinn PJ, Carter ME, Markey B, Carter GR: Clinical Veterinary Microbiology. 1994, Wolfe Publishing, Mosby-Year Book Europe Limited, London. GB, 118-292.
21. Redaelli G, Codazza D: The incidence, pathogenicity and pathology of bacterial and fungal species in the mare's uterus. Folia Veterinaria Latina. 1977, 8: 198-204.
22. Ricketts SW: The barren mare. Diagnosis, prognosis, prophylaxis and treatment for genital abnormality. Part 1. In practice. 1989, 11: 119-125.
23. Ricketts SW, Mackintosh ME: Role of anaerobic bacteria in equine endometritis. J Reprod Fertil. 1987, 35 (Suppl): 343-351.

24. Ricketts SW, Young A, Medici EB: Uterine and clitorial cultures. Equine Reproduction. Edited by: McKinnon AO, Voss JL. 1993, Lea and Febinger, Philadelphia, USA, 234-245.
25. SAS Institute Inc: SAS Procedures Guide, Version 6. 1990, Cary NC: SAS Institute Inc., USA, 325-364. Third
26. Shin SJ, Lein DH, Aronson AL, Nusbaum SR: The bacteriological culture of equine uterine contents, in-vitro sensitivity of organisms isolated and interpretation. J Reprod Fert. 1979, 27 (Suppl): 307-315.
27. STC: Swedish Trotting Association, Section for registration. 1999, Stockholm, Sweden, (Svenska Travsportens Centralförbund)
28. Sternberg S: Antimicrobial resistance in bacteria from pets and horses. Acta vet scand Suppl. 1999, 92: 37-50.
29. SVARM: Swedish Veterinary Antimicrobial Resistance Monitoring. 2001, National Veterinary Institute, Uppsala, Sweden
30. Swartz MN: Use of antimicrobial agents and drug resistance. N Engl J Med. 1997, 337: 491-492. 10.1056/NEJM199708143370709.
31. Waelchli RO, Corboz L, Doebeli M: Streptomycin-resistent Escherichia coli as a marker of vulvovestibular contamination of endometrial culture swabs in the mare. Can J vet Res. 1992, 56: 308-12.

CHAPTER 4

Canine Herpesvirus Seroprevalence and Associated Factors in Dogs of Mexico

Edgar Guillermo Valdivia Lara[1], Jesica Ileana Ángeles Solis[1], Cesar Cuenca Verde[1], Juan Antonio Montaraz Crespo[2], Laura Cobos Marín[3], Juan Carlos Del Río García[4], Guillermo Valdivia Anda[1*]

[1]*Laboratorio de Patogenicidad Microbiana, Unidad de Investigación Multidisciplinaria Facultad de Estudios Superiores Cuautitlán, Universidad Nacional Autónoma de México, Cuautitlan Izcalli, México*

[2]*Laboratorio de Inmunología, Facultad de Estudios Superiores Cuautitlán, Universidad Nacional Autónoma de México, Ciudad de México, México*

[3]*Facultad de Medicina Veterinaria, Universidad Nacional Autónoma de México Ciudad Universitaria, Ciudad de México, México*

[4]*Laboratorio de Micotoxinas, Unidad de Investigación Multidisciplinaria, Facultad de Estudios Superiores Cuautitlán, Universidad Nacional Autónoma de México, Ciudad de México, México*

ABSTRACT

Canine herpesvirus (CHV-1) causes disease associated with high mortality in infected puppies, which represents large financial losses for dog breeders. Since CHV-1 at the time of the study he had not been reported in Mexico, the main objective of this study was to determine the prevalence of antibodies against CHV-1 in canine kennels in the metropolitan area of Mexico City. A commercial enzyme-linked immunosorbent assay (ELISA) was used, and the results were compared to those of a viral neutralization test. The ELISA kit uses the complete viral particle as the antigen. The plaque reduction neutralization test was combined with the immunoperoxidase technique because of the low cytopathic effect of CHV-1. Neutralizing antibodies were also detected in 20 randomly selected samples. The prevalence of CHV-1 with ELISA was 87%. The concordance between ELISA and serum neutralization (SN) was 0.1129, the sensitivity of the ELISA against SN was 1.0 (100%), the positive predictive value was 0.39 (39%), and the negative predictive value was 1 (100%). These results show that ELISA is useful for monitoring the dog population for CHV-1; a positive test result requires confirmation with an SN test, and a negative ELISA result indicates a high probability of being SN-negative. The only variables that were statistically associated with CHV-1 prevalence were breed and kennel. A statistically significant relationship between the degree of ELISA and SN titer was obtained, with a confidence level of 95%. None of the clinical presentation factors was statistically significant. These results suggest that most of the canine population studied in Mexico is in a herpesvirus latency state.

Keywords: Canine Herpesvirus Seroprevalence, Enzyme-Linked Immunosorbent Assay, Serum Neutralization Test

1. INTRODUCTION

There are currently only one reports of herpesvirus disease in dogs in Mexico, and the disease is not found on any of the lists issued by the International Organization for Animal Health (OIE). However, canine herpesvirus type I (CHV-1) has been isolated in various countries worldwide, and recent studies in Europe suggest that the disease is present in the canine population [1] - [4]. CHV-1 is associated with high mortality in infected puppies, which represents large financial losses for dog breeders. Because of the lack of evidence of its prevalence in Mexico, there are currently no available options for diagnosis, treatment, or prophylaxis of CHV-1 infection. It is considered that the virus is poorly immunogenic, and generates neutralizing antibodies that disappear within only a few months after infection [5].

Recent studies have estimated the worldwide population seroprevalence of CHV-1 at 30% - 40%, varying according to the country that reports, with a higher prevalence in canine kennels, considering the size of the farm, hygiene, and kennel cough as potential risk factors. A relationship between seroprevalence, history of abortions, and neonatal mortality has also been established [4] [6]. These results have all been based on detection of immunoglobulin G (IgG) antibodies by enzyme-linked immunosorbent assay (ELISA). Furthermore, inhibition of lytic plaques and reduction in fluorescent spots have been employed in the technical serum neutralization (SN) technique has been mainly used for detection of canine herpesvirus and varicella-zoster virus (VZV, belonging to the same viral genus) antibodies [7] - [11]. Canine herpesvirus, like other members of the genus Varicellovirus, usually remains strongly attached to the cell, resulting in very weak cytolytic effects. Two passages of the viral isolates previously obtained in the laboratory [12] were conducted in the chicken embryo by inoculation at the chorioallantoic membrane. This is a widely used technique for replication of other herpesvirus species such as herpes simplex virus (HSV)-1, HSV-2, HSV-3, VZV, laryngotracheitis virus, and pseudorabies [13]. However, to date, there have been no reports of canine herpesvirus replication in the chicken embryo.

The objectives of the present study were to determine the prevalence by the detection of specific IgG antibodies against CHV-1 in canine kennel breeding centers of the metropolitan area of Mexico City, using a commercial ELISA test, to compare the results of ELISA with those of the viral neutralization test, and to determine factors related to the medical history of the animals with the presence of antibodies.

2. MATERIALS AND METHODS

A serological survey was performed in 8 kennels distributed 1 in Mexico City, 3 in the State of Mexico, 2 Queretaro and 2 in the city of Guadalajara Jalisco in a period from August 2014 to February 2016. To calculate the required sample size, we adopted previously reported statistical methods [12], the sample size calculation is done considering that the prevalence in the country is unknown and that the objective was to compare two diagnostic tests [13]. A sample size of 96.04 individuals was obtained through this method, and therefore we decided to use 100 dogs. Inclusion criteria were: canines inhabiting the kennel, over 1 year and under 7 years of age, and clinically healthy at the time of sampling, were excluded animals showing any clinical disease, animals less than 1 year and animals in which the medical history of the last 3 months was unaware. Three to five milliliters of Blood was collected without anticoagulant, using Vacutainer™ tubes, from the saphenous or jugular vein in and centrifuged at 1200 ×*g*, for 10 min to room temperature in a clinical centrifuge to obtain serum. The serum was dispensed into Eppendorf tubes and frozen at −70°C until use. The number of animals that were allowed to sample per kennel was: Mexico City (8), 3 in the State of Mexico (13, 32 and 3), Queretaro (11 and 7) and in the city of Guadalajara Jalisco (22 and 4). The Kennels and breeding distribution they are shown in Table 1. The kennels were selected from the database of customers that counts to particular veterinary diagnostic laboratory, DIVeT™, in which there was a history of possible illness, owners were asked for their participation and animals that met the inclusion criteria were sampled randomly. All sampled kennels are breeders breeding pure breeds and often are carried to dog shows on the country.

A survey, questionnaire, was performed during sampling in which the following data were collected: kennel, animal, and history of diseases associated with herpesvirus infection in dogs [4] such as reproductive problems, eye secretions , skin injury, and respiratory secretions.

A commercial ELISA kit (EVL™ Woerden, the Netherlands) was used to detect the presence of antibodies against CHV-1. This kit uses the complete viral particle as the antigen. The samples were processed according to the manufacturer recommendations by preparing 1:30, 1:90, 1:270, and 1:810 dilutions of the positive and negative controls and a 1:250 dilution of the serum sample.

Table 1. Calculated titer of anti-CHV-1 (EVL™ ELISA) of animals sampled.

Sample identification	Kennel	Breed	Sex	Age (months)	Calculated titer
001	1	Pug	Female	39	*1069.4
002	1	Pug	Male	46	*295.5
003	1	Pug	Female	14	*376.1
004	1	Pug	Female	14	*854.1
005	1	Pug	Male	27	*1812.3
006	1	Pug	Male	13	*1492.2
007	1	Pug	Female	58	*462.2
008	1	Pug	Female	51	*545.0
009	1	Pug	Female	32	73.6
010	1	Pug	Female	37	-255.4
011	1	Pug	Female	24	*1000.9
012	2	Doberman	Male	13	*1120.2
013	2	Bulldog	Female	42	*760.3
014	3	German shepherd	Female	48	-204.6
015	3	German shepherd	Male	36	-252.1
016	3	German shepherd	Female	72	-483.9
017	3	German shepherd	Female	36	-382.4
018	3	German shepherd	Female	14	*166.3

019	3	Westy	Female	84	*2049.7
020	3	Westy	Female	24	*351.8
021	3	Westy	Female	29	*1864.2
022	3	Westy	Female	24	*225.9
023	3	Westy	Female	29	*2748.5
024	3	German shepherd	Female	60	−336.0
025	6	Neapolitan mastiff	Female	48	*521.8
026	6	Neapolitan mastiff	Macho	45	*1390.6
027	7	German shepherd	Female	39	*240.3
028	7	German shepherd	Female	30	*242.5
029	7	Westy	Male	72	−63.3
030	7	Westy	Female	48	*570.4
031	6	Neapolitan mastiff	Female	48	*2037.6
032	2	Great Danes	Female	18	*1494.4
033	2	Great Danes	Female	30	*2190.9
034	3	German shepherd	Female	42	*140.2
035	7	German shepherd	Male	72	*337.8
036	7	German shepherd	Male	24	*950.4
037	7	German shepherd	Female	30	*1996.7
038	4	Great Danes	Male	12	*2083.9
039	4	Great Dane	Male	12	*963.2
040	4	Great Dane	Male	13	*221.5
041	4	Great Dane	Female	12	*1366.6
042	4	Great Dane	Female	84	*1021.4
043	4	Great Dane	Female	48	*1956.0
044	4	Great Dane	Female	48	*3262.7
045	4	Great Dane	Male	60	*2718.7
046	5	Dachshund	Female	72	*1200.4
047	5	Dachshund	Female	83	*766.8
048	5	Dachshund	Female	83	*972.5
049	5	Dachshund	Male	36	*967.9
050	5	Dachshund	Female	38	*1449.2
051	5	Dachshund	Male	84	*452.9
052	5	Dachshund	Male	60	*1103.9
053	5	Dachshund	Female	13	*1306.2
054	5	Dachshund	Female	12	*1355.0

055	5	Dachshund	Female	84	*2655.9
056	5	Dachshund	Female	38	*1037.6
057	5	Dachshund	Female	83	*1052.7
058	5	Dachshund	Male	12	*1530.5
059	5	Dachshund	Male	81	*605.2
060	5	Dachshund	Female	25	*1915.3
061	5	Dachshund	Male	84	*667.9
062	5	Dachshund	Female	36	*1398.0
063	5	Dachshund	Male	82	*198.3
064	5	Dachshund	Male	48	*264.5
065	5	Dachshund	Female	84	*1670.1
066	5	Dachshund	Male	83	*2361.8
067	5	Doberman	Female	24	*869.3
068	5	Doberman	Female	25	*1253.4
069	5	Doberman	Male	80	*1283.6
070	5	Doberman	Female	12	*1176.8
071	5	Doberman	Male	13	*714.9
072	5	Doberman	Female	13	*886.7
073	5	Doberman	Male	13	*1276.6
074	5	Doberman	Male	13	*218.2
075	5	Golden Retriever	Female	36	*834.5
076	5	Doberman	Male	60	*1333.5
077	8	Dachshund	Female	36	*887.8
078	8	Dachshund	Female	12	*981.8
079	8	Dachshund	Female	12	*484.0
080	8	Dachshund	Male	14	*1347.4
081	8	Dachshund	Female	36	12.8
082	8	Dachshund	Female	12	*130.0

083	8	Dachshund	Male	36	*2575.3
084	8	Dachshund	Female	66	*1786.1
085	8	Dachshund	Female	84	*732.3
086	8	Dachshund	Female	84	*1807.0
087	8	Dachshund	Female	72	*3812.5
088	8	Golden Retriever	Male	30	*550.1
089	8	Chihuahua	Female	60	−142.8
090	8	Chihuahua	Female	84	61.5
091	8	Chihuahua	Female	16	*423.6
092	8	Chihuahua	Female	84	25.5
093	8	Dachshund	Female	12	*1580.7
094	8	Dachshund	Female	80	*125.3
095	8	Dachshund	Male	48	*651.1
096	8	Dachshund	Female	72	*880.9
097	8	Dachshund	Female	84	*1308.0
098	8	Dachshund	Male	12	*1182.6
099	5	Doberman	Female	48	*779.9
100	3	Chihuahua	Female	62	−95.2

*Positive result.

After the procedure was completed, the absorbance of the wells was read at 450 nm in a Multiskan Gen, (Thermo Scientific™) reader. Calculation of the titer was performed based on the results of the absorbance (calculated titer) of each sample, with the assumption that the positive control has a titer of 3600, as indicated by the manufacturer of the kit (EVL™ Woerden).

The data were processed using Statgraphics® software, with generalized linear models, taking the calculated titer as the dependent variable and all the following factors as independent variables: age, sex, breed, kennel, weight, height, country of birth, if the dog had copulated or not, the number of litters, and history of respiratory, eye, skin, or reproductive conditions.

Because the virus used by us at work has a very low cytolytic effect [14], we employ a combination of routine test SN and observe cytopathic effect of the virus by the detection of cell infection by an immunoperoxidase technique, using antibodies against canine herpes virus of dog and anti canine IgG subsequently antibodies conjugated to peroxidase (Sigma™) developed by diaminobenzidine (Sigma™) on wells with cells [7]. Neutralizing antibodies were tested in 20 randomly selected samples

A plaque formation test was used for viral titration. Ten-fold dilutions of the virus isolate (1:10, 1:100, 1:1000, 1:10,000) were made, and 0.5 ml of each was added to each well of a 12-well NUNC™ plate containing a monolayer of Madin-Darby canine kidney MDCK cells (In vitro™, p117) at 90% confluence in minimal essential medium (MEM) (In vitro™) with 5% newborn calf serum (NCS) (In vitro™). The cells were incubated for 1 h at 35°C to allow for viral absorption, with mixing every 15 min. Finally, MEM was added slowly with 1% newborn calf serum and 0.6% agarose (Bioline™) at 45°C to allow for solidification, and then incubated at 35°C for 3 days. A drop of formaldehyde (JTBaker™) was then added to each well so that the agarose could be subsequently removed. The wells were washed carefully with phosphate buffered saline (PBS) and fixed with acetone/methanol (ICR™) for 10 min. Subsequently, 0.5 ml of a 1:80 dilution of positive anti-CHV-1 control serum (EVL™) was added to each well and incubated for 30 min at 37°C. The wells were washed three times with PBS, and anti-canine IgG peroxidase (Sigma Aldrich™) was added and incubated for a further 30 min at 37°C. The plate was washed three times with PBS, and a 0.01% diaminobenzidine solution (Sigma Aldrich™) with hydrogen peroxide (ICR™) was added and held at 15 min at 37°C The plate was washed again with PBS and stained with Harris hematoxylin (Sigma Aldrich™) for 15 s, and the plates were examined microscopically for the presence of plaque forming units, termed peroxidase-positive spotlights (PPS), which were counted only in wells containing 20 - 100 plaques (Figure 1).

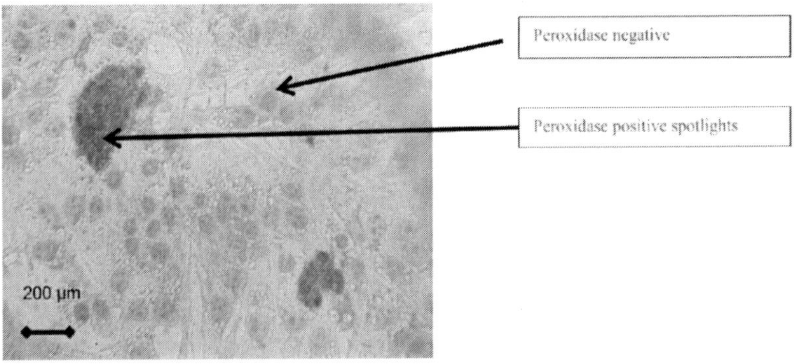

Figure 1. Example of the results taken as positive in the methodology peroxidase positive spotlights (PPS) on cell culture. MDCK cells were infected with canine herpesvirus and incubated for 72 h, and then stained with peroxidase and diaminobenzidine. Magnification, 400×.calculated by ELISA and those obtained from the SN test. For the construction of the model, titers under 8 in the SN and an ELISA result < 100 were considered negative (value of 0) according to published by Gerna, and Chambers [7], for Varicella.

The neutralization assay was performed by incubating equal amounts of virus (50 PPS) with sera at different dilutions (1:4, 1:8, 1:16, 1:32, and 1:64; pretreated for 30 min at 56°C) for 30 min at 37°C. The mixture (0.5 ml) was inoculated to NUNC™ plates containing a monolayer of MDCK cells at 90% confluence. Viral adsorption was allowed for 1 h at 35°C, while redistributing the inoculum every 15 min, and MEM supplemented with 1% calf neonate serum and 0.6% agarose at 45°C was added slowly, allowing for solidification. The plate was incubated at 35°C for three days, and then a drop of 40% formaldehyde was added to each well and the agarose layer was removed the next day. The titer was calculated as the inverse of the highest dilution in which a reduction of at least 50% of the number of plaques was observed. A serum sample was considered negative if the 1:8 dilution of serum did not show a decrease of 50% or more in plaque formation.

Simple regression analysis was used to determine the correlation between the results.

3. RESULTS AND DISCUSSION

The majority of the dogs sampled were females 68% (68/100); the average age was 35.43 months (2.95 years) the standard deviation was 25.7. Overall, 86% of the individuals were born in Mexico, and the remaining 7 were born in the USA (1), Chile (4), Argentina (1), Denmark (1), and Spain (1). Most (85%) of individuals had never left the country, and 61% of the sampled animals had mated, with 32 of the 68 females having produced at least one litter. According to the survey, 26% of the sampled individuals had presented reproductive problems, 14% had a history of infectious tracheobronchitis, 1% had eye infections, and 4% had skin diseases (unrelated to the virus cutaneous presentation). Conventional vaccination schedule was achieved in all the dogs, and none had been previously immunized against CHV-1.

Figure 2 shows the simple linear regression graph of the ELISA results, including the regression equation, slope, and intercept. According to the regression equation, the titer was calculated using the absorbance (**Table 1**). Some of the calculated titers were negative because of the equation, not the actual observation. Besides, the titles relate to some of the analyzed clinical parameters (**Table 2**).

According to the personal recommendations of Dr. Van Herwijnen (EVL™ Woerden, the Netherlands), a result was considered positive if the calculated titer exceeded 100. Of the 100 samples tested, 87 were positive.

The plaque formation test showed a viral titer of 1.72 × 10^4 PFU/ml viral culture. Viral neutralization results were performed for 20 samples from canines selected randomly (**Table 3**). Besides we related to some of the analyzed clinical parameters with ELISA and Serum neutralization test (**Table 4**).

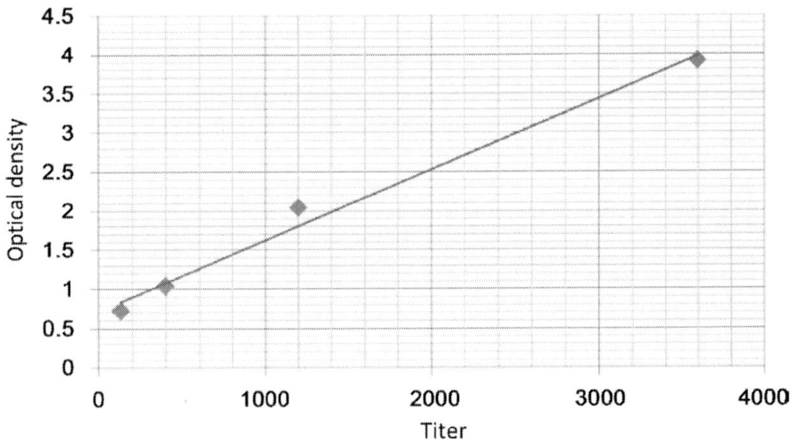

Figure 2. Relationship of the optical density obtained in the ELISA test (EVL™ Woerden, the Netherlands) with the positive control serum titer: $y = 0.0009x + 0.7173$. $R^2 = 0.9924$.

Table 2. Relationship between clinical parameters and results of the ELISA test (EVL™ ELISA).

Parameter	Variable	Elisa negative	Elisa positive
Kennel	1	2 (18.2%)	9 (81.8%)
	2	0 (0%)	4 (100%)
	3	6 (46.2%)	7 (53.8%)
	4	0 (0%)	8 (100%)
	5	0 (0%)	32 (100%)
	6	0 (0%)	3 (100%)
	7	1 (14.3%)	6 (85.7%)
	8	4 (18.2%)	18 (81.8%)
Breed	E. Bulldog	0 (0%)	1 (100%)
	Chihuahua	4 (80%)	1 (20%)
	G. Retriever	0 (0%)	1 (100%)
	Dachshund	1 (2.6%)	37 (97.4%)

	Doberman	0 (0%)	11 (100%)
	Great Dane	0 (0%)	10 (100%)
	L. Retriever	0 (0%)	1 (100%)
	Neapolitan M.	0 (0%)	3 (100%)
	German Shep.	5 (41.7%)	7 (58.3%)
	Pug	2 (18.2%)	9 (81.8%)
	West highland WT	1 (14.3%)	6 (85.7%)
Sex	Male	2 (6.3%)	30 (93.75%)
	Female	11 (16.2%)	57 (83.8%)
Age	Young (24 months old or less)	0 (0%)	30 (100%)
	Adult (more than 24 months old)	13 (18.6%)	57 (81.4 %)
Total		**13 (13%)**	**87 (87%)**

The results for Simple regression analysis are shown in the **Figure 3**. The overall concordance between the ELISA and SN results was 0.1129, which is considered to be low [15], this can be explained, since although both tests measure the levels of antibodies, the detection methods differ. The SN technique detects only neutralizing antibodies that interfere with viral infection directly, whereas the ELISA is unable to distinguish between neutralizing and non-neutralizing antibodies, and therefore has a broader antibody repertoire.

Table 3 Serum neutralization results and ELISA titers in 20 randomly selected dogs

Sample identification	Kennel	Breed	Sex	Age (months)	Calculated titer (ELISA)	Titer (serum neutralization)
002	1	Pug	Male	46	295.5	64
015	3	German shepherd	Male	36	−252.1	<8
016	3	German shepherd	Female	72	−483.9	<8
018	3	German shepherd	Female	14	166.3	<8
022	3	Westy	Female	24	225.9	<8
025	6	Neapolitan mastiff	Female	48	521.8	<8
031	6	Neapolitan mastiff	Female	48	2037.6	8
032	2	Great Dane	Female	18	1494.4	<8
039	4	Great Dane	Male	12	963.2	<8
043	4	Great Dane	Female	48	1956.0	64

045	4	Great Dane	Male	60	2718.7	32
048	5	Dachshund	Female	83	972.5	16
051	5	Dachshund	Male	84	452.9	32
063	5	Dachshund	Male	82	198.3	<8
066	5	Dachshund	Male	83	2361.8	64
071	5	Doberman	Male	13	714.9	<8
079	8	Dachshund	Female	12	484	<8
084	8	Dachshund	Female	66	1786.1	<8
091	8	Chihuahua	Female	16	423.6	<8
094	8	Dachshund	Female	80	125.3	<8

ELISA titers were calculated and the serum neutralization was scored as the reciprocal of the highest dilution that inhibited 50% plaque formation.

Table 4. Agreement between clinical parameters, the ELISA and serum neutralization test.

Parameter	Variable	Elisa test		Serum titer neutralization		Agreement between the two tests
Kennel	1	Positive	1	Positive	1	100%
		Negative	0	Negative	0	100%
	2	Positive	1	Positive	0*	0%
		Negative	0	Negative	1	0%
	3	Positive	2	Positive	0	0%
		Negative	2	Negative	4	50%
	4	Positive	3	Positive	2	66.67%
		Negative	0	Negative	1	0%
	5	Positive	5	Positive	3	60%
		Negative	0	Negative	2	0%
	6	Negative	0	Negative	1	0%
	8	Positive	4	Positive	0	0%
		Negative	0	Negative	4	0%
Breed	Chihuahua	Positive	1	Positive	0	0%
		Negative	0	Negative	1	0%
	Dachshund	Positive	7	Positive	3	42.85%
		Negative	0	Negative	4	0%
	Doberman	Positive	1	Positive	0	0%
		Negative	0	Negative	1	0%
	Great Dane	Positive	4	Positive	2	50%
		Negative	0	Negative	2	0%

	Neapolitan M.	Positive	2	Positive	1	50%
		Negative	0	Negative	1	0%
	German Shep.	Positive	1	Positive	0	0%
		Negative	2	Negative	3	66.67%
	Pug	Positive	1	Positive	1	100%
		Negative	0	Negative	0	100%
	West Highland WT	Positive	1	Positive	0	0%
		Negative	0	Negative	1	0%
Sex	Female	Positive	11	Positive	3	27.27%
		Negative	1	Negative	9	11.11%
	Male	Positive	7	Positive	4	57.14%
		Negative	1	Negative	4	25%
Age	Young (24 months old or less)	Positive	7	Positive	0	0%
		Negative	0	Negative	7	0%
	Adult (more than 24 months old)	Positive	11	Positive	7	63.64%
		Negative	2	Negative	6	33.33%
Total		Positive	18	Positive	7	38.89%
		Negative	2	Negative	13	15.38%

*For 100 animals tested by ELISA, only serum neutralization test is performed on 20 randomly selected animals.

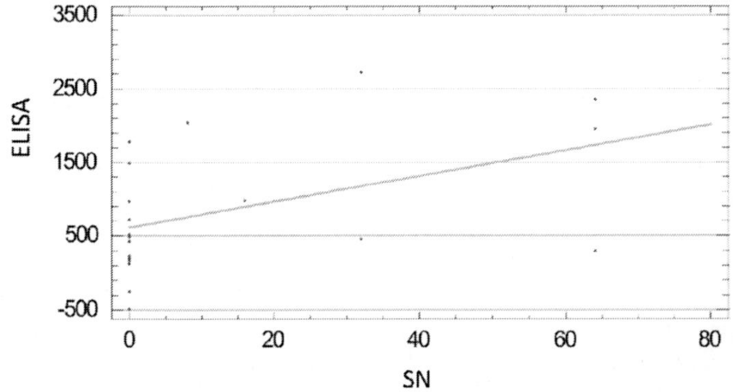

Figure 3. Linear regression of the titers obtained by enzyme-linked immunosorbent assay (ELISA) and the serum neutralization (SN) test. ELISA titers were calculated and SN results were scored as the reciprocal of the highest dilution that inhibited 50% of plaque formation. $Y = 612.6 + 17.54 \times X$; $R^2 = 21.40\%$.

In addition, the ELISA detects antibodies that cross reactive with other proteins, as the test used works with the complete virus but only detects IgG antibodies. Moreover, canine herpesvirus is only weakly immunogenic, and antibodies specific to the virus remain at high titers in the blood for only a short window [11] [16]; therefore, the SN assay might fail to detect low levels of antibodies.

The sensitivity of the ELISA test with respect to the SN result was 1 (100%). In 1989, Takumi *et al.* [9] developed the indirect ELISA technique, and detected a 26.2% positivity rate in contrast to a 5% positive rate obtained by the neutralization technique in 557 dogs. Therefore, ELISA is considered a more sensitive and practical test [17]. Such a highly sensitive diagnostic test is useful in clinical settings where the failure to diagnose a disease (i.e., false negative) would create more problems than overdiagnosis (i.e., false positives). This is the case of a screening test, which is performed by applying a diagnostic test that gives valid and reliable results that are inexpensive, easy to perform, and causes minimal discomfort to the patient.

In this study, the positive predictive value was 39%, indicating that a sample that is deemed positive by the ELISA test has only a 39% chance to be positive by the SN test. This value is relatively low, suggesting that a positive result obtained in a clinical setting or epidemiological study would have to be confirmed by the reference test (SN). Moreover, the negative predictive value was 100%, indicating that if a sample is negative on the ELISA test, it has a 100% chance to be negative in the reference test (SN). This value indicates that a negative result on ELISA can be considered a true negative with confidence.

In the general linear model, breed and kennel were the only variables statistically as-**Table 3**. Serum neutralization results and ELISA titers in 20 randomly selected dogs.

sociated with the calculated titer, with a p-value of 0.0006 and R^2 of 0.3649, indicating that 36.49% of the variability in the calculated titer could be explained by these two variables. None of the factors of the clinical presentation such as infectious tracheobronchitis, skin or eye conditions, or reproductive problems had a statistically significant relationship with the serologically calculated titer, which suggests that most of the canine population studied is in viral latency or that the clinical presentations are due to other etiologies [18].

4. CONCLUSIONS AND IMPLICATIONS

The high prevalence of CHV-1 disease in adult dogs in the central area of Mexico is highly significant when considering that in some areas, our previous research [14] showed that puppies from 1 to 3 months of age were killed by infection, and we were able to isolate strains of the virus. Although the litters belonged to different bitches from those sampled in the present study, our results confirm that canine herpesvirus is present in Mexico, causing clinical infection in newborn puppies and is present in latent form in adults, as has been found in other countries. Therefore, Mexico should adopt epidemic control tactics against the disease.

Overall, these results show that the ELISA test is useful for monitoring and screening of the canine population for CHV-1. A positive test result requires confirmation with the SN test, whereas a negative result in ELISA indicates a high probability of being SN-negative, and further confirmation is not required. Therefore, the ELISA test and SN test developed herein can be used to screen for viruses that have a very low cytolytic effect and are more highly associated with the cells by latency or integration of viral DNA into chromosome cell with or without cell transformation because the literature reports that despite being in viral latency, the body is able to form antibodies against the virus, so it is possible to detect by sensitive tests [16] [17]. This approach is expected to dramatically bring down the cost and time of diagnosis.

ACKNOWLEDGEMENTS

This work was supported by the Project PAPIIT-UNAM IT-202114 "Assessment of canine herpesvirus disease development in Mexico and possible measures for control and prevention".

REFERENCES

1. Lacheretz, A. and Cognard, S. (1998) Epidemiology and Serological Diagnosis of 10 Canine Herpesvirus. Revue de Medecine Veterinaire, 143, 1477-1488.
2. Reading, M.J. and Field, H.J. (1998) A Serological Study of Canine Herpesvirus-1 Infection in the English Dog Population. Archives of Virology, 143, 1477-1488. http://dx.doi.org/10.1007/s007050050391
3. Rijsewijk, F.A.M., Luiten, E.J., Daus, F.J., Van der Heijden, R.W. and Van Oirschot, J. (1999) Prevalence of Antibodies against Canine Herpesvirus 1 in Dogs in the Netherlands in 1997-1998. Veterinary Microbiology, 65, 1-7. http://dx.doi.org/10.1016/S0378-1135(98)00285-5
4. Ronsse, V., Verstegen, J., Onclin, K., Friederic, F. and Poulet, H. (2004) Risk Factors and Reproductive Disorders Associated with Canine Herpesvirus-1 (CHV-1). Theriogenology, 61, 619-636. http://dx.doi.org/10.1016/S0093-691X(03)00249-8
5. Greene, C.E. (2012) Canine Herpesvirus Infection. In: Greene, C.E., Ed., Infectious Diseases of the Dog and Cat, 4th Edition, Saunders, Philadelphia, 48-54.
6. Guigal, P.M., Fontbonne, A., Buff, S., Vincetti, M., Thevenet, F., Pavlowiez, S., et al. (2002) Prevalence of Antibodies against Canine Herpes Virus in French Breeding Kennels. Proceedings of the 3rd EVSSAR European Congress on Reproduction in Companion, Exotic and Laboratory Animals, Liege, 10-12 May 2002, 132; Cited in Theriogenology, 57.

7. Gerna, G. and Chambers, R. (1976) Varicella-Zoster Plaque Assay and Plaque Reduction Neutralization Test by the Immunoperoxidase Technique. Journal of Clinical Microbiology, 4, 437-442.
8. Engels, M., Mayr-Bibrack, B., Ruckstuhl, B. and Metzler, A. (1980) Seroepizootiology of Canine Herpesvirus in Switzerland and Preliminary Trials of a Vaccine. Zentralblatt für Veterinarmedizin Reihe B, 27, 257-267. http://dx.doi.org/10.1111/j.1439-0450.1980.tb01691.x
9. Takumi A., Kusanagi, K., Tuchiya, K., Xuan, X., Azetaka, M. (1989) Serodiagnosis of Canine Herpesvirus infection—Development of an Enzyme-Linked Immunosorbent Assay and Its Comparison with Two Improved Methods of Serum Neutralization Test. Japanese Journal of Veterinary Science, 52, 241-250. http://dx.doi.org/10.1292/jvms1939.52.241
10. Xuan, X., Horimoto, T., Limcumpao, J.A., Takumi, A., Tohya, Y., Takahashi, E. and Mikami, T. (1991) Neutralizing Determinants of Canine Herpesvirus as Defined by Monoclonal Antibodies. Archives of Virology, 116, 185-195. http://dx.doi.org/10.1007/BF01319241
11. Nothling, J.O., Hossy, D., Steckler, D. and Ackermann, M. (2008) Seroprevalence of Canine Herpesvirus in Breeding Kennels in the Gauteng Province of South Africa. Theriogenology, 69, 276-282. http://dx.doi.org/10.1016/j.theriogenology.2007.09.022
12. 1Tilaki, K.H. (2014) Sample Size Estiamation in Diagnostic Test Studies of Biomedical Informatics, Methodological Review. Journal of Biomedical Informatics, 48, 193-204. http://dx.doi.org/10.1016/j.jbi.2014.02.013
13. Bell, M.L., Pinto, A.T., McKenzie, J.E. and Olivier, J. (2014) A Myriad of Methods: Calculated Simple Size for Two Proportions Was Dependent on the Choice of Simple Size Formula and Software. Journal of Clinical Epidemiology, 67, 601-605. http://dx.doi.org/10.1016/j.jclinepi.2013.10.008
14. Valdivia-Lara, E.G., Barron-Romero, B.L., Cobos-Marin, L., Angeles-Solis, J.I., Gonzalez-Gallardo S., Cuenca-Verde, C., et al. (2016) Pathology Isolation and Identification of Canine Herpesvirus (CHV-1) in Mexico. Open Journal of Pathology, 6, 111-121.
15. Watson, P.E. and Petrie, A. (2010) Method Agreement Analysis: A review of Correct Methodology. Theriogenology, 73, 1167-1179. http://dx.doi.org/10.1016/ j.theriogenology.2010.01.003
16. Carmichael, L.E. (1970) Herpesvirus Canis: Aspects of Pathogenesis and Immune Response. Journal of the American Veterinary Medical Association, 156, 1714-1721.
17. Ronsse, V., Verstegen, J., Onclin, K., Guiot, A.L., Aeberle, C. and Nauwynck, H.J. (2002) Seroprevalence of Canine Herpesvirus-1 in the Belgian Dog population in 1997-1998. Reproduction in Domestic Animals, 37, 299-304. http://dx.doi.org/10.1046/j.1439-0531.2002.00363.x
18. Carmichael, L.E. and Grene, C.E. (1998) Canine Herpesvirus Infection. In: Greene, C.E., Ed., Infectious Diseases of the Dog and Cat, WB Saunders, Philadelphia, 28-32.

CHAPTER 5

Pathogenicity study in sheep using reverse-genetics-based reassortant bluetongue viruses

Cristina C. Celma[a], Bishnupriya Bhattacharya[a], Michael Eschbaumer[b,1], Kerstin Wernike[b], Martin Beer[b], Polly Roy[a,*]

[a]*Department of Pathogen Molecular Biology, Faculty of Infectious and Tropical Diseases, London School of Hygiene and Tropical Medicine, United Kingdom*

[b]*Institut fu¨r Virusdiagnostik, Friedrich-Loeffler-Institut, Greifswald-Insel Riems, Germany*

ABSTRACT

Bluetongue (BT) disease, caused by the non-enveloped bluetongue virus (BTV) belonging to the *Reoviridae* family, is an economically important disease that affects a wide range of wild and domestic ruminants. Currently, 26 different serotypes of BTV are recognized in the world, of which BTV-8 has been found to exhibit one of the most virulent manifestations of BT disease in livestock. In recent years incursions of BTV-8 in Europe have resulted in significant morbidity and mortality not only in sheep but also in cattle. The molecular and genetic basis of BTV-8 pathogenesis is not known. To understand the genetic basis of BTV-8 pathogenicity, we generated reassortant viruses by replacing the 3 most variable genes, S2, S6 and S10 of a recent isolate of BTV-8, in different combinations into the backbone of an attenuated strain of BTV-1. The growth profiles of these reassortant viruses were then analyzed in two different ovine cell lines derived from different organs, kidney and thymus. Distinct patterns for each reassortant virus in these two cell lines were observed. To determine the pathogenicity of these reassortant viruses, groups of BTV-susceptible sheep were infected with each of these viruses. The data suggested that the clinical manifestations of these two different serotypes, BTV-1 and BTV-8, were slightly distinct and BTV-1, when comprising all 3 genome segments of BTV-8, behaved differently to BTV-1. Our results also suggested that the molecular basis of BT disease is highly complex.

Keywords: Bluetongue virus, Serotype 8, Reassortment, Non-structural protein NS3, Pathogenicity in sheep

1. INTRODUCTION

Bluetongue (BT), an insect-transmitted, non-contagious viral disease of domestic and wild ruminants is caused by bluetongue virus (BTV). The disease is characterized by inflammation of the mucous membranes, congestion, swelling, hemorrhages and is often accompanied with high mortality in sheep (Erasmus, 1975; Maclachlan et al., 2009). Although cattle and goats usually carry the virus for a certain period of time without showing any apparent clinical signs of disease, they are capable of transmitting the virus to other ruminants via biting *Culicoides* midges. BTV is present in a broad band of countries extending approximately between 40°N and 35°S (Mellor et al., 2000; Purse et al., 2005). Until 15 years ago, Europe was essentially BT-free apart from Cyprus; however, since 1998 at least one of the 26 serotypes of BTV has been active on the continent every year, mainly in the Mediterranean basin (Maclachlan and Guthrie, 2010; Mellor et al., 2008; Purse et al., 2005). In 2006 a highly pathogenic BTV-8 strain emerged for the first time in Northern Europe, spreading very rapidly and affecting thousands of herds. The same serotype re-emerged in 2007 and 2008, causing devastating disease not only in sheep but also in cattle with high morbidity and mortality (Elbers et al., 2008a, 2008b; Wilson and Mellor, 2009). Studies involving molecular epidemiology have also shown that the most severe disease in northern European sheep and cattle was caused by BTV-8 (Dal Pozzo et al., 2013; Martinelle et al., 2013; Purse et al., 2005). The phenotypic differences between BTV-8 compared with less virulent strains suggested that genetic background may be partly responsible. However, the mechanism of pathogenicity is still very poorly understood.

BTV is a member of the *Orbivirus* genus within the *Reoviridae* family. Like other members of the family, BTV has a genome of 10 segmented double-stranded RNA (segments S1–S10) that are enclosed within two capsids. While the inner core is made up of 5 highly conserved proteins (VP1, VP3, VP4, VP6 and VP7), the outer capsid consists of two variable proteins, VP2 (receptor-binding protein and serotype determinant) and VP5 (membrane penetration protein). In addition, BTV also encodes for 4 non-structural proteins (NS1–NS4), of which NS3 encoded by S10 is more variable than NS1 and NS2. NS3 is shown to be involved in virus trafficking and release from the infected host (Beaton et al., 2002; Celma and Roy, 2009). Recently it has been shown that NS3 is also involved in the regulation of the induction of interferon type 1 (Chauveau et al., 2013), suggesting a role in the innate immune response.

In this study, we designed reassortant viruses between BTV-8 and BTV-1 to establish the genetic basis of BTV pathogenicity. The rationale for designing reassortant viruses was based on the two most variable proteins of the outer capsid (VP2 and VP5) and the non-structural protein NS3, which is the most variable within BTV NS proteins. Reassortant viruses were generated using a reverse genetics (RG) system replacing these three RNA segments (S2, S6 and S10) of the low virulent strain, BTV-1, with that of highly virulent BTV-8, either singly or in combinations. The phenotypic characteristics of the disease caused by these reassortant viruses were analyzed by infection of sheep. Our results suggested that all three proteins together are involved in the disease outcome and that the molecular basis of BTV pathogenicity is highly complex.

2. METHODS

2.1. Cells and viruses

BSR cells (BHK-21 subclone) were maintained in Dulbecco modified eagle medium (DMEM, Sigma–Aldrich) supplemented with 5% (v/v) fetal bovine serum (FBS, Invitrogen). PT and SFT-R cells (ovine-derived kidney and thymus cells respectively, Collection of Cell Lines in Veterinary Medicine, Friedrich-Loeffler-Institut, Insel Riems, Germany) were maintained in minimum essential medium eagle (MEM, Sigma–Aldrich) supplemented with 10% (v/v) FBS.

BTV-1 (South African strain) and BTV-8 (Ardennes isolate) viral stocks were generated by infection of BSR cells and kept at 4 °C until use.

2.2. Recovery of reassortant BTV-1/BTV-8 viruses

Segments S2 (VP2), S6 (VP5) and S10 (NS3) (GenBank accession numbers: KJ872780–KJ872782) of BTV-8 were obtained using a sequence-independent cloning system as previously described (Boyce et al., 2008; Maan et al., 2007; Matsuo et al., 2011). Briefly, dsRNAs from purified core particles were ligated to a self-annealing primer before RT-PCR amplification using a specific primer. Each cDNA amplified from segments S2, S6 and S10 of BTV-8 was cloned into pUC19 and fully sequenced (Source Bioscience) before insertion of the T7 promoter at the 5′ end and insertion of a unique restriction enzyme site that generates the correct end of the segment at the 3′ end.

For synthesis of uncapped T7 transcripts for segments S2, S6 and S10, RiboMAX Large-Scale RNA Production System T7 (Promega) or TranscriptAid T7 High Yield Transcription Kit (Thermo Scientific) kits were used according to manufacturer's instructions. Reassortant viruses between BTV-1 and BTV-8 were recovered from confluent monolayers of BSR cells after transfection with a full set of BTV T7 transcripts as described before (Boyce et al., 2008). Individual plaques were picked, amplified and virus stocks were kept at 4 °C.

2.3. Virus growth kinetics and characterization

The genomic dsRNA profile was analyzed for each reassortant. Monolayers of BSR cells were infected with reassortant or parental viruses and upon complete cytopathic effect, infected cells were harvested. The genomic dsRNA was purified with Tri reagent (Sigma) using standard methods. The parental origin of segments S2, S6 and S10 was determined by differential mobility in non-denaturing PAGE or by sequencing.

For virus growth study, monolayers of PT and SFT-R cells were synchronously infected at a multiplicity of infection (MOI) of 1 and samples were collected at 0, 12, 24 and 48 h (hours) post-infection (p.i.). Cells and supernatant were harvested, subjected to two freeze/thaw cycles and the total titer was determined by plaque assay in triplicate and expressed as plaque formation units per ml (PFU/ml) or by tissue culture infective dose 50 ($TCID_{50}$). The mean, standard deviation and the *p* values were also determined by Excel (Microsoft). Viral protein expression was determined by Western blot (WB) using specific antibodies against structural VP5, VP7 and non-structural NS3 proteins. As loading control, an antibody against β-actin (Sigma) was used. Each blotting experiment was repeated twice and the amount of protein expression was quantitated by ImageJ software.

2.4. Pathogenicity studies

Forty-six sheep (East Frisian breed) were segregated in groups of 4 animals and subcutaneously injected with parental BTV-1 (South African strain), BTV-8 (Ardennes isolate) or one of the reassortant viruses at $\sim 1 \times 10^7$ PFU/animal. Control groups received only saline buffer. Whole blood samples were taken at regular intervals before and after infection. For three weeks, a clinical score was assigned daily to each sheep based on a range of clinical signs. These included body temperature, feed intake, attitude, and lameness, appearance of conjunctiva, scleral blood vessels and cornea, appearance of skin and mucous membranes of the nose and mouth, redness and hemorrhaging, nasal and ocular discharge, salivation, respiratory rate and sounds, as well as edema of the tongue, face, ears, periorbital and submandibular regions. Depending on severity, up to four points were awarded per category. All animal procedures were in compliance with European ethical regulations, i.e. directives 91/628, 92/65 and 86/609/EEG, regarding the protection of vertebrate animals used for experimental and/or scientific purposes.

A mixed-design analysis of the variance model (implemented in R, http://www.r-project.org/) was used to test for significant differences in clinical scores between groups. *P* values from post hoc pairwise comparisons were adjusted for multiplicity by the Bonferroni method.

2.5. Viral load

To determine virus replication in animal hosts, RNA was extracted from blood samples at different time points with the NucleoSpin 96 virus core kit (Macherey-Nagel) and analyzed by real-time RT-PCR as described previously (Toussaint et al., 2007) using a LightCycler 480II (Roche Diagnostics). A NS1 (S5) RNA standard was used for absolute quantification.

2.6. Immunological tests

The presence of antibodies against BTV VP7 was measured by ELISA. Serum samples were analyzed by a double-recognition BTV antibody ELISA (PrioCHECK BTV DR, Prionics GmbH). For neutralization tests, serum samples were tested against parental BTV-1 and BTV-8 viruses and the neutralization titer was determined as the highest dilution able to neutralize 100 $TCID_{50}$ of virus in 50% of replicate wells.

3. RESULTS

3.1. Recovery and characterization of reassortant BTV-1 and BTV-8 viruses

Initially BTV-8 (Ardennes isolate) (Le Gal et al., 2008) exact-copy segments S2, S6 and S10, encoding VP2, VP5 and NS3 respectively, were cloned using a sequence-independent method (Boyce et al., 2008; Maan et al., 2007). Subsequently, after ensuring the accuracy of the clones by sequencing, a T7 promoter and a specific restriction enzyme site were introduced by PCR and transcripts were generated using the fully digested plasmids as templates. For virus recovery, BSR cells were transfected with ssRNAs S2, S6 and S10 of BTV-8 in different combinations together with the remaining ssRNAs (either 7 or 8 segments) of BTV-1 (Boyce et al., 2008). Following incubation, reassortant viruses were recovered as described previously (Boyce et al., 2008) and assessment of the plaque morphology established that the recovered BTV1/8VP2.5 (with segments S2 and S6 of BTV-8) and BTV1/8VP2.5.NS3 (with segments S2, S6 and S10 of BTV-8) reassortant viruses had plaque formation similar to the parental viruses (data not shown).

Purification of the genomic dsRNAs from cells infected with independent plaques of BTV1/8VP2.5 and BTV1/8VP2.5.NS3 and their subsequent analysis on a non-denaturing polyacrylamide gel confirmed that the recovered reassortant viruses had the expected dsRNA genome patterns that were equivalent to the parental origin of the segments (Fig. 1A, left panel). In addition, RT-PCR and sequencing also validated the authenticity of the segments that could not be assessed by their mobility in non-denaturing polyacrylamide gels.

Since pathogenicity of BT disease is more pronounced in sheep, ovine cells were evaluated for growth kinetics and protein expression at different times p.i. For this purpose, first an ovine cell line (PT) derived from kidney was infected at MOI of 1 with either wild-type BTV-1 or BTV-8 or reassortant virus BTV-1/8VP2.5 or BTV-1/8VP2.5.NS3. Virus growth kinetics was monitored by plaque assay (Fig. 1A, right panel) of samples taken at 0, 12, 24 and 48 h p.i. Although both reassortant viruses BTV1/8VP2.5 and BTV1/8VP2.5.NS3 consisted of a BTV-1 backbone, the growth characteristics of only the triple reassortant were different to that of BTV-1. Reassortant virus BTV1/8VP2.5 exhibited a similar profile to BTV-1, with only a slight difference at 12 h ($p < 0.05$) but not at ($p > 0.05$) either 24 or 48 h p.i. In contrast, the triple reassortant BTV-1/8VP2.5.NS3 virus titer at 12 h p.i. was significantly different from BTV-1 ($p < 0.05$) but not from BTV-8 at the same time point ($p > 0.05$).

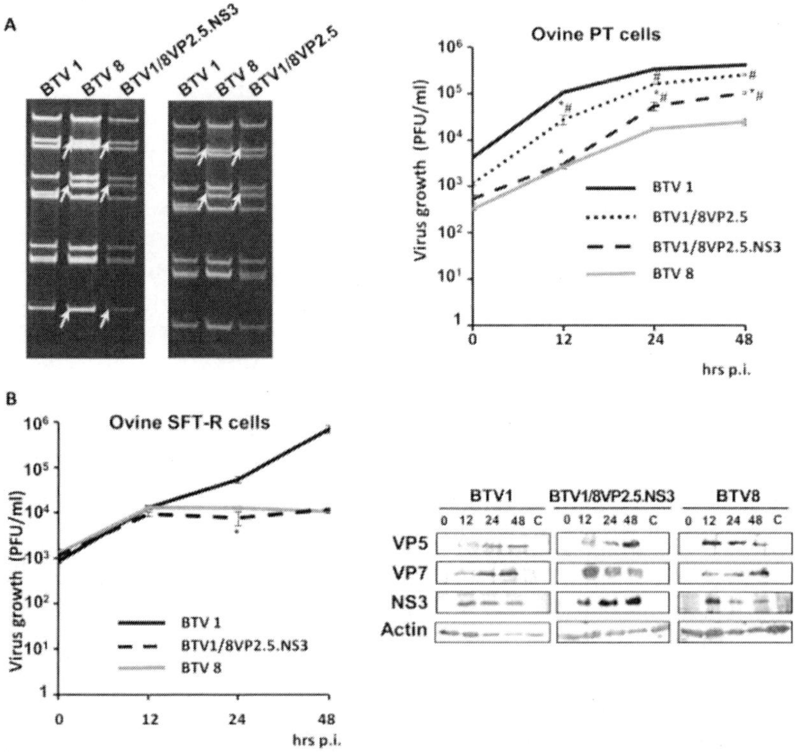

Figure. 1. Characterization of reassortant BTV-1/BTV-8 viruses. (A) Left panel, genomic profile of reassortant viruses generated by RG. DsRNA was extracted from cells infected with either BTV-1, BTV-8, BTV1/8VP2.5.NS3 or BTV1/8VP2.5 and analyzed in a non-denaturing polyacrylamide gel. Arrows indicate reassortant segments S2, S6 or S10 of BTV-8 in each generated virus. Right panel, virus growth profile in the ovine kidney-derived PT cell line. Cells were infected with parental BTV-1 or BTV-8, or reassortant BTV1/8VP2.5.NS3 or BTV1/8VP2.5 and samples were harvested at the indicated times. Titers at each time point were determined in triplicate by plaque assay, expressed as virus growth in plaque formation units per ml (PFU/ml) and plotted in logarithmic scale. Error bars indicate standard deviation. Asterisk and hashtag indicates the significance ($p < 0.05$) of the difference in titers of BTV1/8VP2.5.NS3 or BTV1/8VP2.5 at 12, 24 and 48 h p.i. to WT BTV-1 (*) or BTV-8 (#). (B) Characterization of BTV1/8VP2.5.NS3 in ovine thymus-derived SFT-R cells. Left panel, virus growth. Cells were infected with BTV1/8VP2.5.NS3 or parental BTV-1 or BTV-8 and processed as indicated in (A). The *p* values were designated the same way as in (A). Right panel, protein expression profile from cells infected with BTV1/8VP2.5.NS3 at the indicated time points. Viral proteins were detected by Western blot using specific antibodies against VP5, VP7 or NS3. As loading control, an antibody against cellular actin was used. Protein profile from cells infected with BTV-1 and BTV-8 was used as controls.

However, growth of this virus at later times (24 and 48 h p.i.) exhibited a somewhat different ($p < 0.05$) profile in comparison to both parental viruses, BTV-1 and BTV-8, albeit closer to BTV-8.

Since triple reassortant virus BTV1/8VP2.5.NS3 appeared to have a different growth profile to that of BTV-1, the growth patterns of these two viruses were further analyzed using an alternate ovine cell line, ovine thymus SFT-R cells. As VP2, VP5 and NS3 in the triple reassortant virus BTV1/8VP2.5.NS3 were taken from BTV-8, the latter was also included as a control. Analysis of the growth profiles of the three viruses in SFT-R cells (Fig. 1B, left panel) exhibited that the virus titers of the reassortant virus, BTV-1 and BTV-8 at 12 h p.i. were similar and there was no statistically significant difference ($p > 0.05$) between them. Although virus titers of reassortant and BTV-8 plateaued by 24 h p.i., the only virus that still continued to grow at the later time points was the BTV-1. Further, Western blot analysis of viral protein expression, using antibodies for NS3, VP5 and a major structural protein VP7 (Fig. 1B, right panel), revealed that all three viruses expressed these proteins as early as 12 h p.i. Since the virus titers of the triple reassortant virus, BTV-1 and BTV-8 were not statistically significant from each other at 12 h p.i., the protein expression for all three viruses was measured at 24 and 48 h p.i. Further, densitometric analysis of protein expression at each time point indicated that the increase in expression of VP7 and VP5, but not NS3, was similar to BTV-8; while NS3 was more similar to BTV-1 (results not shown). Altogether, this suggested that exchanging S2, S6 and S10 segments of BTV-1 with that of BTV-8 resulted in a reassortant virus that was more similar to BTV-8 than BTV-1. In addition, the effect of the reassortment was more pronounced in the thymus-derived (SFT-R) than in the kidney (PT) cells.

3.2. Pathogenicity in sheep infected with BTV reassortant viruses

Since in tissue culture the triple reassortant virus BTV1/8VP2.5.NS3 behaved more similarly to BTV-8 and BTV-1/8VP2.5 to BTV-1, experiments were designed to analyze the pathogenicity of these reassortant viruses in animal hosts. For this purpose, four groups containing four sheep in each group were infected with BTV-1 or BTV-8 or BTV1/8VP2.5 or BTV1/8VP2.5.NS3. Two animals injected with saline buffer were used as the control group. The pathogenicity of each virus infection was analyzed by clinical observation of the animals in all groups. A clinical score was assigned daily to each sheep based on a range of clinical symptoms as described in Section 2. Depending on severity, up to four points were awarded per category. Animals infected with the reassortant

BTV1/8VP2.5.NS3 presented a slightly early onset of the highest clinical score compared to BTV1/8VP2.5 or BTV-1 (Fig. 2A).

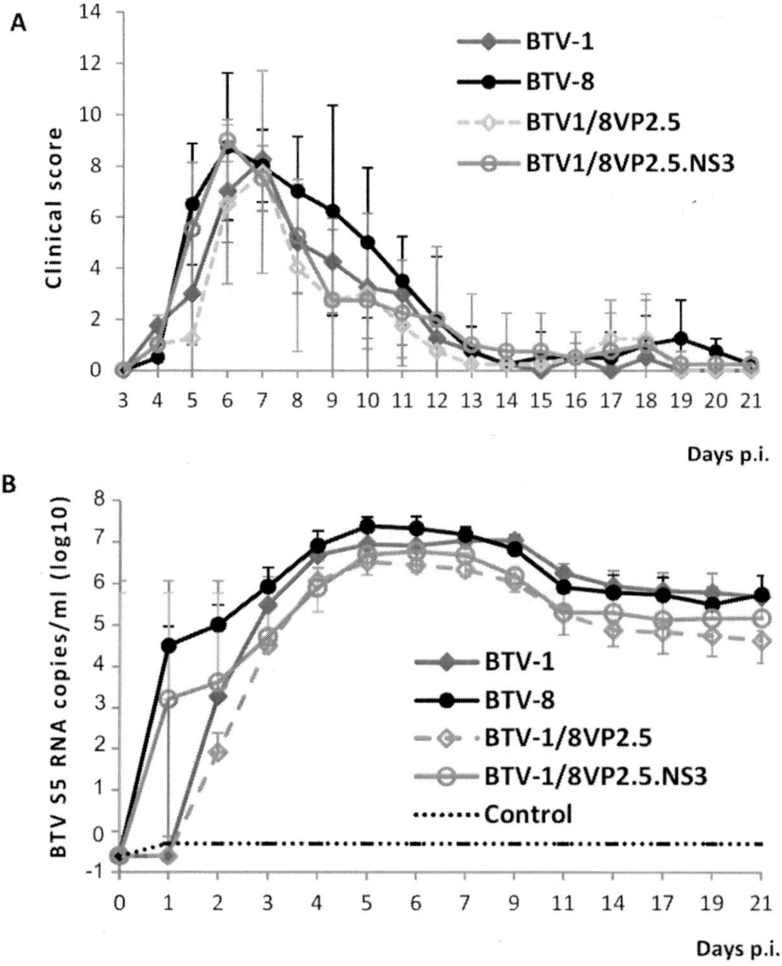

Figure. 2. Clinical manifestation and virus replication in sheep infected with reassortant BTV1/BTV-8 viruses. (A) Animals were inoculated with reassortants BTV1/8VP2.5.NS3, BTV1/8VP2.5 or parental BTV-1 or BTV-8. Infected animals were examined daily. Attitude, body temperature and other clinical signs were recorded and scored as outlined in the text. Means and standard deviation were calculated for each group. (B) Animals were inoculated with virus as above and blood samples for detection of NS1 gene by real-time RT-PCR were taken at the indicated time points. As control, a group of animals injected with saline buffer was included in the study.

As the same pattern was also observed for BTV-8 with a maximum clinical score at 6 days p.i., this data suggested that the NS3 might influence virus pathogenicity and might play an important role in the onset of BT disease. Although the clinical scores in all four infection groups were different from the control group, the observed differences between the infection groups were not statistically significant.

Subsequently, viremia in these infected animals was estimated by analyzing the whole blood samples that were taken before virus inoculation and on days 1 to 7, 9, 11, 14, 17, 19, and 21 p.i. using real-time RT-PCR (Toussaint et al., 2007). Animals infected with the reassortant viruses showed an increase in the number of NS1 gene copies over the first week after infection (Fig. 2B) similar to BTV-1 and BTV-8 viruses, indicating that these viruses can replicate as efficiently as parental viruses in an animal host. In comparison, samples from control sheep were free of BTV RNA. (Fig. 2B, control). While productive infection measured by detecting the presence of antibodies against VP7, the group-specific antigen, by ELISA, also demonstrated that all animals in the infection groups had seroconverted by day 7 after inoculation, no BTV antibody response was detected in any sample from the negative control sheep (data not shown). In addition, the serum neutralization response tested in samples taken at 21 days p.i. confirmed that animals infected with the reassortant viruses BTV1/8VP2.5.NS3 and BTV1/8VP2.5 showed a strong neutralization response against BTV-8 but not BTV-1 (Table 1). In comparison, animals infected with BTV-1 presented a response only against the same serotype. Overall these data indicated that the reassortant viruses are able to replicate in animals and they also generate a strong immune response similar to that of the parental strains. Since the sheep infected with the reassortant BTV1/8VP2.5.NS3 showed a clinical score that was comparable to that of BTV-8, it was hypothesized that NS3 might be playing an important role in the pathogenicity of BT disease.

3.3. Importance of NS3 in bluetongue disease in sheep

To further determine the importance of NS3 in pathogenicity, we generated a reassortant virus in which S10 of BTV-1 was exchanged with S10 of BTV-8, BTV1/8NS3 by RG system (Fig. 3A). In addition, two other reassortants BTV1/8VP2.NS3 and BTV1/8VP5.NS3 were generated to determine the contribution of each segment in the context of NS3. Further, a sequence comparison between the segment S10 of BTV-1 and BTV-8 showed that only 12 residues in the NS3 sequence are different between these two serotypes, of which 5 are located at the putative extracellular domain of NS3.

Table 1. Serum neutralization titer in animals infected with BTV1/8VP2.5.NS3, BTV1/8VP2.5 or parental BTV-1 or BTV-8.

Infection group	**BTV-1 (dilution)**[a]	**BTV-8 (dilution)**[a]
BTV-1 (4 animals)	1/640	Negative
	1/240	Negative
	1/240	Negative
	1/480	Negative
BTV-8 (4 animals)	Negative	1/480
	Negative	1/240
	Negative	1/240
	Negative	1/320
BTV1/8VP2.5 (4 animals)	Negative	1/240
	Negative	1/240
	Negative	1/120
	Negative	1/120
BTV1/8VP2.5.NS3 (4 animals)	Negative	1/480
	Negative	1/320
	Negative	1/480
	Negative	1/120
PBS (2 animals)	Negative	Negative
	Negative	Negative

Negative indicates no detectable neutralizing activity in a 1/10 dilution.

[a]Neutralizing activity in serum samples from sheep 21 days post-infection with reassortant or parental virus was determined as the highest dilution that neutralized 100 $TCID_{50}$ virus in 50% of replicate wells.

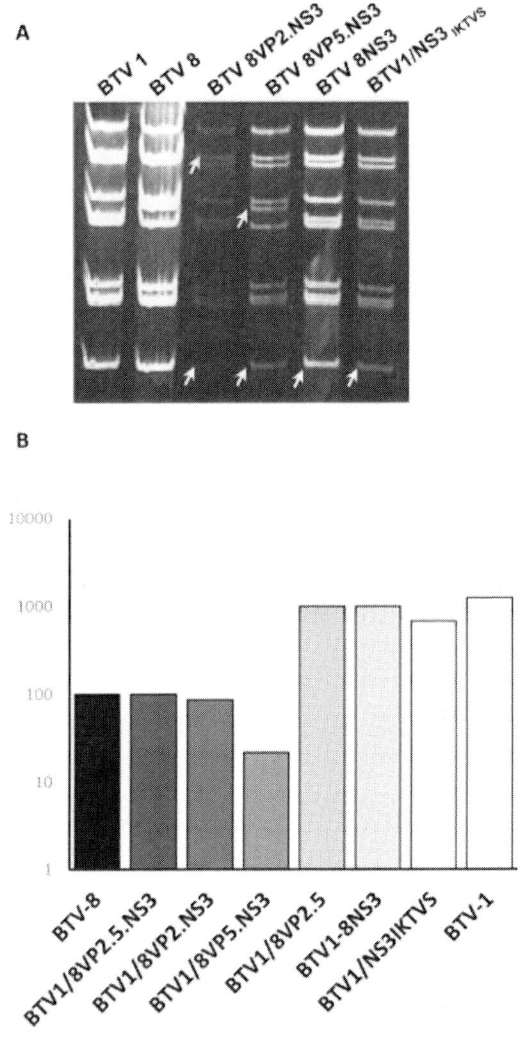

Figure. 3. Characterization of reassortants BTV1/8VP2.NS3, BTV1/8VP5.NS3, BTV1/8NS3 or mutant BTV1/NS3$_{IKTVS}$ viruses. (A) Genomic profile of reassortant viruses generated by reverse genetics. Genomic dsRNA was extracted from cells infected with either reassortant or parental viruses as indicated in each lane and analyzed in a non-denaturing polyacrylamide gel. (B) Virus growth in SFT-R cell line. Cells were infected with parental BTV-1 or BTV-8 or triple reassortant BTV1/8VP2.5.NS3 or dual reassortant BTV1/8VP2.NS3, BTV1/8VP5.NS3 and BTV1/8VP2.5 or single BTV1/8NS3 and BTV1/NS3$_{IKTVS}$ viruses and samples were harvested at 48 h p.i. Total titers were determined by TCID$_{50}$ and expressed as percentage of BTV-8 titer considered as 100%.

To determine if these differences play any role in pathogenicity, we introduced the substitutions $L_{149}I/S_{153}K/A_{154}T/I_{156}V/Q_{158}S$ in the BTV-1 NS3 sequence and generated a mutant virus BTV1/NS3$_{IKTVS}$ that expressed an NS3 protein containing an extracellular domain similar to that of BTV-8. Each reassortant virus was grown in BSR cells and stocks were stored at 4 °C. Although all viruses yielded titers between 1.5×10^7 and 4.6×10^7 PFU/ml, BTV1/8VP5.NS3 yielded consistently lower titers ($\sim 1 \times 10^6$) than the other reassortant viruses.

Further, the growth of each virus was examined in ovine SFT-R cells. Interestingly, both BTV1/8NS3 and BTV1/NS3$_{IKTVS}$ single reassortant viruses behaved like BTV-1, not BTV-8. The other two dual reassortant viruses, BTV1/8VP2.NS3 and BTV1/8VP5.NS3, behaved as an intermediate between two parental viruses as shown in Fig. 3B, suggesting that NS3 might be acting in concert with each of the outer capsid proteins.

For pathogenicity studies, animals were divided into six infection groups and a control group with four sheep in each group. Sheep in the infection groups were inoculated with BTV-1 or BTV-8 or one of four different reassortant BTV-1/BTV-8 viruses. Animals were clinically scored daily and blood samples were taken before and at days 1–11, 14, 17, and 21 days after infection. All the generated viruses (reassortants and mutant) were able to replicate in sheep, as real-time RT-PCR demonstrated an increasing amount of S6 RNA in blood samples over time (data not shown). When the animals were clinically scored, similar to the parental BTV-1, a delay in pathogenicity was observed for the virus BTV1/8VP2.NS3 with a maximum score at 8 days p.i. (Fig. 3B). The reassortants BTV1/8VP5.NS3, BTV1/8NS3, mutant BTV1/NS3I$_{KTVS}$ and BTV-8 reached a maximum score at seven days p.i. (Fig. 4A and B). The clinical scores in all infection groups were significantly different from the control group; however, the observed differences between the infection groups were not statistically significant.

4. DISCUSSION

BTV serotypes, BTV-1 and BTV-8, have displayed a marked ability to spread and severe virulence has been reported during European outbreaks in domestic ruminants, sheep being among the most severely affected livestock species (Allepuz et al., 2010; Elbers et al., 2008b). Detailed investigations in sheep to elucidate the pathogenic mechanisms of BTV-1 under experimental conditions are scarce (Hamblin et al., 1998; Perez de Diego et al., 2011). In comparison, infections carried out with BTV-8 in sheep have been accompanied by a significant variation in results (Darpel et al., 2007; Elbers et al., 2008c; Worwa et al., 2010). Recently, two studies have also reported a comparative study involving pathogenicity of BTV-1 and -8 in sheep and cattle (Dal Pozzo et al., 2013; Sanchez-Cordon et al., 2013).

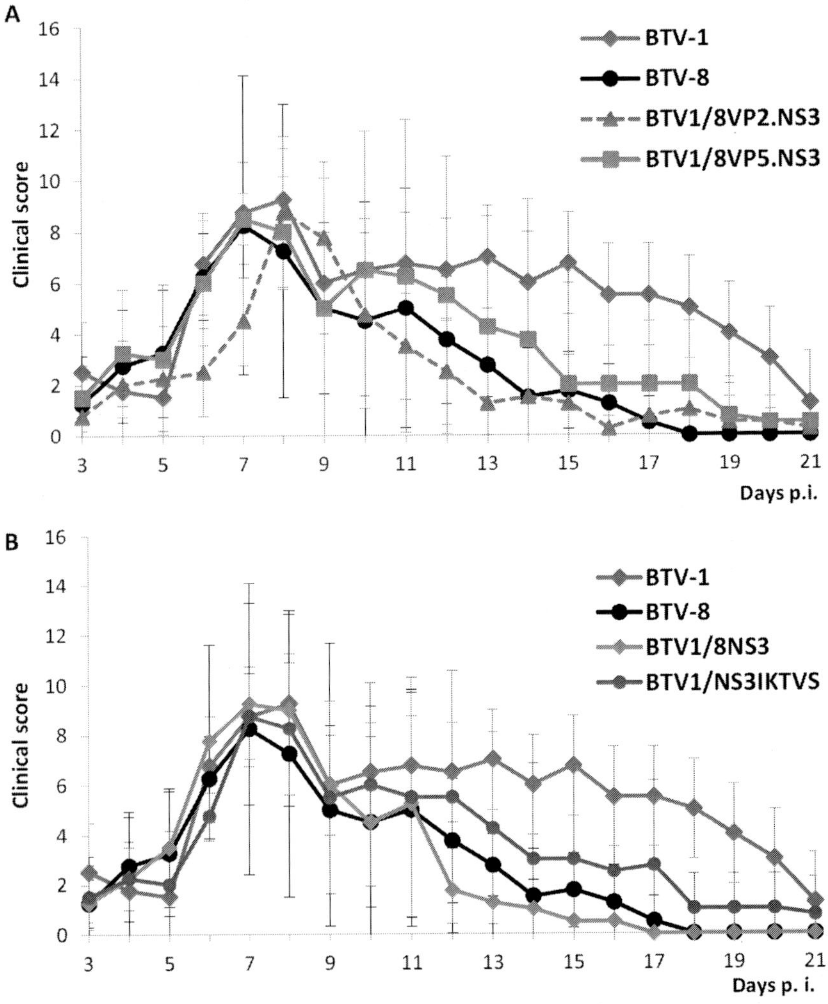

Figure. 4. Pathogenicity of reassortants BTV1/8VP2.NS3, BTV1/8VP5.NS3, BTV1/8NS3 or mutant BTV1/NS3$_{IKTVS}$ viruses. (A, B) Clinical signs were recorded for sheep infected with BTV1/8VP2.NS3 (A), BTV1/8VP5.NS3 (A), BTV1/8NS3 (B) or mutant BTV1/NS3$_{IKTVS}$ (B) viruses using the same scoring described in Fig. 2. Parental BTV-1 and BTV-8 viruses were also included as control (same set of data was plotted in each graph).

This is the first study to investigate how the difference in the clinical manifestations between BTV-1 and BTV-8 is influenced by the genetic origin of the virus. Since there is some evidence in previous reports regarding the involvement of VP2 and VP5, the two most exposed viral capsid proteins as determinants of virus virulence (DeMaula et al., 2000; Huismans and Howell, 1973), we have explored whether variations in pathogenicity are likely to be associated with the two most variable BTV genomic segments, S2 and S6 that encode VP2 and VP5 respectively. In addition, as S10, the gene encoding for NS3 has also been found to be the most variable amongst the segments encoding for the non-structural proteins in BTV infected cells, this protein was also hypothesized to play a potentially important role in BTV pathogenesis. Interestingly, in BTV infected cells it has also been demonstrated that the interaction of VP2 and VP5 with NS3 has an important role in virus assembly and trafficking (Beaton et al., 2002; Bhattacharya and Roy, 2008; Celma and Roy, 2009, 2011; Wirblich et al., 2006).

For this purpose, the BTV RG system was utilized to generate reassortant viruses that consisted of a backbone of BTV-1, with VP2, VP5 and NS3 of BTV-8 in five different combinations (BTV1/8VP2.5, BTV1/8VP2.5.NS3, BTV1/8VP2.NS3, BTV1/8VP5.NS3 and BTV1/8NS3). In each of the reassortants the equivalent genome segment of BTV-1 was replaced by BTV-8. This is the first time that reassortant viruses generated by RG technology have been tested in an animal host.

In order to investigate the influence of tissue specificity in the growth pattern of the reassortant viruses and to compare it with the two parental strains of BTV-1 and BTV-8, sheep kidney (PT) and thymus (SFT-R) cell lines were infected with the respective viruses. Since our primary objective was to assess the importance of the VP2, VP5 and NS3 in BTV pathogenesis, growth curves and virus protein production of reassortant virus BTV1/8VP2.5.NS3 were assessed in the two different ovine cell types. Our results showed that both wild-type viruses as well as the reassortant viruses had different growth profiles in each cell line, indicating that there may be a tissue specificity related to a particular strain that may play a role in virus pathogenicity. In addition, our results also revealed that the growth pattern of BTV1/8VP2.5 and not BTV1/8VP2.5.NS3 was similar to that of BTV-1 in the ovine kidney cell line. Experiments undertaken with BTV1/8VP2.5.NS3, BTV-1 and BTV-8 in ovine thymus cell lines demonstrated a more pronounced similarity in growth profile between BTV1/8VP2.5.NS3 and BTV-8 but not BTV-1. This confirmed that substituting all 3 segments, S2, S6 and S10 of BTV-1, with that of BTV-8 resulted in a reassortant virus that was more similar in behavior to BTV-8 than BTV-1. Since the effect was more pronounced in the ovine thymus SFT-R cell line, this also validated that in cell culture systems the origin of the cultured cells does play a significant role in determining virus growth characteristics. Hence, on the basis of these studies it was postulated that the growth profile of a particular BTV strain was dependent on VP2, VP5 and NS3.

Consequently, experiments designed to analyze the pathogenicity of BTV1/8VP2.5 and BTV1/8VP2.5.NS3 in sheep revealed that parental BTV-8 and reassortant BTV1/8VP2.5.NS3 triggered an earlier onset of clinical disease symptoms than parental BTV-1 and reassortant BTV1/8VP2.5. Interestingly, our results are in direct contrast to another recent investigation undertaken in sheep that have demonstrated that BTV-1 is more pathogenic than BTV-8. The difference in pathogenicity of BTV-1 and BTV-8 between our current study and that undertaken by Sanchez-Cordon et al. (2013) can be attributed to the fact that the BTV-1 strain used to recover our reassortant viruses was a tissue culture adapted strain obtained from South Africa. Subsequently, RT-PCR on blood samples procured from infected sheep and serum neutralization tests also confirmed that BTV1/8VP2.5 and BTV1/8VP2.5.NS3 were able to replicate in animals and that they also were able to generate a strong immune response similar to that of the parental strains. Since sheep infected with reassortant BTV1/8VP2.5.NS3 showed similar clinical scores to that of BTV-8, it led us to further hypothesize that of the 3 proteins, VP2, VP5 and NS3, the non-structural protein NS3 might be an important contributing factor for the development of BT disease. Hence, three more reassortant viruses were generated by RG system in which S10 of BTV-8 was either present singly (BTV1/8NS3) or in a combination with either BTV-8 VP2 (BTV1/8VP2.NS3) or VP5 (BTV1/8VP5.NS3). Sequence comparison of S10 belonging to BTV-1 and BTV-8 highlighted the presence of only 12 amino acid residues that were variable between the two serotypes. Since a cluster containing 5 of these variable residues was located in the putative extracellular domain of NS3, a mutant virus was generated (BTV1/$NS3_{IKTVS}$) that only contained amino acid substitutions ($L_{149}I/S_{153}K/A_{154}T/I_{156}V/Q_{158}S$) in this region of NS3. Consequently, this domain of BTV-1 NS3 became identical to the extracellular domain of BTV-8. When the reassortant and the mutant virus were inoculated in sheep, real-time RT-PCR of blood samples demonstrated that all the newly generated viruses and the two parental strains could replicate in sheep. Although pathogenicity analysis of the reassortant viruses suggested that BTV1/8VP2.NS3 behaved similarly to BTV-1, while the other two reassortant viruses (BTV1/8NS3 and BTV1/8VP5.NS3) and the mutant virus BTV1/$NS3_{IKTVS}$ were similar to BTV-8, the overall clinical scores were not significantly different. Based on our result it was hypothesized that the two outer capsid proteins (VP2 and VP5) and the non-structural protein NS3 are all acting in concert with each other and are to a certain extent responsible for the pathogenicity of a particular strain. Interestingly, earlier reports have demonstrated that the interaction of NS3 with the two outer capsid proteins and cytoplasmic proteins have important consequences for both virus assembly and trafficking in infected cells (Beaton et al., 2002; Bhattacharya and Roy, 2008; Celma and Roy, 2009, 2011; Wirblich et al., 2006). Since mutant virus BTV1/$NS3_{IKTVS}$, consisting of an extracellular loop region identical to that of BTV-8 NS3, did not show a statistical significant difference in virus pathogenicity when compared to the two parental strains, it can be concluded that the loop region of NS3 or its interaction with various cellular partners does not have a significant contribution to BT pathogenesis. Although the role of the remaining seven highly conserved BTV genomic segments cannot be negated, further work on this aspect is beyond the remit of this paper. Clinical experiments undertaken with BTV-1 and BTV-8 showed a different profile in two different sets of experiments (Figs. 2 and 4). Since these two animal experiments were conducted over one year apart, it is possible that in spite of every effort that was made to keep conditions consistent between experiments, some variations occurred. However, it is unlikely that the small variation between experiments invalidates the findings made within each of the experiments. The role of NS3 in regulating interferon response has recently been reported (Chauveau et al., 2013). It will be interesting to investigate interferon responses of the triple reassortant virus versus the two wild-type viruses in the animal. However, these will require a series of experiments and detailed analysis that are beyond the remit of this study.

5. CONCLUSION

The clinical disease and pathogenicity of BTV is a very complex process and cannot be explained by the presence or absence of only one genome segment or its protein product. Much more work needs to be undertaken to clearly understand the genetic basis of BT pathogenesis and to our knowledge, this is the first report in this particular research area.

ACKNOWLEDGEMENTS

This study was funded by the European Union (FP7, ORBIVAC, grant 245266) and partly by the Biotechnology and Biological Sciences Research Council (BBSRC), United Kingdom (grant BB/K015168/1).

REFERENCES

1. Allepuz, A., Garcia-Bocanegra, I., Napp, S., Casal, J., Arenas, A., Saez, M., Gonzalez, M.A., 2010. Monitoring bluetongue disease (BTV-1) epidemic in southern Spain during 2007. Prev. Vet. Med. 96, 263–271.
2. Beaton, A.R., Rodriguez, J., Reddy, Y.K., Roy, P., 2002. The membrane trafficking protein calpactin forms a complex with bluetongue virus protein NS3 and mediates virus release. Proc. Natl. Acad. Sci. U. S. A. 99, 13154–13159.
3. Bhattacharya, B., Roy, P., 2008. Bluetongue virus outer capsid protein VP5 interacts with membrane lipid rafts via a SNARE domain. J. Virol. 82, 10600–10612.
4. Boyce, M., Celma, C.C., Roy, P., 2008. Development of reverse genetics systems for bluetongue virus: recovery of infectious virus from synthetic RNA transcripts. J. Virol. 82, 8339–8348.
5. Celma, C.C., Roy, P., 2009. A viral nonstructural protein regulates bluetongue virus trafficking and release. J. Virol. 83, 6806–6816.
6. Celma, C.C., Roy, P., 2011. Interaction of calpactin light chain (S100A10/ p11) and a viral NS protein is essential for intracellular trafficking of nonenveloped bluetongue virus. J. Virol. 85, 4783–4791.
7. Chauveau, E., Doceul, V., Lara, E., Breard, E., Sailleau, C., Vidalain, P.O., Meurs, E.F., Dabo, S., Schwartz-Cornil, I., Zientara, S., Vitour, D., 2013. NS3 of bluetongue virus interferes with the induction of type I interferon. J. Virol. 87, 8241–8246.
8. Dal Pozzo, F., Martinelle, L., Thys, C., Sarradin, P., De Leeuw, I., Van Campe, W., De Clercq, K., Thiry, E., Saegerman, C., 2013. Experimental coinfections of calves with bluetongue virus serotypes 1 and 8. Vet. Microbiol. 165, 167–172.
9. Darpel, K.E., Batten, C.A., Veronesi, E., Shaw, A.E., Anthony, S., Bachanek- Bankowska, K., Kgosana, L., bin-Tarif, A., Carpenter, S., Muller-Doblies, U.U., Takamatsu, H.H., Mellor, P.S., Mertens, P.P., Oura, C.A., 2007. Clinical signs and pathology shown by British sheep and cattle infected with bluetongue virus serotype 8 derived from the 2006 outbreak in northern Europe. Vet. Rec. 161, 253–261.
10. DeMaula, C.D., Bonneau, K.R., MacLachlan, N.J., 2000. Changes in the outer capsid proteins of bluetongue virus serotype ten that abrogate neutralization by monoclonal antibodies. Virus Res. 67, 59–66.
11. Elbers, A.R., Backx, A., Ekker, H.M., van der Spek, A.N., van Rijn, P.A., 2008a. Performance of clinical signs to detect bluetongue virus serotype 8 outbreaks in cattle and sheep during the 2006-epidemic in The Netherlands. Vet. Microbiol. 129, 156–162.
12. Elbers, A.R., Backx, A., Meroc, E., Gerbier, G., Staubach, C., Hendrickx, G., van der Spek, A., Mintiens, K., 2008b. Field observations during the bluetongue serotype 8 epidemic in 2006. I. Detection of first outbreaks and clinical signs in sheep and cattle in Belgium, France and the Netherlands. Prev. Vet. Med. 87, 21–30.
13. Elbers, A.R., Backx, A., Mintiens, K., Gerbier, G., Staubach, C., Hendrickx, G., van der Spek, A., 2008c. Field observations during the Bluetongue serotype 8 epidemic in 2006. II. Morbidity and mortality rate, case fatality and clinical recovery in sheep and cattle in the Netherlands. Prev. Vet. Med. 87, 31–40.
14. Erasmus, B.J., 1975. The control of bluetongue in an enzootic situation. Aust. Vet. J. 51, 209–210.
15. Hamblin, C., Salt, J.S., Graham, S.D., Hopwood, K., Wade Evans, A.M., 1998. Bluetongue virus serotypes 1 and 3 infection in Poll Dorset sheep. Aust. Vet. J. 76, 622–629.
16. Huismans, H., Howell, P.G., 1973. Molecular hybridization studies on the relationships between different serotypes of bluetongue virus and on the difference between the virulent and attenuated strains of the same serotype. Onderstepoort J. Vet. Res. 40, 93–103.
17. Le Gal, M.C., Dufour, B., Geoffroy, E., Zanella, G., Moutou, F., Millemann, Y., Rieffel, J.N., Pouilly, F., 2008. Bluetongue virus serotype 8 in the Ardennes in 2007. Vet. Rec. 163, 668.

18. Maan, S., Rao, S., Maan, N.S., Anthony, S.J., Attoui, H., Samuel, A.R., Mertens, P.P., 2007. Rapid cDNA synthesis and sequencing techniques for the genetic study of bluetongue and other dsRNA viruses. J. Virol. Methods 143, 132–139.
19. Maclachlan, N.J., Drew, C.P., Darpel, K.E., Worwa, G., 2009. The pathology and pathogenesis of bluetongue. J. Comp. Pathol. 141, 1–16.
20. Maclachlan, N.J., Guthrie, A.J., 2010. Re-emergence of bluetongue, African horse sickness, and other orbivirus diseases. Vet. Res. 41, 35.
21. Martinelle, L., Dal Pozzo, F., Sarradin, P., De Leeuw, I., De Clercq, K., Thys, C., Thiry, E., Saegerman, C., 2013. Pulmonary artery haemorrhage in newborn calves following bluetongue virus serotype 8 experimental infections of pregnant heifers. Vet. Microbiol. 167, 250–259.
22. Matsuo, E., Celma, C.C., Boyce, M., Viarouge, C., Sailleau, C., Dubois, E., Breard, E., Thiery, R., Zientara, S., Roy, P., 2011. Generation of replica-tion-defective virus-based vaccines that confer full protection in sheep against virulent bluetongue virus challenge. J. Virol. 85, 10213–10221.
23. Mellor, P.S., Boorman, J., Baylis, M., 2000. Culicoides biting midges: their role as arbovirus vectors. Annu. Rev. Entomol. 45, 307–340.
24. Mellor, P.S., Carpenter, S., Harrup, L., Baylis, M., Mertens, P.P., 2008. Bluetongue in Europe and the Mediterranean basin: history of occur-rence prior to 2006. Prev. Vet. Med. 87, 4–20.
25. Perez de Diego, A.C., Athmaram, T.N., Stewart, M., Rodriguez-Sanchez, B., Sanchez-Vizcaino, J.M., Noad, R., Roy, P., 2011. Characterization of protection afforded by a bivalent virus-like particle vaccine against bluetongue virus serotypes 1 and 4 in sheep. PLoS ONE 6, e26666.
26. Purse, B.V., Mellor, P.S., Rogers, D.J., Samuel, A.R., Mertens, P.P., Baylis, M., 2005. Climate change and the recent emergence of bluetongue in Europe. Nat. Rev. Microbiol. 3, 171–181.
27. Sanchez-Cordon, P.J., Pleguezuelos, F.J., Perez de Diego, A.C., Gomez- Villamandos, J.C., Sanchez-Vizcaino, J.M., Ceron, J.J., Tecles, F., Garfia, B., Pedrera, M., 2013. Comparative study of clinical courses, gross lesions, acute phase response and coagulation disorders in sheep inoculated with bluetongue virus serotypes 1 and 8. Vet. Microbiol. 166, 184–194.
28. Toussaint, J.F., Sailleau, C., Breard, E., Zientara, S., De Clercq, K., 2007. Bluetongue virus detection by two real-time RT-qPCRs targeting two different genomic segments. J. Virol. Methods 140, 115–123.
29. Wilson, A.J., Mellor, P.S., 2009. Bluetongue in Europe: past, present and future. Philos. Trans. R. Soc. Lond. B: Biol. Sci. 364, 2669–2681.
30. Wirblich, C., Bhattacharya, B., Roy, P., 2006. Nonstructural protein 3 of bluetongue virus assists virus release by recruiting ESCRT-I protein Tsg101. J. Virol. 460–473.
31. Worwa, G., Hilbe, M., Chaignat, V., Hofmann, M.A., Griot, C., Ehrensperger, F., Doherr, M.G., Thur, B., 2010. Virological and pathological findings in bluetongue virus serotype 8 infected sheep. Vet. Microbiol. 144, 264–273.

CHAPTER 6

Defining the "core microbiome" of the microbial communities in the tonsils of healthy pigs

Beth A Lowe[1], Terence L Marsh[1], Natasha Isaacs-Cosgrove[1], Roy N Kirkwood[2,4], Matti Kiupel3 and Martha H Mulks[1*]

[1]*Department of Microbiology and Molecular Genetics and the Center for Microbial Pathogenesis, Michigan State University, East Lansing, MI, USA.*

[2]*Department of Large Animal Clinical Sciences, College of Veterinary Medicine, Michigan State University, East Lansing, MI, USA.*

[3]*Department of Pathobiology and Diagnostic Investigations, College of Veterinary Medicine, Michigan State University, East Lansing, MI, USA.*

[4]*School of Animal and Veterinary Science, University of Adelaide, Adelaide, Australia*

AUTHORS' CONTRIBUTIONS

BAL, TLM, and MHM contributed to the design of the study, performed the data analyses, and wrote the manuscript; NIC prepared and processed the DNA samples; RNK designed the tonsil brushes and performed all veterinary procedures; MK performed necropsies and collection of specimens. All authors read and approved the final manuscript.

ABSTRACT

Background

Porcine tonsils are the colonization site for many pathogenic as well as commensal microorganisms and are the primary lymphoid tissue encountered by organisms entering through the mouth or nares. The goal of this study was to provide an in-depth characterization of the composition and structure of the tonsillar microbial communities and to define the core microbiome in the tonsils of healthy pigs, using high throughput bar-coded 454-FLX pyrosequencing.

Results

Whole tonsils were collected at necropsy from 12 16-week-old finisher pigs from two healthy herds. Tonsil brushes were also used to collect samples from four of these animals. Bacterial DNA was isolated from each sample, amplified by PCR with universal primers specific for the bacterial 16S rRNA genes, and the PCR products sequenced using pyrosequencing. An average of 13,000 sequences were generated from each sample. Microbial community members were identified by sequence comparison to known bacterial 16S rRNA gene sequences.

The microbiomes of these healthy herds showed very strong similarities in the major components as well as distinct differences in minor components. *Pasteurellaceae* dominated the tonsillar microbiome in all animals, comprising ~60% of the total, although the relative proportions of the genera *Actinobacillus, Haemophilus,* and *Pasteurella* varied between the herds. Also found in all animals were the genera *Alkanindiges, Peptostreptococcus, Veillonella, Streptococcus* and *Fusobacterium*, as well as *Enterobacteriaceae* and *Neisseriaceae. Treponema* and *Chlamydia* were unique to Herd 1, while *Arcanobacterium* was unique to Herd 2.

Tonsil brushes yielded similar results to tissue specimens, although *Enterobacteriaceae* and obligate anaerobes were more frequently found in tissue than in brush samples, and *Chlamydia*, an obligately intracellular organism, was not found in brush specimens.

Conclusions

We have extended and supported our previous studies with 16S clone libraries, using 16S rRNA gene pyrosequencing to describe the microbial communities in tonsils of healthy pigs. We have defined a core microbiome, dominated by *Pasteurellaceae*, in tonsil specimens, and have also demonstrated the presence of unique minor components of the tonsillar microbiome present in each herd. We have validated the use of non-invasive tonsil brushes, in comparison to tonsil tissue, which will facilitate future studies.

1. BACKGROUND

The palatine tonsils of pigs are large, flat, follicular structures on the ventral side of the soft palate, at the junction of the oropharynx and nasopharynx, that are constantly exposed to both ingested and inhaled microorganisms [1]. Both the surface of the tonsils and the extensive tubular tonsillar crypts are an important colonization site for many pathogenic and commensal microorganisms, including both bacteria and viruses [1]. Conversely, the tonsils are also the main oropharyngeal lymphoid tissue, and play a key role in surveillance, detection, and initiation of an immune response against organisms that enter through either the mouth or the nares [1, 2].

Asymptomatic carriage in the tonsils provides a reservoir for many bacterial porcine pathogens, such as *Actinobacillus pleuropneumoniae, A. suis, Streptococcus suis, Haemophilus parasuis*, and *Mycoplasma hyopneumoniae*, as well as viral pathogens such as PRRS virus and classical swine fever virus [1-8]. Indeed, the tonsils are a routine sampling site in surveillance of many porcine pathogens [1]. Porcine tonsils are also a reservoir for pathogens capable of causing foodborne infections of humans, including *Salmonella* and *Campylobacter* species, *Escherichia coli, Listeria monocytogenes*, and *Yersinia enterocolitica* [9-13]. The commensal tonsillar microbiota likely interacts with these pathogens to either inhibit or enhance colonization and carriage.

In a previous study, standard aerobic culture and culture-independent construction and analysis of 16 s rRNA gene clone libraries were used to examine the microbial communities in the tonsils of healthy pigs [14]. Gram-positive cocci (*Streptococcus, Staphylococcus*, and *Enterococcus* species) predominated in the culture analysis, but were not detected in the clone libraries. In contrast, *Pasteurellaceae* (*Actinobacillus, Haemophilus*, and *Pasteurella* species) and anaerobic *Fusobacterium, Bacteroides, Porphyromonas*, and *Prevotella* species were the dominant genera found in the 16S rRNA clone libraries. The goal of the current study was to utilize high throughput bar-coded 454-FLX pyrosequencing to provide a more in-depth characterization of the composition and structure of the tonsillar microbial communities and to define the core microbiome in the tonsils of healthy pigs.

2. METHODS

2.1 Animals

The study and all animal procedures were approved by the Michigan State University Institutional Animal Care and Use Committee. Eight 18-20 week old pigs from a high health status herd with no recent history of respiratory disease (Herd 1) and four similar pigs from a currently healthy herd with a history of chronic but undefined respiratory problems (Herd 2) were randomly selected for use in this study. Both herds are farrow-to-finish operations weaning at 21 to 24 days of age, with similar management, located in mid-Michigan. Groups of similarly aged pigs were moved from the nursery to the grow-finish rooms all in-all out, although there was a common airspace via either connecting corridor (Herd 1) or connecting doors (Herd 2). Herd 1 Time 1 contains four Hampshire-Yorkshire crossbred pigs (pigs A-D) that were sampled in June 2007. These four pigs received no vaccinations or in-feed antibiotics. Herd 1 Time 2 contains four purebred Yorkshire pigs (pigs J-M) that were sampled 2 years later (April 2009). These pigs received Tylan (Elanco Animal Health, Indianapolis, IN) in-feed and were vaccinated against PCV2. There were no other significant differences in feed, vaccination, or medication between the two sampling periods for Herd 1. Herd 2 contains four Hampshire-Cambrough crossbred pigs (pigs E-H) that were sampled only once, in July 2007. This herd was also vaccinated against PCV2 and received Tylan in-feed. Additionally, Herd 2 received Pulmotil (Elanco Animal Health) until 8 weeks of age. The standard feed ration for Herd 2 was similar to that for Herd 1.

All pigs were taken off feed at least 3 h prior to collection of specimens. Pigs were anesthetized by intramuscular injection of Telazole (6.6 mg/kg) and Xylazine (3.3 mg/kg) prior to transport from the farms to the necropsy facilities at the Michigan State University Diagnostic Center for Population and Animal Health. Pigs were euthanized within 30 min by overdose with a pentobarbital solution (Fatal-Plus, 100 mg/kg, Vortech Pharmaceutical, Dearborn, MI) delivered intravenously into the vena cava, following standard procedures.

Lung specimens from Herd 1 Time 1 pigs A-D and Herd 2 pigs E-H were aseptically sampled and cultured on blood agar and brain heart infusion agar containing 10 μg/ml NAD. No bacterial isolates were recovered from Herd 1 pigs. *Arcanobacterium pyogenes* was recovered from well-defined lesions in the lungs of 1 pig from Herd 2 (pig F).

2.2 Collection of tonsil tissue

The entire right and left palatine tonsils from all 12 pigs were collected. Tonsils were placed into sterile glass Petri dishes where extraneous connective tissue was removed and tonsils were divided into four quarters using a sterile scalpel. One quarter of the right tonsil was combined with one quarter of the left tonsil and this combined sample was minced, placed into a sterile 15 ml conical tube, and immediately frozen at -20°C for subsequent extraction of community DNA.

2.2 Tonsil brush design and use

For this study, we designed a brush for swabbing porcine palatine tonsils. The tonsil brush consisted of three parts: a 0.5″ diameter soft-bristle test tube brush with approximately 5″ of metal handle (VWR International, West Chester, PA), an 8″ long by 0.5″ diameter wooden dowel used as a handle, and a guard for the brush made from 2 15 ml screw capped centrifuge tubes (Fisher Scientific, Pittsburgh, PA). The brush portion was prepared by cutting the brush head to a length of 1″ and sealing the cut end with a drop of superglue to protect the pig and tissue from injury. Once dry, the brush was soaked for 1 h in 10% hydrochloric acid to destroy any contaminating DNA, rinsed thoroughly with sterile H_2O, and allowed to dry completely. The guard was made by cutting the conical ends off two 15 ml centrifuge tubes, taping the cut ends of the tubes together, and removing one of the caps. The brush was then assembled by inserting the handle of the test tube brush into the end of the dowel and placing the guard over the brush. The guard was secured to the handle with a piece of autoclave tape. Assembled brushes were autoclaved in instant sealing sterilization pouches (Fisher Scientific).

For swabbing porcine tonsils, the cap was removed from the guard and the brush (with the guard still in place) was inserted into the pig's mouth near the tonsil. The guard was then pulled back to expose the brush. Both the right and left tonsils were brushed for approximately 10 s each, with sufficient pressure to allow penetration of tonsillar crypts with the soft brush bristles, and at the same time with care not to cause tissue damage and bleeding. The guard was replaced and brush removed from the pig's mouth. The guard was discarded and the brush removed from the handle and placed into a 50 ml sterile test tube containing 20 ml of ice-cold 80% ethanol. Brushes were then stored at -20°C for subsequent extraction of community DNA.

Tonsil brush specimens were collected from Herd 1 Time 2 pigs (pigs J-M), immediately after euthanasia and prior to removal of the tonsils for tissue collection.

2.3 Isolation of community DNA

Community DNA from pig tonsils was extracted from tonsil tissues using a PowerSoil DNA Isolation Kit (MoBio Laboratories, Carlsbad, CA) as previously described [14]. Community DNA from tonsil brush specimens was extracted using the same kit as follows. The tube containing the tonsil brush and 20 ml of ice-cold 80% ethanol was vortexed vigorously to remove cells and cell debris from the brush. The brush was removed and discarded. The sample in 80% ethanol was divided evenly into 2 sterile Corex® tubes and centrifuged in a refrigerated Sorvall SS-34 rotor at 16,000 × *g* for 30 min. Following centrifugation, supernatants were discarded. One pellet was suspended in 5 ml of ice-cold 80% ethanol and archived at -20°C. The second pellet was suspended in 1 ml phosphate buffered saline (PBS) for DNA extraction. Approximately 0.25 ml of the sample was added to each of 4 MoBio PowerBead tubes. The samples were shaken vigorously in a Bead Beater (BioSpec Products, Bartlesville, OK) for 1.5-2 min at 4°C, and then extracted according to manufacturer's instructions. After purification, the concentration of community DNA was determined spectrophotometrically using a Nanodrop (Thermo Scientific, Wilmington, DE). Fifty percent of the yield was immediately archived at -80°C; the remaining DNA was used for polymerase chain reaction (PCR) amplification and 454 pyrosequencing.

2.4 454 pyrosequencing

For 454 Flx sequencing, community template DNAs were amplified with primers designed by the Ribosomal Database Project (RDP) at Michigan State University [15]. The forward primer contains the Flx-specific terminal sequence (5'-GCCTCCCTCGCGCCATCAG-3') followed by a six base tag and then the 16S rRNA-specific 3' terminus of the composite primer (5'-AYTGGGYDTAAAGNG-3'). The reverse primer was composed of four variants targeting the same 16S rRNA region to maximize coverage of the database (R1 = /5'/TACNVGGGTATCTAATCC; R2 = /5'/TACCRGGGTHT CTAATCC; R3 = /5'/TACCAGAGTATCTAATTC; R4 = /5'/CTACDSRGGTMTC TAATC). The 3' terminus of the forward primer is at *E. coli* position 578 and the 3' terminus of the reverse primer is at position 785. Pilot scale (25 µl) PCR reactions for optimization were followed by 2-3 preparative 50 µl amplification reactions. High fidelity Taq (Invitrogen Platinum) was used with MgSO4 (2.5 mM), the vendor supplied buffer, BSA (0.1 mg/ml), dNTPs (250 µM) and primers (1 µM). A three minute soak at 95°C was followed by 30 cycles of 95°C (45 s), 57°C (45 s) and 72°C (1 min) with a final 4 min extension at 72°C. PCR products were agarose gel purified (2% metaphor in TAE) and bands were extracted with a QIAquick Gel Extraction Kit (Qiagen, Valencia, CA). Gel extracted material was further purified with a Qiagen PCR

Cleanup kit. Quantification of purified PCR product was with PicoGreen using Qubit (Invitrogen, Carlsbad, CA). The PCR products from 20 to 40 samples were combined in equal mass amounts and loaded into a Roche GS Flx system using vendor specified chemistries.

2.5 Sequence analysis tools

All sequences derived from 454 Flx sequencing were processed through the RDP pyrosequencing pipeline [15, 16, 17]. Initial processing included screening and removing short reads (those lacking both primer sequences) and low quality reads (any with errors in the primer sequence). Sequences were sorted based on sequence tags and trimmed of primer and tag sequences. The remaining high quality sequences were taxonomically identified using the Classifier tool at a 60% confidence level. The classifier output was then used for analysis of similarities and difference between herds (Additional files 1, 2, 3, 4, 5). For analysis of the data at the genus level, all genera with fewer than 5 representatives were dropped from the analysis.

To identify members of the family *Pasteurellaceae* and genus *Streptococcus* to the lowest possible phylogenetic level, we obtained all the 138 near full-length type sequences from family *Pasteurellaceae* and genus *Streptococcus* from RDP release 10.22 (August 2010). We also added sequence AF486274 ("*Actinobacillus porcitonsillarum*"). These 139 sequences were aligned by the Infernal aligner [16] trained by RDP [17]. The final reference set contained the region corresponding to the 454 FLX amplicon (*E. coli* position 578 to 784) sliced from the alignment. To determine the nearest neighbor, the 454 FLX sequences passing the RDP Pyro initial filtering were aligned by the Infernal aligner and the distance between each FLX sequence and reference sequences was calculated. The reference sequence with the closest distance was reported. In case of tie, all the reference sequences were reported.

2.6 Statistical analysis

For the statistical analyses of sequences, we used a 0.03% cutoff value for clustering. This is consistent with previous analyses of 454 data [18] as well as the historical value frequently used over the past 15 years [19, 20]. Similarly we used this cutoff in evaluating members of family *Pasteurellaceae* and genus *Streptococcus*. For comparative statistical analyses, aligned sequences were clustered using the RDP Complete Linkage Clustering Tool and the resulting cluster files were used to calculate Jaccard and Sørensen indices [17]. For comparative statistical analyses, aligned sequences were clustered using the RDP Complete

Linkage Clustering Tool and the resulting cluster files were used to calculate Jaccard and Sørensen indices [17]. Cluster files were also reformatted with the EstimateS Formatter Tool through the RDP website. Principle component analysis followed by centroid calculations with a 95% confidence limit were performed in R (version 2.10; http://www.r-project.org/) with Vegan package (http://vegan.r-forge.r-project.org) using the EstimateS formatted files. Chao 1 was calculated using the cluster files derived from each sample and from merged samples for herds using the RDP Pyrosequencing Pipeline. Simpson's Diversity index was calculated with MOTHUR [21].

3. RESULTS

Community DNA was isolated from whole tonsil tissue (Pigs A-M) or tonsil brushings (Pigs J-M) as described in Methods. Tonsil tissue samples were collected in spring 2007 from two different herds, and again in spring 2009 from Herd 1. Tonsil brush samples were collected only in spring 2009 from Herd 1, to test the validity of this rapid and painless sampling technique in comparison to euthanasia and removal of tonsil tissue. Sixteen samples (four groups of four samples) were collected and analyzed by 454 Flx pyrosequencing, and comparisons were made between Herd 1 and Herd 2, between Herd 1 Time 1 and Herd 1 Time 2 tissues, and between tissue and brush samples from Herd 1 Time 2 in these results.

3.1 Bar-coded 16S pyrosequencing

A total of 210,433 quality reads were obtained from the four groups of pigs sampled, with at least 15,000 reads per group. Samples of tonsil tissue from Herd 1 at time 2 yielded the fewest number of quality reads. Table 1 shows the number of reads obtained from each of the four groups of pigs and the percent of those reads that could be taxonomically assigned at a 60% confidence level using the RDP Classifier. Overall, greater than 97% of the total reads could be taxonomically assigned at the phylum, class, and order level. This dropped to 90.5% at the family level and further dropped to 72.3% at the genus level. Taxonomic assignment of reads was consistently lower at all levels for Herd 2 compared to all three groups of samples from Herd 1.

Figure 1 shows the rarefaction plots for the four groups. Herd 1 and Herd 2 plots demonstrate that Herd 2 had significantly more phylotypes and greater unsampled diversity (Figure 1A). Comparison of the three groups of Herd 1 pigs reveals similar trajectories even though the number of reads sampled varied (Figure 1B).

Table 1 Taxonomic characterization of tonsillar microbial communities

	Sample	# Reads[a]	Phylum[b]	Class[b]	Order[b]	Family[b]	Genus[b]
Herd 2	Tissue	99894	95.6%	95.4%	94.8%	82.7%	64.7%
Herd 1 Time 1	Tissue	54932	99.7%	99.6%	99.1%	96.7%	85.0%
Herd 1 Time 2	Tissue	15929	99.8%	99.5%	99.4%	98.7%	70.1%
Herd 1 Time 2	Brush	39678	99.9%	99.5%	99.5%	98.6%	75.0%
Total # reads		210433	205795	205346	204467	190540	152192
Avg % Assigned			**97.8%**	**97.6%**	**97.2%**	**90.5%**	**72.3%**

[a] the sum of all sequences of 4 individuals

[b]% of reads taxonomically assigned at each level

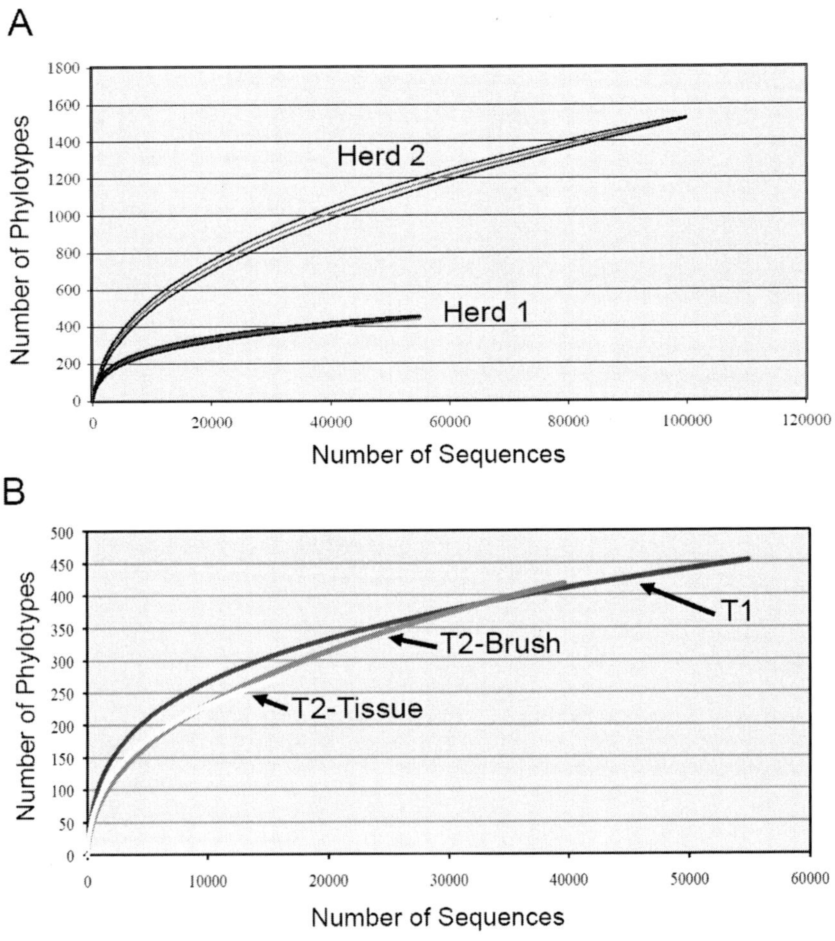

Figure 1 Rarefaction curves computed with the RDP Pyrosequencing Pipeline. Rarefaction curves are presented for each group of samples obtained by 454 pyrosequencing. The curves for herds 1 and 2 at time 1 are shown in panel A, while the curves for all three groups of samples from herd 1 are shown in panel B.

Taken together, this suggests that the microbial community in the tonsils in Herd 2 was more complex at this level of interrogation.

As stated above, a total of 210,433 reads was obtained for the four groups. Table 2 indicates the number of reads made from each individual sample as well as the total for each group. The number of reads for each individual and each group forms the basis of the comparisons for the number of OTUs,

Chao-1 richness, and the Simpson diversity indices. Using a cutoff of 97% identity for species level distinctions, the number of OTUs detected per sample ranged from 57 to 730. In general, the number of OTUs detected increased with the number of reads, though there was no direct proportionality. Herd 2 showed the greatest variability across the four pigs sampled ranging from 197 to 730 OTUs. Pigs from Herd 1, regardless of sampling time or method, had less diversity and variation across pigs (57-251 OTUs). The first sampling that compared Herd 1 with Herd 2 had roughly twice as many sequences in Herd 2 as in Herd 1 (99,894 vs 54,932). However, the number of detected OTUs and Chao 1-estimated OTUs of Herd 2 were 3.3 and 4 fold greater than those of Herd 1. The Shannon and Simpson's Diversity Indices reflected the trends seen with detected and estimated richness, with Herd 2 measurably more diverse than Herd 1. This could be seen in the rank abundance curves (data not shown) where Herd 2 had greater asymmetry (less even) and a longer tail comprised of OTUs with small populations. The Simpson's evenness measurement indicated that all communities were quite uneven (1.0 = perfect evenness) but that the second sampling of Herd 1 derived from extracted tissue was less skewed than other communities.

3.2 Phylum, class, and order level structure of the tonsillar communities

We found members of 17 different phyla of bacteria in one or more tonsil specimens examined (Additional file 1). Microbial communities in all pigs in all four groups of samples were dominated by *Proteobacteria*, which averaged 73.4% of the communities (ranging from 47.0% to 94.5% in individual specimens); *Firmicutes*, which averaged 17.8% (ranging from 3.1% to 45.6%); and *Fusobacteria*, which averaged 5.6% (ranging from 0.6% to 16.3%) of the total reads assigned. Together, the *Proteobacteria, Firmicutes*, and *Fusobacteria* comprised 96.8% (ranging from 87.5% to 99.6% in individual specimens) of the microbial communities in these samples. *Actinobacteria* (1.2%) and *Bacteroidetes* (0.8%) were also found in most pigs in all four groups of samples. These five phyla form the core microbiome of porcine tonsils, and together comprised on average 98.8% (ranging from 89.5% to 100%) of the reads assigned to the phylum level (Table 3). In addition, *Tenericutes* (0.03%) were found in small numbers in at least one pig in each group of samples.

Distribution at the class level followed well from the phylum level data. We found members of 27 different classes of bacteria in at least one of the tonsil specimens (Additional file 2).

Table 2 Diversity and richness of the tonsillar microbial communities

	# Reads	# OTUs[a]	Chao-1[b]	Shannon[c]	Simpson[d]	Simpson evenness[e]
Pig E	43770	582	980	3.14	0.10	0.02
Pig F	11386	197	268	3.40	0.07	0.07
Pig G	16519	485	820	3.73	0.05	0.04
Pig H	28219	730	1224	3.42	0.11	0.01
Herd 2 Time 1	**99894**	**1525**	**2513**	**3.58**	**0.06**	**0.01**
Pig A	12268	128	161	2.37	0.21	0.03
Pig B	14885	190	235	3.17	0.09	0.05
Pig C	9392	182	237	2.81	0.14	0.04
Pig D	18387	135	291	3.23	0.07	0.11
Herd 1 Time 1	**54932**	**453**	**628**	**3.23**	**0.07**	**0.03**
Pig J	5523	122	191	3.26	0.07	0.12
Pig K	2760	67	88	2.70	0.11	0.14
Pig L	6295	167	233	3.12	0.09	0.06

Pig M	1351	57	87	2.45	0.15	0.11
Herd 1 Time 2	**15929**	**273**	**382**	**3.23**	**0.08**	**0.05**
Pig J Brush	13361	155	228	2.04	0.29	0.02
Pig K Brush	5672	102	141	2.38	0.14	0.07
Pig L Brush	9380	251	465	2.35	0.26	0.01
Pig M Brush	11265	136	164	2.83	0.11	0.06
Herd 1 Brush	**39678**	**418**	**650**	**2.53**	**0.18**	**0.01**

[a] number of OTUs (based on 0.03 cut-off) found in each sample or herd
[b] the estimated richness of an environment based on 0.03 cut-off
[c] computed at the RDP Pyrosequencing Pipeline
[d] calculated with MOTHUR [21] using a distance matrix computed at RDP Pyrosequencing Pipeline
[e] derived from Simpson's Index where E = (1/D)/S, D is the Simpson's Index and S is the total number of species (OTUs)

Table 3 The core microbiome of porcine tonsils

Phylum	% of total	Class	% of total	Order	% of total	Family	% of total	Genus	% of total
Proteobacteria	73.4	*Gammaproteobacteria*	69.8	*Pasteurellales*	56.0	*Pasteurellaceae*	60.2	*Actinobacillus*	37.0
								Haemophilus	6.6
								Pasteurella	16.1
				Pseudomonadales	11.8	*Moraxellaceae*	12.3	*Alkanindiges*	12.0
				Enterobacteriales	2.0	*Enterobacteriaceae*	2.2		
		Betaproteobacteria	3.2	*Burkholderiales*	0.3				
				Neisseriales	2.8	*Neisseriaceae*	3.0		
		Alphaproteobacteria	0.3						
Firmicutes	17.8	*Clostridia*	14.3	*Clostridiales*	14.3	*Peptostreptococcaceae*	2.2	*Peptostreptococcus*	2.6
						Veillonellaceae	4.4	*Veillonella*	3.2
		Bacilli	3.5	*Lactobacillales*	3.4	*Streptococcaceae*	0.5	*Streptococcus*	0.6
Fusobacteria	5.6	*Fusobacteria*	5.6	*Fusobacteriales*	5.6	*Fusobacteriaceae*	5.6	*Fusobacterium*	7.0
Actinobacteria	1.2	*Actinobacteria*	1.2	*Actinomycetales*	0.9				
Bacteroidetes	0.8	*Bacteroidia*	0.3	*Bacteroidales*	0.3				
5/17 phyla identified	98.8	8/27 classes identified	98.2	10/34 orders identified	97.4	8/61 families identified	90.4	8/101 genera identified	85.1

NOTE: Almost half of the *Clostridiales* could not be assigned at the family level, and > 92% of the *Neisseriaceae* could not be assigned to a genus.

Classes found in all animals in all four groups of specimens included, in order of prevalence, *Gammaproteobacteria* (69.8% of the total reads taxonomically assigned at the class level), *Clostridia* (14.3%), *Fusobacteria* (5.6%), *Bacilli* (3.5%), and *Betaproteobacteria* (3.2%). *Actinobacteria* (1.2%), *Alphaproteobacteria* (0.3%), and *Bacteroidia* (0.3%) were found in most animals in all groups of samples. These eight classes form the core microbiome of porcine tonsils, and together represent 98.2% (ranging from 89.2% to 99.9% in individual specimens) of the total reads assigned at the class level (Table 3). In addition, *Epsilonproteobacteria* (0.1%), and *Mollicutes* (0.02%) were found at least one animal in each group. Both *Deltaproteobacteria* (0.1%) and *Sphingobacteria* (0.1%) were found in at least one animal in all three groups of tissue specimens but not in the brush specimens.

We found members of 34 different orders of bacteria in at least one tonsil specimen (Additional file 3). The predominant orders found in all four groups of specimens included *Pasteurellales* (56.0%), *Clostridiales* (14.3%), *Pseudomonadales* (11.8%), *Fusobacteriales* (5.6%), *Lactobacillales* (3.4%), *Neisseriales* (2.8%) and *Enterobacteriales* (2.0%). In addition, the *Actinomycetales* (0.9%), *Burkholderiales* (0.3%), and *Bacteroidales* (0.3%) were found in most animals in all groups of specimens. These ten orders form the core microbiome of porcine tonsils, and together represent 97.4% (ranging from 88.0% to 99.7% in individual specimens) of the reads assigned at the order level (Table 3). *Bacillales* (0.14%) and *Campylobacterales* (0.13%) were also found in small numbers in half of the specimens.

3.3 Family and genus level structure of the tonsillar communities

We found members of 61 families (Additional file 4) and 101 genera (Additional file 5) in at least one tonsil specimen. Five families were found in all pigs in all groups of specimens: *Pasteurellaceae* (60.2%), *Moraxellaceae* (12.3%), *Fusobacteriaceae* (5.6%), *Veillonellaceae* (4.4%), and *Neisseriaceae* (3%). In addition, three families, the *Peptostreptococcaceae* (2.2%), *Enterobacteriaceae* (2.2%), and *Streptococcaceae* (0.5%), were found in most animals in all groups of specimens. These eight families form the core microbiome in porcine tonsils, and represent 90.4% (ranging from 73.5% to 99.0% in individual specimens) of the reads assigned at the family level (Table 3). It should be noted that almost half (46.8%) of the *Clostridiales* could not be assigned at the family level.

Of the 101 genera identified in these samples, 49 were found in both herds (Additional file 5). Thirty-seven genera represented at least 0.1% of the total reads from all specimens (Figure 2). Of these 37 genera, 13 were found in all 4 groups of specimens, 2 were found only in Herd 1, 1 was found only in Herd 2, and 8 were found in tissue specimens but not in brush specimens.

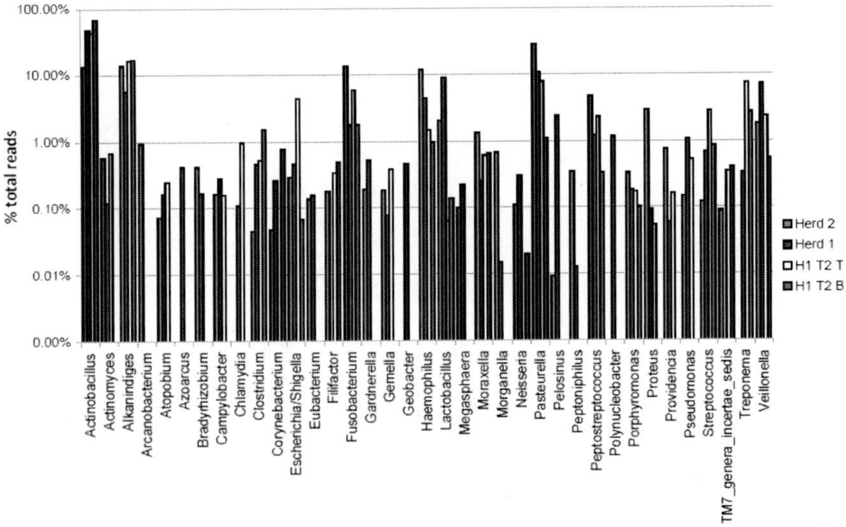

Figure 2 Taxonomic characterization of the four groups of samples obtained by 454 pyrosequencing. Bars illustrate the proportion of reads classified into particular genera. Only genera that contain at least 0.01% of the total number of reads are shown

The relative distribution of the top ten genera found in these specimens is shown in Figure 3. These 10 genera comprised on average 88.3% (ranging from 67.2% to 98.8%) of the total genera in the microbial communities in these specimens. *Actinobacillus (Pasteurellaceae), Alkanindiges (Moraxellaceae*), *Fusobacterium (Fusobacteriaceae*), and *Haemophilus (Pasteurellaceae*) were found in all pigs in all groups of specimens. *Pasteurella (Pasteurellaceae*), *Veillonella (Veillonellaceae*), *Peptostrep-tococcus (Peptostreptococcaceae)*, and *Streptococcus (Streptococcaceae)* were found in almost all pigs in all groups of specimens. These eight genera form the core microbiome in porcine tonsils, and represent 85.1% of the reads assigned to the genus level (Table 3). In addition, *Moraxella (Moraxellaceae)* and *Lactobacillus (Lactobacillaceae)* were found in more than half of the samples tested, and are included in the top ten genera shown in Figure 3. It should be noted that most (> 92%) of the *Neisseriaceae* could not be assigned at the genus level.

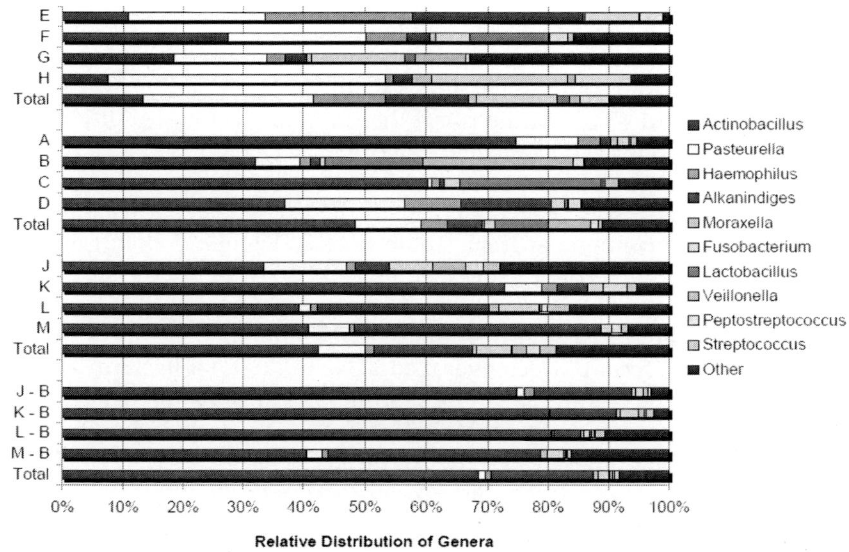

Figure 3 Relative distribution of the ten most abundant genera identified. The distribution of genera in each individual pig, as well as the group totals are shown.

3.4 Species level structure of tonsillar communities

We utilized a pairwise distances program to compare the 454 16S sequences from each pig to the V4 (variable region 4) regions of the type strains for species in the families *Pasteurellaceae* and *Streptococcaceae*. Using a 97% cutoff, we determined the closest affiliation for each sequence.

Sequences with closest affiliations to *Actinobacillus indolicus, A. minor, "A. porcitonsillarum"*, and *Haemophilus parasuis* were found in all samples. Sequences with closest affiliation to *A. porcinus, A. rossii, H. felis, Pasteurella aerogenes, P. canis, P. multocida*, and *Streptococcus suis* were found in most samples. Finally, sequences with closest affiliation to *S. plurextorum, A. lignieresii*, and *A. seminis* were found in small numbers in 40% of the samples.

3.5 Comparison of Herd 1 time 1 and time 2 communities

To determine whether the microbial communities in a given swine herd change over time, we compared the communities in tonsil tissue from pigs from Herd 1 sampled two years apart, in 2007 (time 1) and 2009 (time 2). Overall, the core microbiome of the two groups of samples remained quite similar at the phylum, class, order, and family levels, with the exception that *Neisseriales* were more frequently identified at time 2 (10.1% of the total) than time 1 (0.6%) (Additional file 3) and *Lactobacillaceae* were more common at time 1 (7.8% of the total) than time 2 (0.04%) (Additional file 4). Both were dominated by *Pasteurellaceae*, which comprised 64.2% of the total at time 1 and 50.3% at time 2 (Additional file 4). The distribution of the top ten genera was very similar, with the exception that *Lactobacillus* was much more common at time 1 than time 2 (Figure 3). Both groups of samples also contained the genera *Treponema* (phylum *Spirochaetes*) and *Chlamydia* (phylum *Chlamydiae*), with higher numbers of both seen at time 2. In addition, all Herd 1 time 1 samples also contained the genus *Pelosinus* (family *Veillonellaceae*), which averaged 2.3% of the total in Herd 1 time 1 but was not found at time 2 (Additional file 5). No genus present in most animals in the sample were identified as unique to Herd 1 time 2. There were no significant differences between the clusters at a 97% cutoff aligned to species of *Pasteurellaceae* and *Streptococcaceae* identified in the two groups of Herd 1 samples.

There were a variety of organisms associated with soil and water, such as *Polynucleobacter, Geobacter*, and *Azoarcus*, that were found only in Herd 1 at time 1, and generally only in one or two animals (Additional file 5). Indeed, Herd 1 time 1 samples contained members of 6 phyla, *Chloroflexi, Cyanobacteria, Gemmatimonadetes, Planctomycetes, Synergistetes*, and *Verrucomicrobia*, which together comprised 0.5% of the total tonsillar communities in Herd 1 time 1 samples but were not found in time 2 samples from herd 1 (Additional file 1).

3.6 Comparison of Herd 1 time 1 and Herd 2 communities

The microbial communities of Herd 1 time 1 and Herd 2 tissue samples showed strong similarities in the core microbiomes as well as distinct differences. In both herds, the tonsillar microbiomes were dominated by *Pasteurellaceae* (64.2% of Herd 1, 57.4% of Herd 2). However, the distribution of genera within that family varied between the herds; 75% of the *Pasteurellaceae* in Herd 1 were identified as genus *Actinobacillus*, while in Herd 2, 50% of the *Pasteurellaceae* were identified as genus *Pasteurella* (Figure 3 and Additional file 5). Reads identified as genus *Fusobacterium* formed a larger percentage of the total in Herd 2 (13.3%) than Herd 1 (1.7%). Distribution of the remaining major genera in the core microbiome was similar in the two herds (Figure 3).

Of the 101 genera identified, 41 were unique to Herd 1 and 11 were unique to Herd 2 (Additional file 5). Of those genera unique to Herd 1, only 2, *Treponema* (phylum *Spirochaetes)* and *Chlamydia* (phylum *Chlamydiae*) were found in most pigs from Herd 1. Reads identified as *Treponema* were found in all three groups of Herd 1 samples, although in smaller numbers at time 1 (0.3% of the total) than at time 2 (an average of 3.9% from tissue and brush samples), but were not found in Herd 2 (Figure 2). Reads identified as *Chlamydia* comprised on average 0.3% of the total reads in both groups of Herd 1 tissue specimens but were not found in brush specimens (Figure 2).

Of the 11 genera unique to Herd 2, only *Arcanobacterium* (phylum *Actinobacteria*, family *Actinomycetaceae)* was found in all animals. Reads identified as *Arcanobacterium* comprised 0.9% of Herd 2, but were not found in any Herd 1 specimen (Figure 2). This was the only genus unique to Herd 2 that was found in most animals and represented ≥ 0.1% of the total genera identified in all specimens. In addition, reads assigned to proposed phylum SR1 comprised 0.05% of Herd 2 but were not found in Herd 1.

At the 97% cutoff, both Herd 1 and Herd 2 contained the same core clusters of *Pasteurellaceae* and *Streptococcaceae*, although the relative proportions varied between the two groups of samples. For example, Herd 1 contained a higher fraction of sequences most closely affiliated with *A. minor* and fewer affiliated with *A. porcitonsillarum* than Herd 2. Furthermore, sequences most closely related to *Streptococcus plurextorum* and *S. thermophilus* were found in most samples from Herd 1, but not Herd 2.

3.7 Comparison of communities in Herd 1 time 2 tissue and brush samples

The predominant families (with ≥ 1% of the total assigned reads) and genera in both tissue and brush specimens from Herd 1 time 2 are the *Pasteurellaceae* (*Actinobacillus, Pasteurella, Haemophilus*); *Moraxellaceae (Alkanindiges, Moraxella); Neisseriaceae* (not assigned at the genus level); *Spirochaetaceae (Treponema); Veillonellaceae* (*Veillonella*); *Fusobacteriaceae* (*Fusobacterium*); *Streptococcaceae (Streptococcus)* and *Peptostreptococcaceae (Peptostreptococcus, Filifactor)* (Figure 2).

Enterobacteriaceae (several different species) and obligate anaerobes were more frequently found in tissue than in brush samples (Figures 2 and 3; Additional files 4 and 5). *Chlamydia*, an obligately intracellular organism, comprised 0.95% of the reads assigned to the genus level in the tissue specimens, but was not found in the brush specimens (Additional file 5). Other differences generally reflect either a very small number of reads or reads from only 1-2 samples (Additional files).

3.8 Statistical comparison of communities

Figure 4 shows the Jaccard analysis of the clustered sequences from each tonsil community. The samples from Herd 1 and Herd 2 from the same year (Time 1, 2007) are clearly distinguishable. Samples from Herd 1 taken 2 years later (Time 2, 2009) group with samples taken in time 1 from Herd 1, but are distant from Herd 2. The Jaccard indices of the time 2 sampling from Herd 1 where community samples were derived from both tonsil tissue and brushed tonsils indicate high similarity between these two sampling methods. Some variability exists within the Herd 1 Time 2 samples, as indicated by Pigs K and J from the brush samples where substantial similarities exist with at least two pigs from Herd 2 (lower left of Panel A).

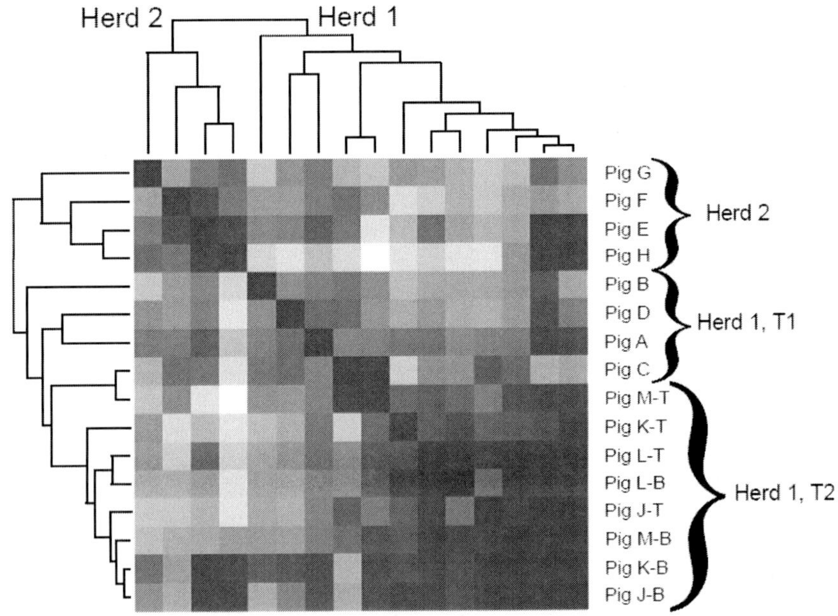

Figure 4 Jaccard indices of pig tonsil communities. Indices are presented clustered and plotted in heat map format where light to dark indicates increasing similarity.

Principle component analysis (PCA), using the first two factors (PC1 and PC2) was performed using communities from each pig sampled (Figure 5). Each point represents one tonsil community while the colored areas represent the 95% confidence limit of each group. Using the first two components explains 63% of the total variation among the individual samples. This demonstrated that the microbial communities were distinguishable from one another, but relatively close in phylogenetic space as judged by the range of eigenvalues.

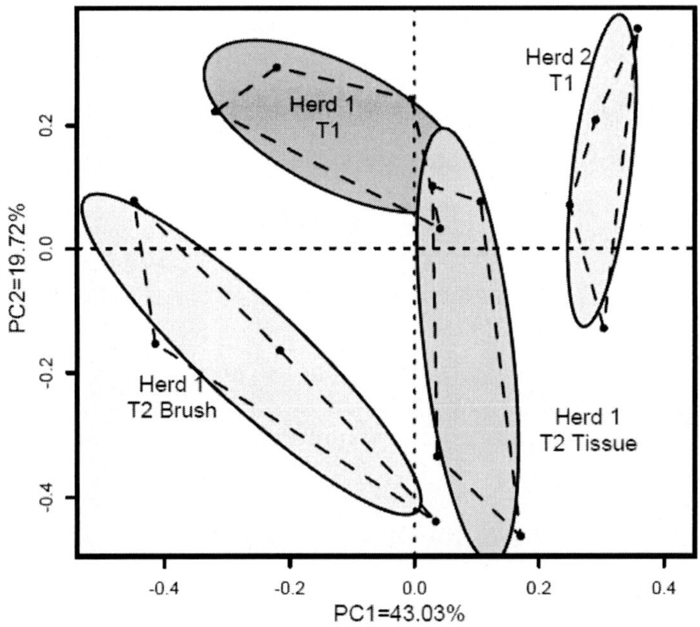

Figure 5 Principle Component Analysis (PCA) results on all individuals sampled. PCA was performed at the level of OTUs, clustering sequences at a 3% difference. The PCA plot of tonsillar communities shows PCA analysis using the first two components, accounting for 62.75% of the sample variation. Each point represents the tonsillar community of one individual pig. Colored circles represent the 95% confidence limit for each group of samples.

4. DISCUSSION

We have previously reported the first culture-independent analysis of the microbial communities of the tonsils of healthy pigs [14]. In the previous study, we analyzed 831 16S rRNA gene sequences from clone libraries constructed from samples from eight pigs from two healthy herds. In the current study, we analyzed the same samples (Herd 1 Time 1 and Herd 2 samples), as well as additional samples collected from Herd 1 two years after the first set by two different collection methods (Herd 1 Time 2 tissue and brush samples), using high through-put next generation sequencing methods to generate 210,433 sequences to extend our initial community analysis to greater depth.

Overall, we found strong similarities between the four groups of samples as well as minor unique differences. We identified a "core microbiome" for porcine tonsils that includes eight core genera from six core families (*Pasteurellaceae, Moraxellaceae, Fusobacteriaceae, Veillonellaceae, Peptostreptococcaceae*, and *Streptococaceae*) as well as members of the *Enterobacteriaceae*, which varied in genera found from sample to sample, and *Neisseriaceae*, which could not be identified to the genus level (Table 3). Two additional genera, *Moraxella* and *Lactobacillus*, that are included in the ten most abundant genera identified (Figure 3) were found less consistently, and in particular were missing from most of the Herd 1 Time 2 tissue specimens, and therefore are not included in the core microbiome that we have defined as "found in most animals in all groups".

As in the previous study [14], *Pasteurellaceae (Actinobacillus, Haemophilus,* and *Pasteurella* species) dominated the tonsillar microbial communities in all pigs examined, comprising on average 60.2% of the total reads, and ranging from 39.2% to 87.0% in individual pigs. The distribution of genera within the family *Pasteurellaceae* - with *Actinobacillus* predominate in Herd 1 samples and *Pasteurella* in Herd 2-also compares well with the previous study.

However, a major difference between the results of the two studies is the glaring lack of *Bacteroidetes* in the current data. In the previous study [14], sequences identified as belonging to the order *Bacteroidales* (genera *Bacteroides, Prevotella*, and *Porphyromonas*) comprised the second most dominant group (30% of the sequenced clones) after the *Pasteurellales*, and were found in almost all animals. Three additional species of *Porphyromonadaceae* (*Dysgonomonas, Parabacteroides*, and *Tannerella)* were found in a few animals, particularly from Herd 2. In contrast, *Bacteroidales* comprised 0.3% of the sequence reads in the current study, including among the Herd 1 time 1 and Herd 2 samples that were the identical samples used in the previous study. An unexpectedly low abundance of *Bacteroidetes* has been found in other studies using high-throughput bar-coded pyrosequencing [22-24]. One potential explanation cited is variation in the samples analyzed [22-24], which is not the case in our study. These were the same DNA samples used in the previous study [14]. A second explanation would be partial degradation of these samples, resulting in loss of *Bacteroidetes* DNA. However, these same samples have also recently been analyzed with 454-Titanium primers and shown to still contain *Bacteroidetes* DNA (M. H. Mulks & T. L. Marsh, unpublished observations). A more likely explanation is amplification bias during the

PCR reactions. However, all of the primer sets used in these studies, which targeted three different variable regions of the 16S gene-the V4 region in the current study, V5 [22], and V6 regions [23, 24]-were shown in silico to cover the *Bacteroidetes* species, and the V4 primers were tested experimentally against genomic DNA from known *Bacteroides* isolates and shown to amplify 16 s rDNA. It is likely that members of the *Bacteroidetes* are also part of the core microbiome of porcine tonsils, despite the lack of evidence in our current data.

While there were clear and strong similarities between the core microbiomes of all of the groups examined, there were also unique differences in minor genera found or missing from particular groups. These differences can not readily be explained by differences in overall herd management or antibiotic usage in the groups (no antibiotics in Herd 1 time 1, Tylan in Herd 1 time 2, and Tylan plus Pulmotil in Herd 2). For example, reads identified as *Arcanobacterium* were found in all Herd 2 samples, and comprised 0.93% of the reads from that herd, but were not found in any Herd 1 sample. In contrast, reads identified as *Treponema* were found in all but one sample from Herd 1, but not in any sample from Herd 2, and *Chlamydia* were found in Herd 1 tissue samples but not in Herd 2 samples. *Lactobacillus* was abundant in most samples from both Herd 1 time 1 and Herd 2, but was rare in Herd 1 time 2 samples. *Pelosinus* was abundant only in Herd 1 time 1, not Herd 2 or Herd 1 time 2 samples. There were many other genera found in small numbers in 1-2 animals per group that were unique to that group, such as *Polynucleobacter* and *Geobacter* in Pig D from Herd 1 time 1 (Additional file 5), but no others that could be found in most animals in one group but not in animals of another group. These results indicate that, despite the small sample number, we can identify differences in the minor genera found in the two different herds.

One goal of this project was to test tonsil brushes as an alternative, non-invasive method to collect tonsil samples, eliminating the need to euthanize animals to collect tonsil tissue. The Jaccard analysis (Figure 4) clearly indicated that all samples from the second sampling of Herd 1 were more similar to each other than to samples from Herd 1 and 2. We could detect differences between the brush and tissue extraction procedures as indicated in Figure 5, but the difference was small based on the range of eigenvalues. The detected statistical differences were a consequence of an increase in the percentage of reads identified as *Actinobacillus*, fewer sequences of *Fusobacterium, Veillonella*, and *Peptostreptococcus*), and no detectable sequences from the obligate intracellular pathogen *Chlamydia* in the brush specimens. These results indicate that use of tonsil brushes provides a slightly different but sufficiently representative picture of the microbial community that is suitable for within or across herd comparisons when compared to direct extraction of tonsil tissue.

A wide range of bacterial and viral porcine pathogens are routinely isolated from the tonsils [1]. In this study, we identified large numbers of sequences whose closest affiliate in the database were *Haemophilus parasuis* and *Pasteurella multocida* (on average 12.8% and 9.3%, respectively, of reads identified at the 97% cutoff), as well as small numbers of sequences closest to *Streptococcus suis* (on average 0.4% of reads identified) from almost all samples. However, we did not find sequences affiliated with *Actinobacillus pleuropneumoniae* or *A. suis*, which have been reported to be found in most swine herds in Ontario, Canada [7]. Small numbers of sequences closest to *Mycoplasma* were found in a few pigs, but these were not identified beyond the Classifier function of the RDP. Herd 1 has been regularly tested and found to be free of *A. pleuropneumoniae, A. suis*, and *Mycoplasma*, which is substantiated by these results. We were surprised not to find sequences consistent with the presence of pathogenic *Actinobacillus* species in Herd 2, which has had a history of chronic but undefined respiratory problems. It is possible that these chronic problems are related to the higher numbers of *Pasteurella sequences* found in Herd 2, or to the presence of another known respiratory pathogen, *Arcanobacterium*, found in Herd 2 but not Herd 1.

In addition to porcine pathogens, many bacterial agents of foodborne infections of humans have been isolated from pig tonsils, including members of the *Enterobacteriaceae* such as *Salmonella* species, *Escherichia coli*, and *Yersinia enterocolitica* as well as *Campylobacter* species and *Listeria monocytogenes* [9-13]. We found low numbers of *Campylobacter* (0.17% of total reads) and *Escherichia* (0.59% of total reads) in most of the pig tonsils in this study. In addition, we found other *Enterobacteriaceae* (1.9% of the total) that are rarely associated with human foodborne illness, including *Citrobacter, Enterobacter, Morganella, Proteus*, and *Providencia*, in one or more pigs. We did not find *Salmonella, Yersinia*, or *Listeria* in these tonsil samples from healthy pigs.

The only other mammalian system where the tonsillar microbiota has been reported is in humans. Culture-based studies of human tonsils have identified *Streptococcus pyogenes; S. pneumoniae;* Group C, F, and G β-hemolytic streptococci; several α-hemolytic and non-hemolytic streptococci; *Staphylococcus aureus; Haemophilus influenzae; H. parainfluenzae;* and *Moraxella catarrhalis* in aerobic cultures [25-31]. Many species of the *Bacteroides-Prevotella-Porphyromonas* group, *Fusobacterium, Lactobacillus, Peptostreptococcus*, and *Veillonella* have also been isolated using anaerobic cultures. Culture-independent analyses using 16S rRNA gene sequences, by clone libraries [31, 32], PhyloChip microarray [32], and pyrosequencing [22], have all supported the predominance of members of the phylum *Firmicutes* in human tonsils/oropharynx, with *Streptococcus mitis, S. australis, S. parasanguinis, S. intermedius, Gemella sanguinis, G. haemolysans, Granulicatella adjacens*, and *Gr. elegans* most commonly found [31, 33]. *Proteobacteria* and *Bacteroidetes* were also commonly identified by 16S sequencing techniques [31-33]. In strong contrast to our results with porcine tonsils, *Pasteurellaceae* were identified as a very small percent of the human tonsillar or oropharyngeal community by culture-independent methods [31, 32].

We recognize that the short reads derived from 454-Flx limit the phylogenetic information. However previous workers have reported that short reads suffice (at some level) for community analyses [18, 34]. As next generation sequencing improves and the read lengths grow, short reads will become less of an issue.

We also recognize that this study is limited to four animals from one herd and eight from a second herd, all healthy animals from two farms that are geographically close and have similar management styles. Clearly this is a preliminary "core microbiome" that will likely evolve as samples from more diverse herds are analyzed. With this limitation, the strong similarities seen between samples suggest that next generation sequencing will help to develop a robust phylogenetic view of the tonsil community across geographically distant herds and commercially relevant species of pigs and management styles, and allow comparison of communities in healthy animals to those in animals with disease as well as asymptomatic carriers of pathogens.

5. CONCLUSIONS

The 16S rRNA gene pyrosequencing results reported here extend and support our previous studies using 16S clone libraries to describe the microbial communities in tonsils of healthy pigs. Our results have defined a core microbiome found in tonsil specimens from two herds and at two time points from the same herd, and have also demonstrated the presence of minor components of the tonsillar microbiome unique to each herd. How the normal microbiota of the tonsils varies with and affects acquisition and carriage of pathogens, both porcine pathogens and those associated with foodborne illness in humans, is the subject of ongoing studies.

ACKNOWLEDGEMENTS

We thank Rhiannon LeVeque and Sargurunathan Subashchandrabose for assistance with collection of specimens, and Fan Yang for assistance with the statistical analysis. This research was supported by grants from the National Pork Board and the Michigan State University Center for Microbial Pathogenesis.

REFERENCES

1. Horter DC, Yoon KJ, Zimmerman JJ: A review of porcine tonsils in immunity and disease. Anim Health Res Rev 2003, 4(2):143-155.
2. Belz GT, Heath TJ: Tonsils of the soft palate of young pigs: crypt structure and lymphoepithelium. Anat Rec 1996, 245(1):102-113.
3. Arends JP, Hartwig N, Rudolphy M, Zanen HC: Carrier rate of Streptococcus sui capsular type 2 in palatine tonsils of slaughtered pigs. J Clin Microbiol 1984, 20(5):945-947.
4. Chiers K, Donne E, Van Overbeke I, Ducatelle R, Haesebrouck F: Actinobacillus pleuropneumonia infections in closed swine herds: infection patterns and serological profiles. Vet Microbiol 2002, 85(4):343-352.
5. Horter DC, Pogranichniy RM, Chang CC, Evans RB, Yoon KJ, Zimmerman JJ: Characterization of the carrier state in porcine reproductive and respiratory syndrome virus infection. Vet Microbiol 2002, 86(3):213-228.
6. Cheville NF, Mengeling WL: The pathogenesis of chronic hog cholera (swine fever). Histologic, immunofluorescent, and electron microscopic studies. Lab Invest 1969, 20(3):261-274.
7. MacInnes JI, Gottschalk M, Lone AG, Metcalf DS, Ojha S, Rosendal T, Watson SB, Friendship RM: Prevalence of Actinobacillus pleuropneumoniae, Actinobacillus suis, Haemophilus parasuis, Pasteurella multocid, and Streptococcus sui in representative Ontario swine herds. Can J Vet Res 2008, 72(3):242-248.
8. Marois C, Cariolet R, Morvan H, Kobisch M: Transmission of pathogenic respiratory bacteria to specific pathogen free pigs at slaughter. Vet Microbiol 2008, 129(3-4):325-332.
9. Bucher M, Meyer C, Grotzbach B, Wacheck S, Stolle A, Fredriksson-Ahomaa M: Epidemiological data on pathogenic Yersinia enterocolitic in Southern Germany during 2000-2006. Foodborne Pathog Dis 2008, 5(3):273-280.
10. Autio T, Markkula A, Hellstrom S, Niskanen T, Lunden J, Korkeala H: Prevalence and genetic diversity of Listeria monocytogene in the tonsils of pigs. J Food Prot 2004, 67(4):805-808.
11. Fredriksson-Ahomaa M, Gerhardt M, Stolle A: High bacterial contamination of pig tonsils at slaughter. Meat Sci 2009, 83:334-336.
12. Fedorka-Cray PJ, Kelley LC, Stabel TJ, Gray JT, Laufer JA: Alternate routes of invasion may affect pathogenesis of Salmonella typhimuriu in swine. Infect Immun 1995, 63(7):2658-2664.
13. Swanenburg M, van der Wolf PJ, Urlings HA, Snijders JM, van Knapen F: Salmonella in slaughter pigs: the effect of logistic slaughter procedures of pigs on the prevalence of Salmonell in pork. Int J Food Microbiol 2001, 70(3):231-242.
14. Lowe BA, Marsh TL, Isaacs-Cosgrove N, Kirkwood RN, Kiupel M, Mulks MH:Microbial communities in the tonsils of healthy pigs. Vet Microbiol 2011, 147(3-4):346-357.
15. Cole JR, Chai B, Farris RJ, Wang Q, Kulam SA, McGarrell DM, Garrity GM, Tiedje JM: The Ribosomal Database Project (RDP-II): sequences and tools for high-throughput rRNA analysis. Nucleic Acids Res 2005, 33 Database:D294-296.
16. Nawrocki E, Kolbe D, Eddy S: Infernal 1.0: inference of RNA alignments. Bioinformatics 2009, 25:1335-1337.
17. Cole J, Wang Q, Cardenas E, Fish J, Chai B, Farris R, Kulam-Syed-Mohideen A, McGarrell D, Marsh T, Garrity G, et al: The Ribosomal Database Project: improved alignments and new tools for rRNA analysis. Nucleic Acids Res 2009, 37:D141-D145.
18. Huse SM, Welch DM, Morrison HG, Sogin ML: Ironing out the wrinkles in the rare biosphere through improved OUT clustering. Environ Microbiol 2010, 12(7):1889-1898.
19. Stackebrandt E, Goebel BM: Taxonomic note: a place for DNA:DNA reassociation and 16S rRNA sequence analysis in the present species definition in Bacteriology. Int J Syst Bacteriol 1994, 44:846-849.

20. Rossello-Mora R, Amann R: The species concept for prokaryotes. FEMS Microbiol Rev 2001, 25(1):39-67.
21. Schloss PD, Westcott SL, Ryabin T, Hall JR, Hartmann M, Hollister EB, Lesniewski RA, Oakley BB, Parks DH, Robinson CJ, et al: Introducing mothur: open-source, platform-independent, community-supported software for describing and comparing microbial communities. Appl Environ Microbiol 2009, 75(23):7537-7541.
22. Laserevic V, Whiteson K, Huse S, Hernandez D, Farinelli L, Ostera M, Schrenzel J, Francois P: Metagenomic study of the oral microbiota by Illumina high-throughput sequencing. J Microbiol Meth 2009, 79:266-271.
23. Andersson AF, Lindberg M, Jakobsson H, Backhed F, Nyren P, Engstrand L: Comparative analysis of human gut microbiota by bar-coded pyrosequencing. PLoS One 2008, 3:e2836.
24. Jakobsson HE, Jernberg C, Andersson AF, Sjolund-Karlsson M, Jansson JK, Engstrand L: Short-term antibiotic treatment has differing long-term impacts on the human throat and gut microbiome. PLoS One 2010, 5: e9836.
25. Toner JG, Stewart TJ, Campbell JB, Hunter J: Tonsil flora in the very young tonsillectomy patient. Clin Otolaryngol 1986, 11:171-174.
26. Gaffney RJ, Timon CI, Freeman DJ, Walsh MA, Cafferkey MT: Bacteriology of tonsil and adenoid and sampling techniques of adenoidal bacteriology. Respir Med 1993, 87:303-308.
27. Gaffney RJ, Freeman DJ, Walsh MA, Cafferkey MT: Differences in tonsil core bacteriology in adults and children: a prospective study of 262 patients. Respir Med 1991, 85:383-388.
28. Gaffney RJ, Cafferkey M: Bacteriology of normal and diseased tonsils assessed by fine-needle aspiration: Haemophilus influenza and the pathogenesis of recurrent acute tonsillitis. Clin Otolaryngol 1998, 23:181-185.
29. Cafferkey MT, Timon CI, O'Regan M, Walsh M: Effect of pre-operative antibiotic treatment on the bacterial content of the tonsil. Clin Otolaryngol 1993, 18:512-516.
30. Brook I, Foote PA: Microbiology of normal tonsils. Ann Otol Rhinol Laryngol 1990, 99:980-983.
31. Aas JA, Paster BJ, Stokes LN, Olsen I, Dewhirst FE: Defining the normal bacterial flora of the oral cavity. J Clin Microbiol 2005, 43:5721-5732.
32. Lemon KP, Klepac-Ceraj V, Schiffer HK, Brodie EL, Lynch SV, Kolter R: Comparative analyses of the bacterial microbiota of the human nostril and oropharynx. mBio 2010, 1:e00129-00110.
33. Paster BJ, Olsen I, Aas JA, Dewhirst FE: The breadth of bacterial diversity in the human periodontal pocket and other oral sites. Periodontology 2006, 42:80-87.
34. Liu ZZ, Lozupone C, Hamady M, Bushman FD, Knight R: Short pyosequencing reads suffice for accurate microbial community analysis. Nucleic Acids Res 2007, 35:e120.

CHAPTER 7

Identification of Bacterial Fish Pathogens in Brazil by Direct Colony PCR and 16S rRNA Gene Sequencing

F. A. Sebastião, L. R. Furlan, D. T. Hashimoto, F. Pilarski*

Aquaculture Center of São Paulo State University—CAUNESP, Universidade Estadual Paulista, Jaboticabal, Brazil

ABSTRACT

Intensive fish farming systems in Brazil have increased the disease incidence, mainly of bacterial origin, due to higher stocking density, high organic matter levels and poor quality of the aquatic environment that causes high mortality rates during outbreaks. The identification of pathogenic species using a fast and reliable method of diagnosis is essential for successful epidemiological studies and disease control. The present study evaluated the use of direct colony PCR in combination with 16S rRNA gene sequencing to diagnose fish bacterial diseases, with the goal of reducing the costs and time necessary for bacterial identification. The method was successful for all 178 isolates tested and produced bands with the same intensity as the standard PCR performed using pure DNA. In conclusion, the genetics methods allowed detecting the most common and important pathogens in Aquaculture, including 12 species of occurrence in Brazilian fish farms. The results of the present study constitute an advance in the available diagnostic methods for bacterial pathogens in fish farms.

Keywords: Direct Colony PCR, 16S rRNA Sequencing, Bacterial Fish Pathogens

1. INTRODUCTION

Due to its high water availability and favorable climate conditions, Brazil displays high potential for the development of fish farming, which is an activity that has been growing substantially over the last few years. According to the Food and Agriculture Organization of the United Nations [1], Brazil is the second largest aquaculture producer in Latin America and the Caribbean, and freshwater aquaculture (tilapia, carp, and native fish) represented 87% (545,300 ton) of the total aquaculture production in 2011.

The growing interest in this activity and, consequently, the search for higher profitability, have been leading producers to adopt super-intensive production systems. However, the high density of confined fish, inadequate farming management practices, and water contamination by toxic products cause chronic stress and immunosuppression in farmed animals. These effects lead to the occurrence of diseases and epizootic outbreaks caused by pathogens that would not have high expression in natural environments [2]. Knowledge about the etiological agents, pathogenesis, biochemistry, antigenicity, epizootiology, and inter-relationship of stress and environmental factors of bacterial infections affecting fish is essential to avoid and control diseases. However, these factors have not been well studied, especially because fish farming is a recent activity, with its intensification beginning in the 1990s [3].

Gram-negative bacteria such as Aeromonas, Flavobacterium, Pseudomonas, and Francisella and gram-posi- tive bacteria from the genera Streptococcus and Lactococcus [4] - [11] are some of the pathogens responsible for economic losses in Brazil. They can cause high fish mortality rates up to 72 h after infection [12].

Although the number of studies focusing on the diagnosis of bacterial etiological agents has increased over the last few years, there are still few available alternatives for the control of fish bacterial infections in Brazil. Therefore, quicker and more effective diagnostic alternatives are necessary, which would help control diseases before they lead to irreversible clinical consequences and high mortality rates. Molecular diagnostic methods use reduced volumes of sample material and exhibit high sensitivity, specificity, and accuracy in pathogen detection [13].

Methods that do not require purified DNA extraction, such as direct colony PCR, are quicker and less expensive and may greatly aid in the early detection of fish pathogens [14]. In addition, because not all microorganism sequences are catalogued in current databases, the use of universal and degenerate primers is a wise strategy. For this reason, methods based on 16S rRNA ribosomal gene amplification and sequencing have been widely explored [13].

The use of universal PCR primers is based on the hypothesis that the primers used are complementary to conserved regions of genes in the environment, resulting in amplification; in turn, heterogeneity is found inside of the fragments flanked by the primers, in hypervariable regions [15] [16]. This method has been revolutionizing microbial ecology, from studies of non-cultivable bacteria to the correct identification of pathogens for accurate diagnoses.

The aim of the present study was to evaluate the direct colony PCR combined with 16S rRNA gene sequencing as a faster and less expensive method to identify fish bacterial pathogens, compared to the classic PCR protocol. Moreover, we have used these methods to demonstrate the efficiency of genetic approaches for the practical evaluation of the diagnosis of aquaculture diseases in Brazilian fish farms.

2. MATERIALS AND METHODS

2.1. Bacterial Strains and Culture Conditions

178 bacterial isolates were obtained between 2010 and 2014 from the following hosts (n = number of assessed fish): tilapia (Oreochromis niloticus n = 93), tambaqui (Colossoma macropomum n = 10), carp (Cyprinus carpio n = 3), cachara (Pseudoplatystoma reticulatum n = 34), and pacu (Piaractus mesopotamicus n = 8).

The hosts exhibited clinical signs of bacterial diseases, such as skin ulcerative lesions, hemorrhagic septicemia, meningoencephalitis, fin rot, exophthalmia, and were collected at fish farms in different regions of Brazil:Dourados (Mato Grosso do Sul State, 22°13'16"S, 54°48'20"W, n = 38), Rio de Janeiro (Rio de Janeiro State, 22°54'S, 43°10'W, n = 4), Itambaracá (Paraná State, 23°0'49"S, 50°24'7"W, n = 10), Itaju (22°25'37"S, 45°27'11"W), Arealva (22°1'38"S, 48°54'36"W), Porto Ferreira (21°51'18"S, 47°28'45"W), Guaíra (20°19'5"S, 48°18'42"W), Santa Fé do Sul (20°12'43"S, 50°55'38"W), Palmital (22°47'30"S, 50°12'18"W) and Jaboticabal (21°15'19"S, 48°19'21"W—São Paulo State, n = 123).

For the isolation of bacteria, scrapings were performed using sterile swabs on fish kidneys and brain. Gram- negative colonies were plated on TSA

(Tryptic Soy Agar-Biolife), and TSB (Tryptic Soy Broth-Biolife) and incubated for 24 h in bacteriological incubator adjusted to 28°C. While gram-positive colonies were seeded in Columbia blood agar (Difco) incubated for 24 - 72 h at 30°C and subcultured in BHI (Brain Heart Infusion Broth, Himedia).

The strains of Palmital (SP) were obtained directly from the Laboratory of Aquatic Animal Disease, APTA, Votuporanga, SP.

2.2. Molecular Identification of Isolates

Two methods of molecular diagnosis were compared in this study aiming to evaluate the efficiency of direct colony PCR (time and cost effectiveness) in relation to the PCR amplification of purified DNA by extraction, both combined with gene sequencing (Table 1).

The standard PCR of purified DNA method followed the steps below.

2.2.1. DNA Extraction

One colony of each isolated was transferred to a tube containing appropriate liquid culture medium (TSB for gram-negative and BHI for gram-positive) and incubated at 28°C until the OD_{600} was between 1 and 1.5. Following incubation, 1.0 mL of the bacteria culture was centrifuged at 12,000×g for 1 min, the supernatant was discarded, and the pellet was frozen at −20°C until DNA extraction. The Axyprep® miniprep kit for bacterial genomic DNA was used according to the manufacturer's instructions (Axygen Biosciences, Union City, CA, USA). DNA was quantified by fluorometry using a Qubit 2.0 fluorometer (Life Technologies, NY, USA).

2.2.2. Standard PCR

PCR was performed in a 25 μL final volume, containing 2.5 μL of 10X buffer (10 mM Tris-HCl, 50 mM KCl), 0.2 μL of 25 mM dNTP, 1.0 μL of 50 mM MgSO4, 0.2 μL of Taq High Fidelity (Platinum®Taq DNA Polymerase, Life Technologies, NY, USA), 2.0 μL of each primer (10 pmol), 25 ng of DNA template, and Milli-Q water up to the final volume. The PCR program consisted of 94°C for 2 min; 35 cycles of 94°C for 30 seconds, 55°C for 30 s, and 68°C for 1.5 min; and final extension at 68°C for 10 min. We used the primers 8F/907R (Table 2), specific for the 16S rRNA bacterial gene [15] [17] [18]. The resulting amplicons of approximately 900 bp (base pair) were analyzed by electrophoresis in 1.5% agarose gel stained with ethidium bromide, according to Sambrook et at. [19].

2.2.3. Purification of PCR Products and Gene Sequencing

PCR products were purified using a MinElute Kit (Qiagen, Crawley, West Sussex, UK) according to the manufacturer's instructions. Purified PCR products were quantified using a Qubit 2.0 fluorometer, and gene sequencing was performed using 50 ng/μL per sample. Sequencing was performed according to Sanger [20]. PCR products were amplified using AmpliTaq polymerase and BigDye Terminator (Applied Biosystems) according to the manufacturer's instructions, using the primer 907R. Sequencing was performed using an ABI PRISM 3730 DNA analyzer (Applied Biosystems).

Table 1. Steps of the two methods compared in this study: Standard PCR of purified DNA and direct colony PCR, both combined with the 16S rRNA gene sequencing.

Steps	Standard PCR	Direct colony PCR
Bacterial isolation	X	X
Replication in broth	X	-
DNA extraction	X	-
PCR	X	X
Electrophoresis	X	X
PCR product purification	X	-
16S rRNA sequencing	X	X
Nucleotide analysis	X	X

Table 2. Sequence of primers used for amplification of the 16S rRNA gene.

Primer	Target sequence 5'- 3'
8F	AGA GTT TGA TYM TGG CTC AG
907R	CCG TCA ATT CMT TTR AGT TT

2.3. Direct Colony PCR

This method allows PCR to be performed on colonies isolated from Petri dishes, without the step of DNA extraction. Colonies (1 - 2 mm diameter) were inoculated by placing a sterile tooth pick at the bottom of a PCR tube (0.2 mL) and incubated at −20°C overnight. The following solution was then added in the PCR tube: 2.0 μL of 10× buffer (10 mM Tris-HCl, 50 mM KCl), 1.2 μL of 50 mM $MgCl_2$, 0.2 μL of 25 mM dNTP, 0.7 μL of each primer 8F/907R (10 pmol/μL), 0.2 μL of Taq DNA polymerase (2.5 U), and Milli-Q water up to 20 μL. The PCR program consisted of 95°C for 5 min; 30 cycles of 95°C for 1 min, 54°C for 1.5 min, and 72°C for 1 min; and final extension at 72°C for 5 min. The amplified PCR products, at 50 ng/μL mean concentration, were analyzed by electrophoresis in a 1.5% agarose gel stained with ethidium bromide [19]. The gels were visualized under UV light, using a ChemiDoc MP imaging system (Bio-Rad). Samples were quantified by fluorometry using a Qubit 2.0 fluorometer and sequenced as described above.

After sequencing, samples of both methods had their nucleotides analyzed.

2.4. Analysis of Nucleotide Sequences

The obtained sequences were visualized using the Bio Edit Sequence Alignment Editor software (v. 7.1.11). Phred quality of sequences was determined. The initial and final portions of the sequences were then removed, keeping only the high-quality fragment.

After trim, sequences were exported in FASTA format and compared with the GenBank database (http://www.ncbi.nlm.nih.gov/genbank/) using the Eztaxon algorithm (http://www.ezbiocloud.net/eztaxon/identify). 100% coverage and identity ≥ 98% were considered for specific identification. Sequences were also submitted to Ribosomal Database Project II (http://rdp.cme.msu.edu) for comparison and identification.

The sequences obtained in the present study were deposited at NCBI GenBank under accession numbers KJ560937 to KJ561113. The complete list of species identified, accession numbers, place of origin, fish species, season and year of collection, and size of amplified PCR products were included as Supplementary Material.

The Brazilian isolates tested were S. agalactiae (n = 23), S. iniae (n = 4), Lactococcus lactis (n = 11), L. raffinolactis (n = 2), L. garvieae (n = 16), Enterococcus casseliflavus (n = 16), E. durans (n = 2), E. faecalis (n = 11), Edwardsiella tarda (n = 5), Aeromonas hydrophila (n = 16), A. jandaei (n = 2), A. veronii (n = 15), Pseudomonas sp. (n = 15).

A phylogenetic diagram was constructed for validation of the sequencing data, using the 138 isolates listed above from the 178 of the present study. In addition, we used as reference 16 sequences originated from different countries (Table 3), obtained from GenBank database.

The 154 FASTA sequences were aligned using the ClustalW Multiple Alignment tool (BioEdit Sequence Alignment Editor software, v. 7.1.11). The data were then entered in Mega software (v. 5.05) to determine the best substitution model. As a result of the preliminary analysis, a maximum-likelihood phylogenetic diagram was constructed, using the Kimura 2-parameter model, with a gamma-shape parameter with 5 categories, the nearest-neighbor-interchange tree inference option. The stability of internal nodes was assessed by bootstrap analysis with 1000 replicates.

3. RESULTS AND DISCUSSION

3.1. The Comparison of the Two Methods: Direct Colony PCR and Classic PCR Protocol

We found no difference in the band size in agarose electrophoresis, nor in the peaks pattern of electrophero- grams in the two methods evaluated.

Bands resulting from the direct colony PCR exhibited the same intensity as those of the standard PCR of purified DNA, for all 178 isolates tested (Figure 1).

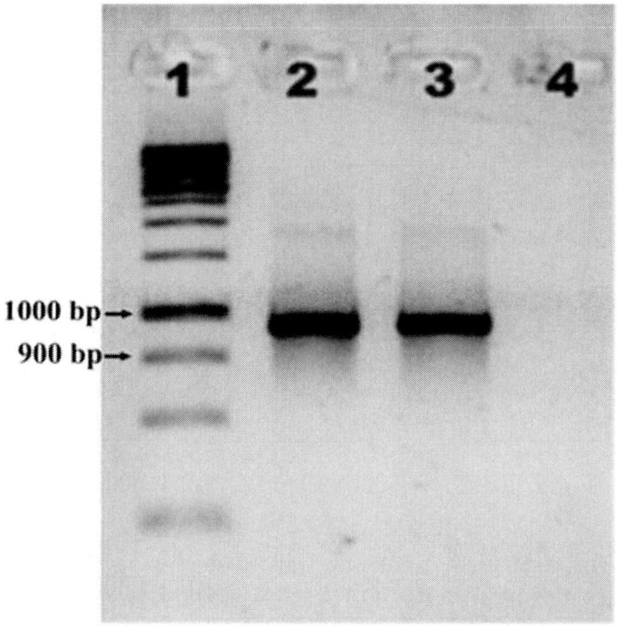

Figure 1. Electropherogram of 1.5% agarose gel stained with ethidium bromide, showing amplification of 16S rRNA gene (primers 8F/907R). Lane 1, Marker 1kb. Lane 2, direct colony PCR. Lane 3, standard PCR. Lane 4, negative control.

Table 3. Reference strains used for the maximum-likelihood phylogenetic analysis and their places of origin.

Genbank access number	Identification	Origin
JX861241	*Aeromonas veronii*	India
ATCC35624	*Aeromonas veronii*	Japan
NR_074841	*Aeromonas hydrophila*	USA
JN644061	*Aeromonas jandaei*	China
ATCC49568	*Aeromonas jandaei*	India
NBRC_105688	*Edwardsiella tarda*	Japan
EU239205	*Pseudomonas fulva*	Korea
KC210866	*Enterococcus casseliflavus*	China
AB530699	*Enterococcus faecalis*	Thailand
NR_036922	*Enterococcus durans*	Germany
KC176716	*Streptococcus agalactiae*	China
NR_027517	*Streptococcus dysgalactiae*	Japan
NR_025148	*Streptococcus iniae*	Israel
NR_044359	*Lactococcus raffinolactis*	South Korea
KC429785	*Lactococcus lactis*	China
ATCC49156	*Lactococcus garvieae*	USA

Electropherograms resulting from the sequencing of both methods exhibited Phred quality scores ≥20. All the isolates had the same results of bacterial identification for both techniques (direct colony PCR and standard PCR of purified DNA). Thus, direct colony PCR was a less expensive and faster diagnostic method, as shown on Table 4. There were 51% savings in cost analysis per sample for direct colony PCR compared to Standard PCR of purified DNA. Moreover, direct colony PCR reduces 2 days in time to issue the final report. After the installation of a bacterial outbreak, fish shoals can be decimated by up to 72 hours. Therefore, rapid diagnosis in aquaculture is a critical point in the production chain, which can be assessed by the genetic tools of the present study.

A faster diagnosis is important, since the one based on classical microbiology techniques (isolation, platting and biochemical tests) can exceed the time for treatment in seven to 15 days and, in many cases, ending up inconclusive. The molecular diagnosis, on the other hand, can provide a faster, low cost, conclusive diagnosis, which is essential to determine the best treatment in fish farming (Table 5).

Besides, in an attempt to control disease outbreaks, in classical scenery in Brazil, producers use multiple antibiotics indiscriminately, selecting resistant strains, contaminating fish, water and raising the risks to consumer health, endorsing the need for rapid and effective diagnosis [21].

A maximum-likelihood phylogenetic tree was built to validate the sequencing data (Figure 2). The bacterial isolates of the same species or phylogenetic related were correctly grouped into a common branch, as expected. The principle of maximum likelihood for phylogenetic inference evaluates the probability of a given model of evolutionary changes explaining the origin of the data observed. In this method, the initial tree is constructed using the neighbor-joining method, and the length of each branch is adjusted to maximize the likelihood that the information will produce the topology of the tree for the desired evolutionary model [22].

These results confirm and validate the direct colony PCR method to be applied as a reliable tool for the identification of bacterial fish pathogens in aquaculture. Although this method has already been used in previous studies for different purposes [14] [23], the present study represents the first practical application for the diagnosis of aquaculture diseases, a field lacking in terms of technological advancement.

Table 4. Cost analysis per sample for bacterial identification, performed in university laboratory already equipped.

	Standard PCR		**Direct colony PCR**	
Steps	US$	Time	US$	Time
Bacterial isolation on plate	0.31	24 - 72 h	0.31	24 - 72 h
Replication in broth	0.41	24 - 48 h	-	-
DNA extraction	5.6	1 - 3 h	-	-
PCR	2.52	3 h	2.52	3 h
Electrophoresis	0.27	1.5 h	0.27	1.5 h
PCR product purification	2.54	1 h	-	-
16S rRNA sequencing	5.0	24 h	5.0	24 h
Nucleotide analysis	-	1 h	-	1 h
Overall	**16.65**	**4 - 7 days**	**8.1**	**3 - 5 days**

*Isolation times vary depending on the species being cultured.

Table 5. Advantages and disadvantages of each method for aquaculture diagnosis.

	Classical microbiology	**Standard PCR**	**Colony PCR**
Advantages	Less technicization	Conclusive diagnosis, faster than the classical Microbiology	Conclusive diagnosis; 51% more economical; 24 - 48 h faster than standard PCR.
Disadvantages	Inconclusive and time consuming diagnosis	More costly; It depends on bacterial culture	1% to 3% can fail; Still depends bacterial culture

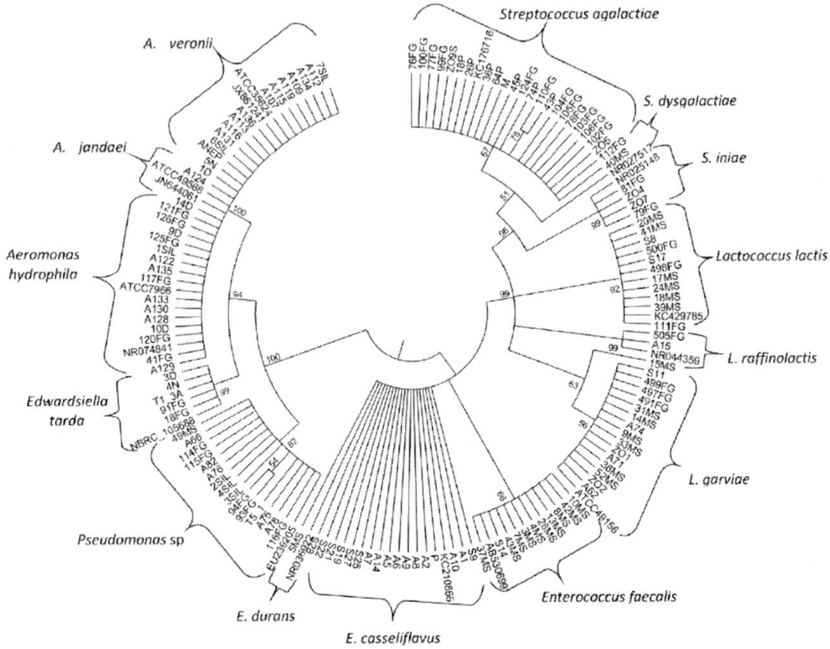

Figure 2. Relationship among different bacteria species using 16S rRNA gene sequences, inferred by maximum-likelihood method. The phylogenetic diagram shows the correct clustering of related fish bacteria isolated in the present study.

3.2. The Analysis of the Common Bacterial Fish Pathogens

Direct colony PCR, combined with gene sequencing, was able to detect the most common and important pathogens in aquaculture, such as Aeromonas hydrophila, Aeromonas veronii, Aeromonas jandaei, Streptococcus agalactiae, Streptococcus iniae, Streptococcus dysgalactiae, Edwardsiella tarda, Pseudomonas sp., Lactococcus garvieae, Citrobacter freundii, Plesiomonas shigelloides, and Enterococcus sp.

As shown in Figure 3, genera related to pathogenic bacteria and with higher frequency among 178 bacterial isolates of this study were Aeromonas (31%), Lactococcus (23%), Enterococcus (22%), Streptococcus (20%), Pseudomonas (11%), Citrobacter (6%), Edwardsiella (5%), Acinetobacter (3%), Enterobacter (2%), Plesiomonas (1%) and Weissela (1%).

Of the 43 Aeromonas isolates, 53% were identified as A. hydrophila by 16S rRNA gene sequencing. This result is in accordance with previous reports that found this species to be predominant [24]. In turn, A. veronii corresponded to 40% of the isolates. The seasonality was also observed in the present study: at higher temperatures (Spring/Summer) there were higher isolation rates of these pathogens [25], which causes hemorrhagic septicemia, characterized by small superficial lesions, focal hemorrhages, ulcers, abscesses, and abdominal distension. Internally, there can be ascitic fluid accumulation, anemia, and lesions in the liver and kidneys [26].

For the genus Lactococcus, the emerging species L. garvieae corresponded to 52% of the total 29 isolates of this genus, followed by L. lactis with 41% incidence in fish originating from the states included in the present study, with higher incidence in P. reticulatum. The species L. garvieae has been isolated from several fish species worldwide, namely in Japan [27], South Africa [28], Europe [29], and Brazil. Its first outbreak was reported in 2009 [7]. Fish with lactococcal infection exhibit lethargy, anorexia, skin darkening and swim closer to the water surface [30], resulting in considerable economic losses, especially during the summer months when the water temperature increases [4]. Few studies report L. lactis as an opportunistic pathogen. However, L. lactis subsp. lactis has been responsible for a 100% loss of hybrid sturgeons (Huso huso × Acipenser ruthenus) in a fish farm in Taiwan, China [31].

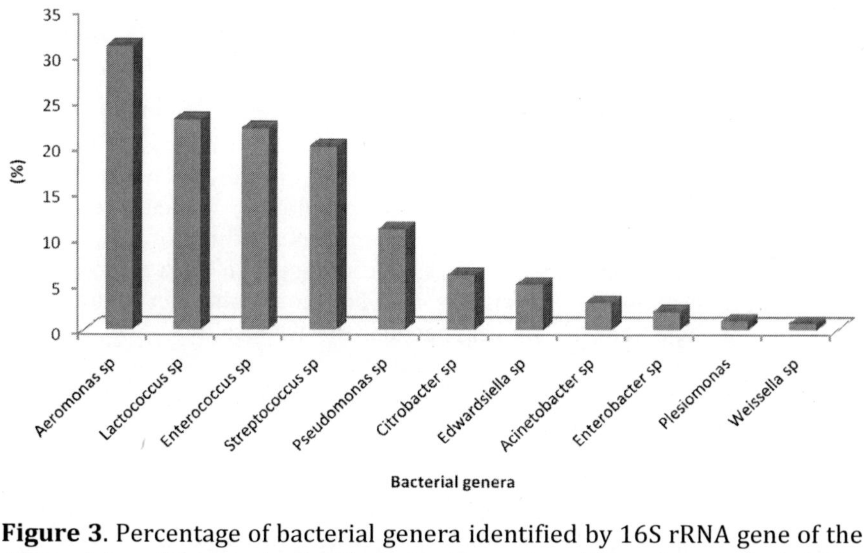

Figure 3. Percentage of bacterial genera identified by 16S rRNA gene of the 178 isolates of this study distributed in the states of Mato Grosso do Sul, São Paulo, Paraná and Rio de Janeiro.

In the present study, 31 Enterococcus strains were isolated from skin and kidney samples. Of these, 55% were E. casseliflavus, 36% E. faecalis, 6% E.durans, and 3% E. sulfureus. The predominance of E. casseliflavus has also been observed among isolates from water and sediment, accounting for 66.7% of a total of 410 Enterococcus sp. isolates in Thailand [32].

Of the 27 Streptococcus strains originating from the states of Mato Grosso do Sul, Paraná, and São Paulo, 89% corresponded to S. agalactiae; this was previously observed by Netto et al. [33] and Figueiredo et al. [10]. Although infection by S. agalactiae is the main cause of losses in tilapia farming worldwide, this pathogen has also been isolated from "cachara" originating from Mato Grosso do Sul. S. agalactiae has been identified in several other fish species, such as Sparus auratus, Liza klunzingeri [34], and Pampusargenteus [35]. Infected fish have meningoencephalitis, exophthalmia, erratic swimming, mainly.

The species P. putida (27%) and P. fulva (20%) were the predominant Pseudomonas species observed (n = 15). Eissa et al. [36] observed an incidence of 30.83% of Pseudomonas species in Nile tilapia in Egypt. Hussain [37] and Zorrilla et al. [38] reported 13.5%, and 9.7% incidence, respectively, of Pseudomonas species in marine fish, values that are similar to the 11% incidence found in the present study. P. fluorescens, P. angulliseptica, P. aeruginosa and P. putida were identified in various species of fish as causative agents of Pseudomonas septicemia. The disease is characterized by petechial hemorrhage, darkness of the skin, detached scales, abdominal ascitis and exophthalmia [39].

As the number of isolates from each region was dissimilar and low, it would not be advisable to determine a frequency profile of pathogens by location, neither the prevalence of bacterial genera by fish species, but we emphasize the importance of drawing a regional profile in aquaculture health monitoring programs and preventive management, therefore, in case of disease outbreak, treatment measures are different in each region, since factors such as light, water quality and soil contamination, quantity of parasites, management, etc are also peculiar to each locality.

4. CONCLUSION

Direct colony PCR combined with 16S rRNA gene sequencing constitutes an efficient alternative for diagnosing bacterial fish diseases, with decreased cost and time compared with the classical methods used in Brazil, such as isolation, biochemical tests, and conventional PCR.

ACKNOWLEDGEMENTS

F. A. Sebastião, L. R. Furlan, D. T. Hashimoto, F. Pilarski (FAPESP-Process 2011/07951-5) for the financial support; the Aquaculture Center (CAUNESP/UNESP, Jaboticabal) and the Laboratory of Microbial and Plant Biochemistry, Technology Department (Laboratório de Bioquímica de Microrganismos e Plantas, FCAV/UNESP Jaboticabal), UNESP, for technical support; and Dr. Fabiana Garcia for donating the bacterial strains.

REFERENCES

1. FAO Fishery and Aquaculture Country Profiles. Brazil (2010) Country Profile Fact Sheets. FAO Fisheries and Aquaculture Department [online]. Rome. Updated 1 June 2010.
2. Dash, S.S., Dasi, B.K., Pattnaik, P., Samal, S.K., Sahu, S. and Ghosh, S. (2009) Biochemical and Serological Characte- rization of Flavobacterium columnare from Freshwater Fishes of Eastern India. Journal of World Aquaculture Society, 40, 236-247. http://dx.doi.org/10.1111/j.1749-7345.2009.00246.x
3. Shama, S., Brandão, D.A., Vargas, A.C., Costa, M.M. and Pedrozo, A.F. (2000) Bactérias com potencial patogênico nos rins e lesões externas de jundiás (Rhamdia quelen) cultivados em sistema semi-intensivo. Ciência Rural, 30, 293- 298. http://dx.doi.org/10.1590/S0103-84782000000200016
4. Vendrell, D., Balcazar, J.L., Ruiz-Zarzuela, I., Blas, I.D., Girones, O. and Muzquiz, J.L. (2006) Lactococcus garvieae in Fish: A Review. Comparative Immunology, Microbiology and Infectious Diseases, 29, 177-198. http://dx.doi.org/10.1016/j.cimid.2006.06.003
5. Olivares-Fuster, O., Klesius, P.H., Evans, J. and Arias, C.R. (2008) Molecular Typing of Streptococcus agalactiae Isolates from Fish. Journal of Fish Diseases, 31, 277-283. http://dx.doi.org/10.1111/j.1365-2761.2007.00900.x
6. Staroscik, A.M., Hunnicutt, D.W., Archibald, K.E. and Nelson, D.R. (2008) Development of Methods for the Genetic Manipulation of Flavobacterium columnare. BMC Microbiology, 8, 115. http://www.biomedcentral.com/1471-2180/8/115
7. Evans, J.J., Klesius, P.H. and Shoemaker, C.A. (2009) First Isolation and Characterization of Lactococcus garvieae from Brazilian Nile Tilapia, Oreochromis niloticus (L.), and Pintado, Pseudoplathystoma corruscans (Spix & Agassiz). Journal of Fish Diseases, 32, 943-951. http://dx.doi.org/10.1111/j.1365-2761.2009.01075.x
8. Birkbeck, T.H., Feist, S.W. and Verner-Jeffreys, D.W. (2011) Francisella Infections in Fish and Shellfish. Journal of Fish Diseases, 34, 173-187. http://dx.doi.org/10.1111/j.1365-2761.2010.01226.x
9. Burr, S.E., Goldschmidt-Clermont, E., Kuhnert, P. and Frey, J. (2012) Heterogeneity of Aeromonas Populations in Wild and Farmed Perch, Perca fluviatilis L. Journal of Fish Diseases, 35, 607-613. http://dx.doi.org/10.1111/j.1365-2761.2012.01388.x
10. Figueiredo, H.C.P., Nobrega-Netto, L., Leal, C.A.G., Pereira, U.P. and Mian, G.F. (2012) Streptococcus iniae Outbreaks in Brazilian Nile Tilapia (Oreochromis niloticus L.) Farms. Brazilian Journal of Microbiology, 43, 576-580. http://dx.doi.org/10.1590/S1517-83822012000200019
11. Beaz-Hidalgo, R. and Figueras, M.J. (2012) Molecular Detection and Characterization of Furunculosis and Other Aeromonas Fish Infections. In: Carvalho, E., Ed., Health and Environment in Aquaculture, InTech Open Access Publisher, 97-132. http://dx.doi.org/10.5772/29901
12. Silva, B.C., Mouriño, J.L.P., Vieira, F.N., Jatobá, A., Seiffert, W.Q. and Martins, M.L. (2012) Haemorrhagic Septicaemia in the Hybrid Surubim (Pseudoplatystoma corruscans ×

Pseudoplatystoma fasciatum) Caused by Aeromonas hydrophila. Aquaculture Research, 43, 908-916. http://dx.doi.org/10.1111/j.1365-2109.2011.02905.x
13. Janda, J.M. and Abbott, S.L. (2007) 16S rRNA Gene Sequencing for Bacterial Identification in the Diagnostic Laboratory: Pluses, Perils, and Pitfalls. Journal of Clinical Microbiology, 45, 2761-2764. http://dx.doi.org/10.1128/JCM.01228-07
14. Coton, E. and Coton, M. (2005) Multiplex PCR for Colony Direct Detection of Gram-Positive Histamine- and Tyramine-Producing Bacteria. Journal of Microbiological Methods, 63, 296-304. http://dx.doi.org/10.1016/j.mimet.2005.04.001
15. Ben-Dov, E., Shapiro, O.H., Siboni, N. and Kushmaro, A. (2006) Advantage of Using Inosine at the 3' Termini of 16S rRNA Gene Universal Primers for the Study of Microbial Diversity. Applied and Environmental Microbiology, 72, 6902-6906. http://dx.doi.org/10.1128/AEM.00849-06
16. Claesson, M.J., Wang, Q., O'Sullivan, O., Greene-Diniz, R., Cole, J.R., Ross, R.P. and O'Toole, P.W. (2010) Comparison of Two Next-Generation Sequencing Technologies for Resolving Highly Complex Microbiota Composition Using Tandem Variable 16S rRNA Gene Regions. Nucleic Acids Research, 38, e200. http://dx.doi.org/10.1093/nar/gkq873
17. Lane, D.J., Pace, B., Olsen, G.J., Stahlt, D.A., Sogint, M.L. and Pace, N.R. (1985) Rapid Determination of 16S Ribosomal RNA Sequences for Phylogenetic Analyses. Proceedings of the National Academy of Sciences of the United States of America, 82, 6955-6959. http://dx.doi.org/10.1073/pnas.82.20.6955
18. Felske, A., Rheims, H., Wolterink, A., Stackebrandt, E. and Akkermans, A.D. (1997) Ribosome Analysis Reveals Prominent Activity of an Uncultured Member of the Class Actinobacteria in Grassland Soils. Microbiology, 143, 2983- 2989. http://dx.doi.org/10.1099/00221287-143-9-2983
19. Sambrook, J. and Russel, D.W. (2001) Molecular Cloning. 3rd Edition, Cold Spring Harbor Laboratory Press, New York.
20. Sanger, F., Nicklen, S. and Coulson, A.R. (1977) DNA Sequencing with Chain-Terminating Inhibitors. Proceedings of the National Academy of Sciences of United States of America, 74, 5463-5467. http://dx.doi.org/10.1073/pnas.74.12.5463
21. Meireles, M.A.O.M. (2008) Uso de antimicrobianos e resistência bacteriana: Aspectos socioeconômicos e comporta- mentais e seu impacto clínico e ecológico. Monograph (Microbiology Expert). Universidade Federal de Minas Gerais, Belo Horizonte.
22. Tamura, K., Peterson, D., Peterson, N., Stecher, G., Nei, M. and Kumar, S. (2011) MEGA5: Molecular Evolutionary Genetics Analysis Using Maximum Likelihood, Evolutionary Distance, and Maximum Parsimony Methods. Molecular Biology and Evolution, 10, 2731-2739. http://dx.doi.org/10.1093/molbev/msr121
23. Kong, P., Richardson, P.A. and Hong, C.X. (2005) Direct Colony PCR-SSCP for Detection of Multiple Pythiaceous Oomycetes in Environmental Samples. Journal of Microbiological Methods, 61, 25-32. http://dx.doi.org/10.1016/j.mimet.2004.10.019
24. Belem-Costa, A. and Cyrino, J.E.P. (2006) Antibiotic Resistence of Aeromonas hydrophila Isolated from Piaractus mesopotamicus (Holmberg, 1887) and Oreochromis niloticus (Linnaeus, 1758). Scientia Agricola, 63, 281-284. http://dx.doi.org/10.1590/S0103-90162006000300011
25. Pereira, C.S., Amorim, S.D., Santos, A.F.M., Reis, C.M.F., Theophilo, G.N.D. and Rodrigues, D.P. (2008) Characterization of Aeromonas spp. Isolates from Newborns Hospitalized. Revista da Sociedade Brasileira de Medicina Tropical, 41, 179-182. http://dx.doi.org/10.1590/S0037-86822008000200009
26. Garcia, F., Pilarski, F., Onaka, E.M., Moraes, F.R. and Martins, M.L. (2007) Hematology of Piaractus mesopotamicus Fed Diets Supplemented with Vitamins C and E, Challenged by Aeromonas hydrophila. Aquaculture, 271, 39-46. http://dx.doi.org/10. 1016/j.aquaculture.2007.06.021
27. Nishiki, I., Furukawa, M., Matui, S., Itami, T., Nakai, T. and Yoshida, T. (2011) Epidemiological Study on Lactococcus garvieae Isolates from Fish in Japan. Fisheries Science, 77, 367-373. http://dx.doi.org/10.1007/s12562-011-0332-0

28. Bekker, A., Hugo, C., Albertyn, J., Boucher, C.E. and Bragg, R.R. (2011) Pathogenic Gram-Positive Cocci in South African Rainbow Trout, Oncorhynchus mykiss (Walbaum). Journal of Fish Diseases, 34, 483-487. http://dx.doi.org/10.1111/j.1365-2761.2011.01259.x
29. Eyngor, M., Zlotkin, A., Ghittino, C., Prearo, M., Douet, D.G., Chilmonczyk, S. and Eldar, A. (2004) Clonality and Diversity of the Fish Pathogen Lactococcus garvieae in Mediterranean Countries. Applied and Environmental Microbiology, 70, 5132-5137. http://dxdoi.org/10.1128/AEM.70.9.5132-5137.2004
30. Avci, H., Aydoğan, A., Tanrikul, T.T. and Birincioğlu, S.S. (2010) Pathological and Microbiological Investigations in Rainbow Trout (Oncorhynchus mykiss Walbaum, 1792) Naturally Infected with Lactococcus garvieae. Kafkas Üniversitesi Veteriner Fakültesi Dergisi, 16, S313-S318.
31. Chen, M.H., Hung, S.W., Shyu, C.L., Lin, C.C., Liu, P.C., Chang, C.H., Shia, W.Y., Cheng, C.F., Lin, S.L., Tu, C.Y., Lin, Y.H. and Wang, W.S. (2012) Lactococcus lactis Subsp. Lactis Infection in Bester Sturgeon, a Cultured Hybrid of Huso huso × Acipenser ruthenus, in Taiwan. Research in Veterinary Science, 93, 581-588. http://dx.doi.org/10.1016/ j. rvsc.2011.10.007
32. Petersen, A. and Dalsgaard, A. (2003) Species Composition and Antimicrobial Resistance Genes of Enterococcus spp., Isolated from Integrated and Traditional Fish Farms in Thailand. Environmental Microbiology, 5, 395-402. http://dx.doi.org/10.1046/j.1462-2920.2003.00430.x
33. Netto, L.N., Leal, C.A.G. and Figueiredo, H.C.P. (2011) Streptococcus dysgalactiae as an Agent of Septicaemia in Nile Tilapia, Oreochromis niloticus (L.). Journal of Fish Diseases, 34, 251-254. http://dx.doi.org/10.1111/j.1365-2761.2010.01220.x
34. Evans, J.J., Wiedenmayer, A.A. and Klesius, P.H. (2002) A Transport System for Maintenance of Viability of Acinetobacter calcoaceticus, Streptococcus iniae, and Streptococcus agalactiae over Varying Time Periods. Bulletin of the European Association of Fish Pathologists, 22, 238-246.
35. Duremdez, R., Al-Marzouk, A. and Qasem, J.A. (2004) Isolation of Streptococcus agalactiae from Cultured Silver Pomfret, Pampus argenteus (Euphrasen), in Kuwait. Journal of Fish Diseases, 27, 307-310. http://dx.doi.org/10.1111/j.1365-2761.2004.00538.x
36. Eissa, N.M.E., Abou, E.E.N., Shaheen, A.A. and Abbass, A. (2010) Characterization of Pseudomonas Species Isolated from Tilapia "Oreochromis niloticus" in Qaroun and Wadi-El-Rayan Lakes, Egypt. Global Veterinaria, 5, 116-121.
37. Hussain, R.A. (2002) Studies on Some Bacterial Infections Affecting Certain Marine Fishes in the Arabian Gulf of Kingodom of Saudi Arabia. Ph.D. Dissertation., Faculty of Veterinary Medicine and Animal Resources, King Faisal University, Al-Ahsa.
38. Zorrilla, I., Chabrillón, M., Arijo, S., Díaz-Rosales, P., Martínez-Manzanares, E., Balebona, M.C. and Moriñigo, M.A. (2003) Bacteria Recovered from Diseased Cultured Gilthead Sea Bream (Sparus aurata L.) in Southwestern Spain. Aquaculture, 218, 11-20. http://dx.doi.org/10.1016/S0044-8486(02)00309-5
39. Austin, B. and Austin, D.A. (2007) Bacterial Fish Pathogens. Diseases of Farmed and Wild Fish. Springer-Praxis Publishing, Ltd., Chichester.

CHAPTER 8

Immunization with a dominant-negative recombinant Herpes Simplex Virus (HSV) type 1 protects against HSV-2 genital disease in guinea pigs

Richard Brans†[1,2] **and Feng Yao***[1]

[1]Department of Surgery, Brigham and Women's Hospital, and Harvard Medical School, Boston, MA 02115, USA and 2Department of Dermatology and Allergology, University Hospital of the RWTH Aachen, 52074 Aachen, Germany

AUTHORS' CONTRIBUTIONS

RB participated in designing the experiments, carried out the animal studies, cell culture work, virus assays, and drafted the manuscript. FY developed the HSV-1 recombinant CJ9-gD, designed the experiments, and participated in their coordination and drafting the manuscript. Both authors read and approved the final manuscript.

ABSTRACT

Background

CJ9-gD is a novel dominant-negative recombinant herpes simplex virus type 1 (HSV-1) that is completely replication-defective, cannot establish detectable latent infection in vivo, and expresses high levels of the major HSV-1 antigen glycoprotein D immediately following infection. In the present study, CJ9-gD was evaluated as a vaccine against HSV-2 genital infection in guinea pigs.

Results

Animals immunized with CJ9-gD developed at least 700-fold higher titers of HSV-2-specific neutralization antibodies than mock-immunized controls. After challenge with wild-type HSV-2, all 10 control guinea pigs developed multiple genital lesions with an average of 21 lesions per animal. In contrast, only 2 minor lesions were found in 2 of 8 CJ9-gD-immunized animals, representing a 40-fold reduction on the incidence of primary genital lesions in immunized animals ($p < 0.0001$). Immunization significantly reduced the amount and duration of viral shedding and provided complete protection against neurological symptoms, while 90% of mock-immunized animals succumbed due to the severity of disease. Importantly, immunized animals showed no signs of recurrent disease or viral shedding during a 60-days observation period after recovery from primary infection, and carried 50-fold less latent viral DNA load in their dorsal root ganglia than the surviving mock-vaccinated controls ($p < 0.0001$).

Conclusions

Collectively, we demonstrate that vaccination with the HSV-1 recombinant CJ9-gD elicits strong and protective immune responses against primary and recurrent HSV-2 genital disease and significantly reduces the extent of latent infection.

1. BACKGROUND

Genital herpes is the main cause of genital ulcer disease worldwide and is due to infections with herpes simplex virus (HSV) [1, 2]. HSV-2 accounts for most cases of genital herpes [3]. Recent studies indicate that in developed countries HSV-1 has become the main causative agent for primary genital herpes, especially among adolescents, women, and homosexual men [4, 5, 6, 7]. The prevalence of HSV-2 in the general population ranges from 10%-60%, indicating that genital herpes is one of the most common sexually transmitted diseases [2, 8].

After primary genital infection, HSV establishes latent infection in dorsal root ganglia with lifelong persistence, subsequently giving rise to intermittent reactivation and recurrent disease [9]. As the clinical appearance of genital HSV infection varies from unspecific symptoms to typically painful lesions [10], only 10-25% of people who are seropositive for HSV-2 are aware that they have genital herpes. HSV is intermittently shed from the genital mucosa in the absence of symptoms causing subconscious transmission of disease [11]. Vertical transmission of HSV to neonates is associated with a high mortality rate and a high incidence of neurological sequelae in survivors [12]. In addition, genital herpes has been linked to an increased risk of sexually acquiring and transmitting human immunodeficiency virus (HIV), which can be strongly reduced by HSV antiviral therapy [13, 14]. To date, the treatment and prevention of primary and recurrent disease is limited [15]. Experimental vaccine approaches against genital herpes have included peptides, proteins, killed virus, DNA vaccines, heterologous replicating viral vectors, replication-defective viruses, and attenuated replication-competent viruses [16, 17]. Considering the general impact of HSV-1 diseases and rising importance of primary genital herpes caused by HSV-1, a desirable vaccine should be capable of offering effective protective immunity against both HSV subtypes.

A main target for subunit vaccine development has been HSV glycoprotein D (gD), a major antigen on the viral envelope [17]. Subunit vaccines containing gD in combination with an adjuvant appeared to be safe and effective against genital herpes in guinea pigs [18, 19, 20], but failed to provide general protection in clinical trials [21, 22]. Replication-defective viruses lacking functions essential for viral replication or assembly of progeny virus particles have a broad antigenic spectrum and are more efficient than subunit vaccines in eliciting protective immune responses against genital HSV in mice and guinea pigs [23]. However, the use of replication-defective viruses, particularly when used in latently infected individuals, imposes certain risks, as they might regain replication competence in the presence of wild-type virus or reactivate latent wild-type virus infections [24].

To minimize these safety concerns, using the T-REx™ gene switch technology (Invitrogen, Carlsbad, CA) developed in our laboratory and the dominant-negative mutant polypeptide UL9-C535C of HSV-1 origin binding protein UL9, we generated a novel class of replication-defective HSV-1 recombinant, CJ83193, which can prevent its own viral DNA replication as well as that of wild-type HSV-1 and HSV-2 in co-infected cells [25, 26]. To increase its safety and vaccine efficacy against HSV infections, we recently constructed a CJ83193-derived HSV-1 recombinant CJ9-gD by replacing the essential UL9 gene with an extra copy of the HSV-1 gD (gD1) gene under the control of the tetO-bearing hCMV major immediate-early promoter [27]. We demonstrated that unlike the gD gene controlled by the endogenous promoter whose expression is dependent on viral replication [28], CJ9-gD expresses high-levels of gD at the immediate-early phase of HSV infection. Moreover, CJ9-gD is completely replication-defective and causes no detectable infection in trigeminal ganglia after ocular or nasal infection in mice [27]. In mice, CJ9-gD induces strong and long-lasting humoral and Th1-associated cellular immune responses against HSV-1 and HSV-2 [27, 29]. Immunization with CJ9-gD protects mice against HSV-1 ocular keratitis and guinea pigs against HSV-1 skin disease [27, 30] as well as genital herpetic disease caused by wild-type HSV-1 and HSV-2 in mice [29]. Previously, we have shown further that CJ9-gD is a safer and more effective vaccine than non-gD-expressing parental CJ83193 virus against HSV-1 infection [27, 29].

The guinea pig model of HSV-2 genital infection offers a unique advantage over the mouse model to investigate the efficacy of candidate HSV vaccine in protection against primary and recurrent HSV-2 genital infection and disease. Specifically, following primary intravaginal infection with HSV-2, guinea pigs develop vesicular lesions resembling those in humans, including development, appearance, and duration of disease. In contrast to mice in which spontaneous reactivation from latent infection rarely occurs in the vaginal tract, guinea pigs undergo episodic spontaneous recurrent infection and disease after recovering from initial genital disease [31, 32]. In the current report, we investigate whether CJ9-gD can serve as an effective vaccine in protection against both primary and recurrent HSV-2 genital infection and disease in guinea pigs following intravaginal challenge with wild-type HSV-2.

2. RESULTS

2.1 Induction of HSV-2-specific neutralization antibodies

The ability of CJ9-gD to elicit HSV-2-specific neutralizing antibodies was determined (Fig. 1). The HSV-2-specific neutralization antibody titer was detected in serum from all immunized guinea pigs and increased significantly from the first to the second vaccination ($p < 0.005$) with a peak titer 3 weeks after the second vaccination of 1400. No HSV-2-specific neutralization antibody was detected in serum from mock-immunized animals at 1:2-dilution before challenge. After challenge with the wild-type HSV-2, the neutralization antibody titer in immunized animals increased 2-fold ($p > 0.05$) and was 1.5-fold higher than that in mock-immunized controls following challenge.

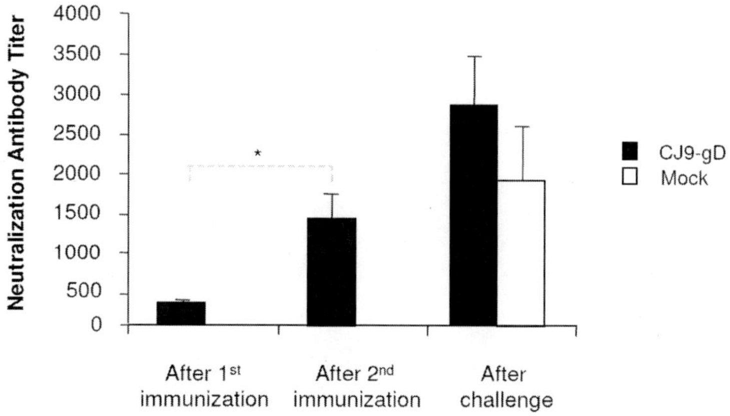

Figure 1 Induction of HSV-2-specific neutralizing antibodies in immunized guinea pigs. Two sets of guinea pigs (n = 8; n = 10) were injected s.c. with 5 × 10^6 PFU/animal of CJ9-gD or with DMEM and boosted after 3 weeks. Blood was taken 3 weeks after each immunization and 5 weeks after challenge. After heat inactivation, serum from each animal was assayed separately for HSV-2-specific neutralizing antibody titers on Vero cell monolayers. The results represent average titers ± SEM. P-value was assessed by Student's t-test (* $p < 0.005$).

2.2 Reduction of intravaginal replication of challenge HSV-2 in immunized guinea pigs

Six weeks after the initial immunization, two sets of guinea pigs (n = 8; n = 10), which were either mock-immunized (n = 10) or immunized with 5 × 10^6 PFU of CJ9-gD (n = 8), were challenged intravaginally with wild-type HSV-2 strain MS. To examine the effectiveness of immunization with CJ9-gD against intravaginal replication of challenge HSV-2, vaginal swabs were taken on days 1, 2, 3, 5, 7 and 9 after challenge. As shown in Fig. 2A, the yields of challenge virus were significantly lower in immunized guinea pigs compared with those in mock-immunized controls from days 1 to 7 (p-values for days 1, 2, and 3 < 0.05, p-value for days 5 and 7 < 0.005), with a reduction of 207-fold on day 1 ($p = 0.036$) and 220-fold on day 2 ($p = 0.012$).

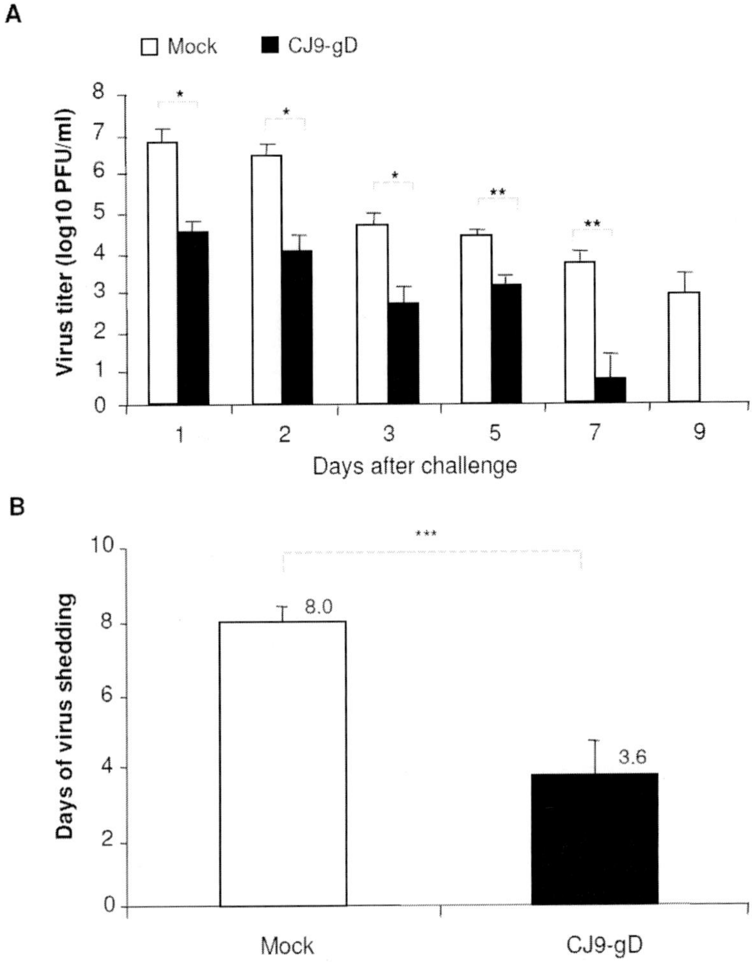

Figure 2 Reduction of challenge HSV-2 vaginal replication in guinea pigs immunized with CJ9-gD. One set of 8 and one set of 10 guinea pigs were inoculated s.c. with either 5×10^6 PFU/animal of CJ9-gD or DMEM and boosted after 3 weeks. At 6 weeks guinea pigs were challenged intravaginally with 5×10^5 PFU of HSV-2 strain MS. Vaginal swabs were taken on days 1, 2, 3, 5, 7, and 9 post-challenge. Infectious virus in swab materials was assessed by standard plaque assay on Vero cell monolayers. Viral titers are expressed as the mean ± SEM in individual vaginal swabs (**A**). The duration of viral shedding is represented as the mean number of days during which infectious virus was detected in swab materials following challenge ± SEM (**B**). P-values were assessed by Student's t-test (* $p < 0.05$, ** $p < 0.005$, *** $p < 0.0005$)

By day 9 no challenge virus was detected in CJ9-gD-immunized guinea pigs, whereas 50% of mock-immunized animals continued to shed virus at an overall average yield of more than 7.1×10^2 PFU/ml. Compared with mock-immunized controls, the average duration of viral shedding in immunized guinea pigs decreased markedly from more than 8 days to 3.6 days (Fig. 2B, $p < 0.0005$).

2.3 Protection against primary HSV-2 genital disease in immunized guinea pigs

After intravaginal challenge with wild-type HSV-2, animals were monitored daily for signs of disease. The development and clinical appearance of lesions caused by challenge virus in mock-vaccinated guinea pigs was consistent with previous observations.

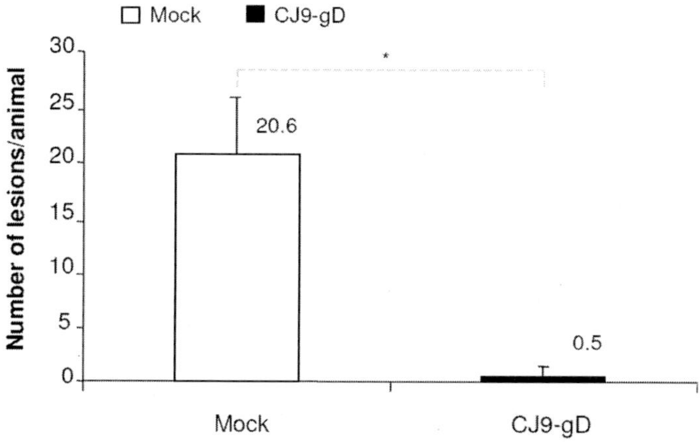

Figure 3 Prevention of primary HSV-2 genital lesions in guinea pigs immunized with CJ9-gD. Mock-immunized and CJ9-gD-immunized guinea pigs described in Fig. 2 were monitored daily for clinical symptoms following challenge with wild-type HSV-2. The average number of lesions per immunized animals was compared with that found in mock-immunized guinea pigs. The indicated values represent the mean number of lesions ± SD on day 6 post-challenge. P-value was assessed by Student's t-test (* $p < 0.0001$).

The impact of immunization with CJ9-gD on the incidence of skin lesions is summarized in Fig. 3. All 10 mock-immunized guinea pigs (100%) developed multiple genital herpes lesions following challenge with wild-type HSV-2. In contrast, only 2 of 8 animals immunized with 5×10^6 PFU of CJ9-gD exhibited two mild herpetiform lesions, resulting in an average of 0.5 lesions per immunized animal. In the corresponding control group, an average of 20.6 lesions per mock-vaccinated animal was detected on day 6 post-challenge ($p < 0.0001$). Thus, the overall incidence of primary herpetic skin lesions in immunized animals was reduced 40-fold compared to mock-immunized controls. Moreover, the two lesions found in the above mentioned immunized guinea pigs were less pronounced and healed off within 2 to 6 days, whereas in mock-immunized animals lesions turned into macerating ulcers with prolonged healing (Fig. 4A) and associated with systemic spreading of virus. All immunized guinea pigs survived the study and showed no signs of neurological illness, whereas 5 of 10 mock-immunized animals (50%) were sacrificed by day 14 after challenge due to hind limb paralysis and severity of disease. The mortality rate in this group increased to 90% by day 41 after challenge (Fig. 4B).

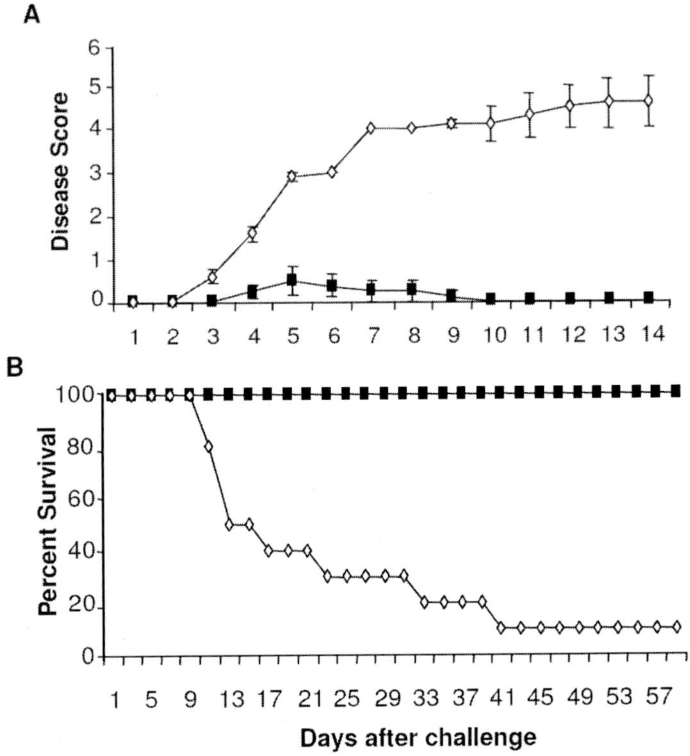

Figure 4 Prevention of primary HSV-2 disease in guinea pigs immunized with CJ9-gD. After challenge with wild-type HSV-2, individual guinea pigs described in legend of Fig. 3 were observed during a 60-day

follow-up period for the incidence of genital and disseminated HSV-2 disease using the following score: 0 = no disease; 1 = redness or swelling; 2 = a few small vesicles; 3 = several large vesicles; 4 = several large ulcers with maceration; 5 = paralysis; and 6 = death. Presented is the disease score for the first 15 days after challenge (A.) and the percentage of survival until day 60 after challenge (B.).

2.4 Protection against recurrent HSV-2 infection in immunized guinea pigs

After recovery from intravaginal challenge with wild-type HSV-2, surviving animals were monitored daily from day 30 to day 60 for signs of recurrent disease. In addition, vaginal swabs were taken daily and assayed for the presence of infectious virus. All immunized animals, and 3 of the 10 mock-immunized controls that survived the first 30 days following challenge with wild-type HSV-2 were monitored for recurrent HSV-2 infection. Two of the mock-immunized animals had recurrent viral shedding between days 30 and 60, whereas one had recurrent lesions. In contrast, no lesions or recurrent viral shedding were detected in immunized guinea pigs (Table 1).

Table 1 Prevention of recurrent HSV-2 infection in guinea pigs immunized with CJ9-gD

	Mock (n = 3)	**CJ9-gD (n = 8)**
Recurrency[1]	3/3	0/8
Recurrent lesions[2]	1/3	0/8
Recurrent shedding[3]	2/3	0/8

[1] Overall number of guinea pigs with recurrent lesions and/or recurrent shedding between days 30 and 60 after challenge.
[2] Number of guinea pigs with recurrent genital lesion between days 30 and 60 after challenge.
[3] Number of guinea pigs from which virus was detected in vaginal swab material by plaque assay on Vero cell monolayers between days 30 and 60 after challenge.

2.5 Protection from latent viral infection in immunized guinea pigs

At day 60 after challenge, 12 lower lumbar and sacral dorsal root ganglia per guinea pig were harvested from all 8 animals immunized with CJ9-gD and the 2 surviving mock-immunized controls. The whole DNA was extracted and tested for latent viral DNA using quantitative real-time PCR. The limit of detection was 5 DNA copies per reaction (correlation coefficient +/- SD: 0.96 +/- 0.016). As shown in Fig. 5, the average amount of latent HSV DNA per guinea pig was 50-fold greater in mock-vaccinated controls than in immunized animals (261486 DNA copies vs. 5229 DNA copies, $p \leq 0.0001$).

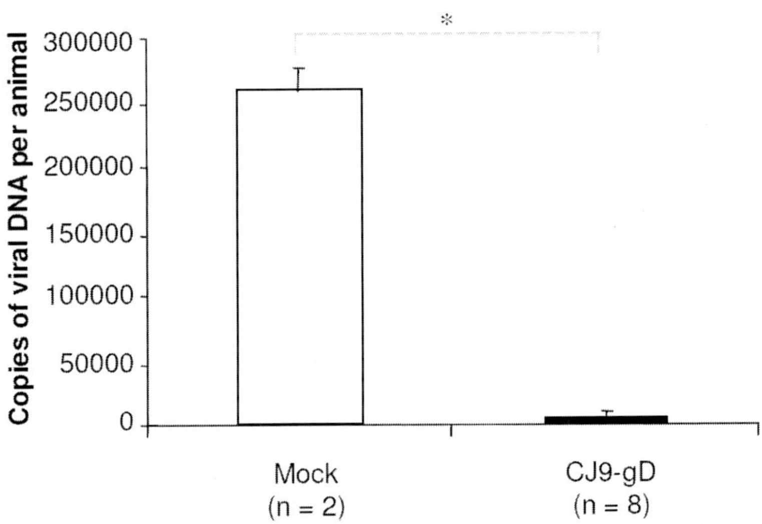

Figure 5 Protection from latent viral infection in guinea pigs immunized with CJ9-gD. Sixty days after challenge, 12 lower lumbar and sacral dorsal root ganglia (DRG) per guinea pig were harvested from all 8 immunized guinea pigs and the 2 surviving mock-immunized controls. The whole DNA was extracted and quantified for the presence of latent viral DNA using quantitative real-time PCR. The amount of viral DNA per guinea pig (A) was determined. The results are indicated as mean values ± SEM. P-value was assessed by Student's t-test (* $p < 0.0001$).

3. DISCUSSION

Although to date no vaccine capable of completely preventing HSV infection has been reported, it is believed that great benefits can be obtained from developing a vaccine that prevents disease with or without partial protection from infection as demonstrated with pertussis and influenza virus vaccines [34]. Our earlier studies demonstrate that immunization with CJ9-gD induces strong and long-lasting HSV-1- as well as HSV-2-specific humoral and Th1- cellular immune responses in mice, leading to a significant reduction in the amount and duration of acute replication of wild-type HSV-1 and HSV-2 after vaginal challenge compared with mock-immunized controls. At an immunization dose of 2×10^6 PFU of CJ9-gD, mice were completely protected from HSV-1 and HSV-2 disease [29]. We were, however, unable to evaluate whether immunization with CJ9-gD is effective in protection against recurrent HSV genital infection and disease in mice. Therefore, in the present report we used guinea pigs to explore the efficacy of immunization with CJ9-gD against HSV-2 primary as well as recurrent genital infection and disease.

We demonstrate that immunization with CJ9-gD at a dose of 5×10^6 PFU elicits high levels of neutralizing antibodies against HSV-2 in guinea pigs. Titers increased significantly from the first to the second vaccination, indicating a boosting effect. Like in mice [29], immunization with CJ9-gD induced about 7-fold higher neutralization antibody titers in guinea pigs against HSV-1 than HSV-2 ($p < 0.0001$) (data not shown). In the present study cellular immune responses were not tested due to the lack of sufficient immunological reagents specific for guinea pigs. We did, however, demonstrate in mice that immunization with CJ9-gD elicits strong HSV-specific CD4+ and CD8+ T-cell responses against HSV-1 and in a lesser extent against HSV-2 at levels similar or comparable to those induced by wild-type HSV-1 [27, 29]. Given the demonstrated importance for both HSV-specific CD4+ and CD8+ T cells in clearance of virus from infected epithelium and in neural tissues in the mouse model of HSV infections [35,40], it is not surprising that HSV-2 cross-specific-T cell response elicited in CJ9-gD-immunized guinea pigs plays a critical role in limiting the replication of challenge virus during both primary infection and reactivation from latency.

Immunization with CJ9-gD significantly reduced the amount and duration of wild-type virus replication as well as the number of genital lesions after vaginal challenge with HSV-2 compared with that in mock-immunized guinea pigs. Only 2 of 8 immunized animals developed 2 mild and fast healing herpetiform lesions with no signs of systemic involvement. Morbidity was quite extensive in mock-vaccinated animals with an average of 20.6 lesions per animal, a high incidence of systemic involvement, and a mortality rate of 90%. High mortality rates in mock-vaccinated animals after vaginal challenge with wild-type HSV-2 have been reported by other groups [19, 41] and limit the evaluation of recurrences. The extent of disease might be influenced by the viral strain or stock, the viral titer and by the inoculation method used. Despite the extensive disease in mock-vaccinated animals, CJ9-gD provided good protection against genital challenge with wild-type HSV-2 in immunized guinea pigs. Therefore, it is reasonable to anticipate that protection would be more effective should a lower dose of challenge virus or a more gentle inoculation be selected.

In accordance with the protection against primary disease, neither recurrent vaginal shedding of infectious virus nor recurrent genital lesions were found in CJ9-gD-immunized animals. Quantitative PCR analysis shows that the amount of latent HSV DNA in dorsal root ganglia was 50-fold lower in immunized guinea pigs compared with the 2 mock-immunized guinea pigs that survived following challenge with wild-type HSV-2 ($p < 0.0001$). Recall that CJ9-gD cannot establish detectable latent infection in sensory ganglia in mice following ocular or intranasal infection [27] nor in dorsal root ganglia after subcutaneous immunization [29]. The viral DNA detected in dorsal root ganglia of CJ9-gD-immunized guinea pigs after vaginal challenge should be primarily the challenge wild-type HSV-2 viral DNA. Taken together, these results are consistent with observations that a reduced latent infection is associated with a lower incidence of reactivation and recurrent disease [20, 41, 42].

Several vaccine candidates have been tested in guinea pigs against genital HSV-2 infection. The subunit vaccine gD2/AS04, which contains the HSV-2 major antigen glycoprotein D (gD2) in combination with the adjuvant aluminium hydroxide and 3-O-deacylated-monophosphoryl lipid A (MPL), was effective in prevention of primary and recurrent genital disease in immunized animals following challenge with wild-type HSV-2 [19, 20]. In phase 3 clinical trials, however, it only provided 73-74% efficacy in protecting against development of genital herpes disease in HSV-seronegative women [21]. A recent study showed that the replication-defective HSV-2 recombinant dl 5-29 was more effective than the HSV-2-gD-based subunit vaccine in inducing HSV-2-specific neutralizing antibodies and CD8+ T-cell response in mice [43]. CJ9-gD is an HSV-1 recombinant defective at level of viral DNA replication, and therefore, similar to dl5-29, capable of expressing a broad spectrum of viral antigens. In addition, it has a unique dominant-negative effect on viral replication (UL9-C535C expression) and expresses high levels of the major HSV-1 antigen gD at the immediate-early phase of infection [27]. Immunization with CJ9-gD led to 220-fold reduction in the yield of challenge wild-type HSV-2 in genital swabs materials on day 2 post-challenge compared with mock-immunized controls. Noting that immunization with gD2/AS04 resulted in less than 14-fold challenge wild-type HSV-2 (strain MS) viral replication compared with mock-immunized controls on day 2 post-challenge, and all mock-immunized animals survived after recovery from primary disease caused by challenge virus [20], our study suggests that CJ9-gD could potentially be more efficacious than gD2 subunit vaccine against HSV-2 genital disease. It will be interesting to test the vaccine efficacy of gD2/AS04 and CJ9-gD in protecting against HSV-2 genital herpes in the same experimental settings. Moreover, in light of that CJ9-gD expresses high-level of gD, and induction of both effective mucosal and systemic immune responses is likely required for an optimal protection against HSV genital infection, it would be of great interest to investigate the effectiveness of CJ9-gD in induction of humoral and T-cell immunity following different routes of immunization and whether the efficacy of CJ9-gD in eliciting mucosal immune response can be enhanced by gD subunit prime/CJ9-gD boost regimen involving combination of mucosal and systemic immunization [44, 45, 46].

Many type-common and type-specific antibodies as well as T cell epitopes have been identified against various HSV-1 and HSV-2 proteins. Mice immunized with CJ9-gD develop stronger humoral and cellular immune responses against HSV-1 than against HSV-2, and are significantly better protected against genital infection with HSV-1 than with HSV-2 [29]. These findings are in agreement with the previous reports that in rodents HSV vaccines are generally less effective in prevention of heterotypic HSV infection than homotypic infection [47, 48]. Combined with observations that humans who were previously infected with HSV-2 are less likely to experience re-infection with a heterologous strain of HSV-2 than individuals with prior HSV-1 infection [49, 50, 51, 52, 53], it is reasonable to believe, that a CJ9-gD-like dominant-negative HSV-2 recombinant would be more effective in prevention of genital HSV-2 infection than the HSV-1 recombinant CJ9-gD. This HSV-2 recombinant is currently being devised.

4. CONCLUSIONS

We demonstrate that immunization with a replication-defective and dominant-negative HSV-1 recombinant CJ9-gD expressing high levels of gD can induce strong cross-protective immunity against primary and recurrent HSV-2 genital infection and disease in guinea pigs. We show further that the latent viral load of challenge wild-type HSV-2 is significantly reduced in immunized guinea pigs compared with the mock-immunized controls. Collectively, CJ9-gD represents a new class of HSV-1 recombinant, which is avirulent, unable to establish detectable latent infection in vivo, and serves as an effective vaccine against genital HSV infection and disease in both mice and guinea pigs.

5. METHODS

5.1 Animals

Female Hartley guinea pigs (300-350 g) were obtained from Charles River Breeding Laboratories (Wilmington, MA). The described animal experiments were conducted according to the protocols approved by the Harvard Medical Area Standing Committee on Animals and the American Veterinary Medical Association. The Harvard Medical School animal management program is accredited by the Association for Assessment and Accreditation of Laboratory Animal Care (AAALAC) and meets National Institutes of Health standards as set forth in "The Guide for the Care and Use of Laboratory Animals" (National Academy Press, 1996).

5.2 Cells and viruses

African Green Monkey Kidney (Vero) cells and RUL9-8 cells, a cell line derived from U2OS cells expressing UL9 and the tetracycline repressor (tetR), were grown and maintained in DMEM growth medium as previously described [33].

Wild-type HSV-2 MS strain (ATCC, Manassas, VA) was propagated and plaque assayed on Vero cells. CJ9-gD was derived from CJ83193 by replacing the essential UL9 gene with the HSV-1 gD gene driven by the tetO-containing hCMV major immediate-early promoter [27]. CJ83193 is a replication-defective virus, in which both copies of the HSV-1 ICP0 gene were replaced by DNA sequences encoding the dominant-negative HSV-1 polypeptide UL9-C535C under control of the tetO-bearing hCMV major immediate-early promoter [25]. CJ9-gD was propagated and plaque assayed in RUL9-8 cells.

5.3 Immunization and challenge

One set of 8 guinea pigs and one set of 10 guinea pigs were randomly assigned to 2 groups. Animals were either mock-immunized with DMEM (n = 10) or immunized with 5×10^6 PFU of CJ9-gD (n = 8) in a volume of 50 μl s.c. in the right and left upper flank per guinea pig. On day 21 after primary immunization, animals were boosted. At the same time and one day prior to viral challenge, serum was obtained from saphenous veins and stored at -80°C. Six weeks after the initial immunization, the animals were preswabbed with a moist sterile calcium alginate swab (Fisher Scientific, Waltham, MA) and inoculated intravaginally with 100 μl of culture medium containing 5×10^5 PFU of HSV-2 strain MS. Animals were kept on their backs with their rear part elevated under the influence of anesthesia for 30-45 min post-infection.

5.4 Detection of HSV-2-specific neutralization antibody titers

Blood was obtained from the saphenous veins and neutralization antibody titers were determined in the presence of complement as described previously [28, 30].

5.5 Clinical observations

After challenge with wild-type HSV-2 strain MS, the animals were monitored daily until day 60. The number of lesions were counted and the progress of disease was scored using a modified method [31]: 0 = no disease; 1 = redness or swelling; 2 = a few small vesicles; 3 = several large vesicles; 4 = several large ulcers with maceration; 5 = paralysis; and 6 = death.

5.6 Assay of acute and recurrent vaginal shedding of challenge virus

After challenge with wild-type HSV-2 strain MS, vaginal mucosae were swabbed with a moist calcium alginate swab (Fisher Scientific, Waltham, MA) on days 1, 2, 3, 5, 7 and 9. From days 30 to 60 post challenge swabs were taken daily. Swabs were kept in 1 ml DMEM and stored at -80°C until assayed for infectious virus by standard plaque assay on Vero cell monolayers.

5.7 Quantitative real-time PCR

At day 60 after intravaginal challenge with HSV-2 strain MS, 12 lower lumbar and sacral dorsal root ganglia were collected from each of the surviving guinea pigs. Dorsal root ganglia were kept separately in 0.5 ml of normal growth medium and stored at -80°C for further processing. DNA was isolated from each dorsal root ganglion and assayed for viral DNA by quantitative real-time PCR as described previously [27].

5.8 Statistical analysis

For statistical analysis unpaired Student's t-tests were performed.

ACKNOWLEDGEMENTS

This work was supported by Public Health Service Grant 5RO1AI05088 from the National Institutes of Health.

REFERENCES

1. Paz-Bailey G, Ramaswamy M, Hawkes SJ, Geretti AM: Herpes simplex virus type 2: epidemiology and management options in developing countries. Sex Transm Infect. 2007, 83 (1): 16-22. 10.1136/sti.2006.020966.
2. Xu F, Sternberg MR, Kottiri BJ, McQuillan GM, Lee FK, Nahmias AJ, Berman SM, Markowitz LE: Trends in herpes simplex virus type 1 and type 2 seroprevalence in the United States. Jama. 2006, 296 (8): 964-973. 10.1001/jama.296.8.964.
3. Whitley RJ: Herpes simplex encephalitis: adolescents and adults. Antiviral Res. 2006, 71 (2-3): 141-148. 10.1016/j.antiviral.2006.04.002.
4. Lafferty WE, Downey L, Celum C, Wald A: Herpes simplex virus type 1 as a cause of genital herpes: impact on surveillance and prevention. J Infect Dis. 2000, 181 (4): 1454-1457. 10.1086/315395.
5. Jin F, Prestage GP, Mao L, Kippax SC, Pell CM, Donovan B, Templeton DJ, Taylor J, Mindel A, Kaldor JM: Transmission of herpes simplex virus types 1 and 2 in a prospective cohort of HIV-negative gay men: the health in men study. J Infect Dis. 2006, 194 (5): 561-570. 10.1086/506455.
6. Roberts CM, Pfister JR, Spear SJ: Increasing proportion of herpes simplex virus type 1 as a cause of genital herpes infection in college students. Sex Transm Dis. 2003, 30 (10): 797-800. 10.1097/01.OLQ.0000092387.58746.C7.
7. Nieuwenhuis RF, van Doornum GJ, Mulder PG, Neumann HA, van der Meijden WI: Importance of herpes simplex virus type-1 (HSV-1) in primary genital herpes. Acta Derm Venereol. 2006, 86 (2): 129-134.
8. Malvy D, Halioua B, Lancon F, Rezvani A, Bertrais S, Chanzy B, Daniloski M, Ezzedine K, Malkin JE, Morand P: Epidemiology of genital herpes simplex virus infections in a community-based sample in France: results of the HERPIMAX study. Sex Transm Dis. 2005, 32 (8): 499-505. 10.1097/01.olq.0000161188.82905.b7.
9. Whitley RJ: Herpes simplex virus infection. Semin Pediatr Infect Dis. 2002, 13 (1): 6-11. 10.1053/spid.2002.29752.
10. Corey L, Adams HG, Brown ZA, Holmes KK: Genital herpes simplex virus infections: clinical manifestations, course, and complications. Ann Intern Med. 1983, 98 (6): 958-972.
11. Wald A, Zeh J, Selke S, Warren T, Ryncarz AJ, Ashley R, Krieger JN, Corey L: Reactivation of genital herpes simplex virus type 2 infection in asymptomatic seropositive persons. N Engl J Med. 2000, 342 (12): 844-850. 10.1056/NEJM200003233421203.
12. Jones CA: Vertical transmission of genital herpes: prevention and treatment options. Drugs. 2009, 69 (4): 421-434. 10.2165/00003495-200969040-00003.
13. Abu-Raddad LJ, Magaret AS, Celum C, Wald A, Longini IM, Self SG, Corey L: Genital herpes has played a more important role than any other sexually transmitted infection in driving HIV prevalence in Africa. PLoS ONE. 2008, 3 (5): e2230-10.1371/journal.pone.0002230.
14. Freeman EE, Orroth KK, White RG, Glynn JR, Bakker R, Boily MC, Habbema D, Buve A, Hayes R: Proportion of new HIV infections attributable to herpes simplex 2 increases over time: simulations of the changing role of sexually transmitted infections in sub-Saharan African HIV epidemics. Sex Transm Infect. 2007, 83 (Suppl 1): i17-24. 10.1136/sti.2006.023549.

15. Whitley RJ, Roizman B: Herpes simplex virus infections. Lancet. 2001, 357 (9267): 1513-1518. 10.1016/S0140-6736(00)04638-9.
16. Koelle DM, Corey L: Recent progress in herpes simplex virus immunobiology and vaccine research. Clin Microbiol Rev. 2003, 16 (1): 96-113. 10.1128/CMR.16.1.96-113.2003.
17. Stanberry LR: Clinical trials of prophylactic and therapeutic herpes simplex virus vaccines. Herpes. 2004, 11 (Suppl 3): 161A-169A.
18. Stanberry LR, Bernstein DI, Burke RL, Pachl C, Myers MG: Vaccination with recombinant herpes simplex virus glycoproteins: protection against initial and recurrent genital herpes. J Infect Dis. 1987, 155 (5): 914-920.
19. Bourne N, Bravo FJ, Francotte M, Bernstein DI, Myers MG, Slaoui M, Stanberry LR: Herpes simplex virus (HSV) type 2 glycoprotein D subunit vaccines and protection against genital HSV-1 or HSV-2 disease in guinea pigs. J Infect Dis. 2003, 187 (4): 542-549. 10.1086/374002.
20. Bourne N, Milligan GN, Stanberry LR, Stegall R, Pyles RB: Impact of immunization with glycoprotein D2/AS04 on herpes simplex virus type 2 shedding into the genital tract in guinea pigs that become infected. J Infect Dis. 2005, 192 (12): 2117-2123. 10.1086/498247.
21. Stanberry LR, Spruance SL, Cunningham AL, Bernstein DI, Mindel A, Sacks S, Tyring S, Aoki FY, Slaoui M, Denis M: Glycoprotein-D-adjuvant vaccine to prevent genital herpes. N Engl J Med. 2002, 347 (21): 1652-1661. 10.1056/NEJMoa011915.
22. Corey L, Langenberg AG, Ashley R, Sekulovich RE, Izu AE, Douglas JM, Handsfield HH, Warren T, Marr L, Tyring S: Recombinant glycoprotein vaccine for the prevention of genital HSV-2 infection: two randomized controlled trials. Chiron HSV Vaccine Study Group. Jama. 1999, 282 (4): 331-340. 10.1001/jama.282.4.331.
23. Dudek T, Knipe DM: Replication-defective viruses as vaccines and vaccine vectors. Virology. 2006, 344 (1): 230-239. 10.1016/j.virol.2005.09.020.
24. Koelle DM, Ghiasi H: Prospects for developing an effective vaccine against ocular herpes simplex virus infection. Curr Eye Res. 2005, 30 (11): 929-942. 10.1080/02713680500313153.
25. Yao F, Eriksson E: A novel anti-herpes simplex virus type 1-specific herpes simplex virus type 1 recombinant. Hum Gene Ther. 1999, 10 (11): 1811-1818. 10.1089/10430349950017491.
26. Yao F, Eriksson E: Inhibition of herpes simplex virus type 2 (HSV-2) viral replication by the dominant negative mutant polypeptide of HSV-1 origin binding protein. Antiviral Res. 2002, 53 (2): 127-133. 10.1016/S0166-3542(01)00207-8.
27. Lu Z, Brans R, Akhrameyeva NV, Murakami N, Xu X, Yao F: High-level expression of glycoprotein D by a dominant-negative HSV-1 virus augments its efficacy as a vaccine against HSV-1 infection. J Invest Dermatol. 2009, 129 (5): 1174-1184. 10.1038/jid.2008.349.
28. Augustinova H, Hoeller D, Yao F: The dominant-negative herpes simplex virus type 1 (HSV-1) recombinant CJ83193 can serve as an effective vaccine against wild-type HSV-1 infection in mice. J Virol. 2004, 78 (11): 5756-5765. 10.1128/JVI.78.11.5756-5765.2004.
29. Brans R, Akhrameyeva NV, Yao F: Prevention of genital herpes simplex virus type 1 and 2 disease in mice immunized with a gD-expressing dominant-negative recombinant HSV-1. J Invest Dermatol. 2009, 129 (10): 2470-2479. 10.1038/jid.2009.86.
30. Brans R, Eriksson E, Yao F: Immunization with a dominant-negative recombinant HSV type 1 protects against HSV-1 skin disease in guinea pigs. J Invest Dermatol. 2008, 128 (12): 2825-2832. 10.1038/jid.2008.142.
31. Stanberry LR, Kern ER, Richards JT, Abbott TM, Overall JC: Genital herpes in guinea pigs: pathogenesis of the primary infection and description of recurrent disease. J Infect Dis. 1982, 146 (3): 397-404.
32. Stanberry LR, Kern ER, Richards JT, Overall JC: Recurrent genital herpes simplex virus infection in guinea pigs. Intervirology. 1985, 24 (4): 226-231. 10.1159/000149647.
33. Yao F, Theopold C, Hoeller D, Bleiziffer O, Lu Z: Highly efficient regulation of gene expression by tetracycline in a replication-defective herpes simplex viral vector. Mol Ther. 2006, 13 (6): 1133-1141. 10.1016/j.ymthe.2006.01.009.

34. Stanberry LR, Cunningham AL, Mindel A, Scott LL, Spruance SL, Aoki FY, Lacey CJ: Prospects for control of herpes simplex virus disease through immunization. Clin Infect Dis. 2000, 30 (3): 549-566. 10.1086/313687.
35. Kuklin NA, Daheshia M, Marconi PC, Krisky DM, Rouse RJ, Glorioso JC, Manican E, Rouse BT: Modulation of mucosal and systemic immunity by enteric administration of nonreplicating herpes simplex virus expressing cytokines. Virology. 1998, 240 (2): 245-253. 10.1006/viro.1997.8926.
36. Ghiasi H, Perng G, Nesburn AB, Wechsler SL: Either a CD4(+)or CD8(+)T cell function is sufficient for clearance of infectious virus from trigeminal ganglia and establishment of herpes simplex virus type 1 latency in mice. Microb Pathog. 1999, 27 (6): 387-394. 10.1006/mpat.1999.0314.
37. Wakim LM, Jones CM, Gebhardt T, Preston CM, Carbone FR: CD8(+) T-cell attenuation of cutaneous herpes simplex virus infection reduces the average viral copy number of the ensuing latent infection. Immunol Cell Biol. 2008, 86 (8): 666-675. 10.1038/icb.2008.47.
38. van Lint A, Ayers M, Brooks AG, Coles RM, Heath WR, Carbone FR: Herpes simplex virus-specific CD8 + T cells can clear established lytic infections from skin and nerves and can partially limit the early spread of virus after cutaneous inoculation. J Immunol. 2004, 172 (1): 392-397.
39. Johnson AJ, Chu CF, Milligan GN: Effector CD4 + T-cell involvement in clearance of infectious herpes simplex virus type 1 from sensory ganglia and spinal cords. J Virol. 2008, 82 (19): 9678-9688. 10.1128/JVI.01159-08.
40. Zhang X, Chentoufi AA, Dasgupta G, Nesburn AB, Wu M, Zhu X, Carpenter D, Wechsler SL, You S, BenMohamed L: A genital tract peptide epitope vaccine targeting TLR-2 efficiently induces local and systemic CD8 + T cells and protects against herpes simplex virus type 2 challenge. Mucosal Immunol. 2009, 2 (2): 129-143. 10.1038/mi.2008.81.
41. Lekstrom-Himes JA, Pesnicak L, Straus SE: The quantity of latent viral DNA correlates with the relative rates at which herpes simplex virus types 1 and 2 cause recurrent genital herpes outbreaks. J Virol. 1998, 72 (4): 2760-2764.
42. Sawtell NM: The probability of in vivo reactivation of herpes simplex virus type 1 increases with the number of latently infected neurons in the ganglia. J Virol. 1998, 72 (8): 6888-6892.
43. Hoshino Y, Dalai SK, Wang K, Pesnicak L, Lau TY, Knipe DM, Cohen JI, Straus SE: Comparative efficacy and immunogenicity of replication-defective, recombinant glycoprotein, and DNA vaccines for herpes simplex virus 2 infections in mice and guinea pigs. J Virol. 2005, 79 (1): 410-418. 10.1128/JVI.79.1.410-418.2005.
44. Neutra MR, Kozlowski PA: Mucosal vaccines: the promise and the challenge. Nat Rev Immunol. 2006, 6 (2): 148-158. 10.1038/nri1777.
45. Gallichan WS, Woolstencroft RN, Guarasci T, McCluskie MJ, Davis HL, Rosenthal KL: Intranasal immunization with CpG oligodeoxynucleotides as an adjuvant dramatically increases IgA and protection against herpes simplex virus-2 in the genital tract. J Immunol. 2001, 166 (5): 3451-3457.
46. Tengvall S, Lundqvist A, Eisenberg RJ, Cohen GH, Harandi AM: Mucosal administration of CpG oligodeoxynucleotide elicits strong CC and CXC chemokine responses in the vagina and serves as a potent Th1-tilting adjuvant for recombinant gD2 protein vaccination against genital herpes. J Virol. 2006, 80 (11): 5283-5291. 10.1128/JVI.02013-05.
47. Boursnell ME, Entwisle C, Blakeley D, Roberts C, Duncan IA, Chisholm SE, Martin GM, Jennings R, Ni Challanain D, Sobek I: A genetically inactivated herpes simplex virus type 2 (HSV-2) vaccine provides effective protection against primary and recurrent HSV-2 disease. J Infect Dis. 1997, 175 (1): 16-25.
48. Da Costa XJ, Morrison LA, Knipe DM: Comparison of different forms of herpes simplex replication-defective mutant viruses as vaccines in a mouse model of HSV-2 genital infection. Virology. 2001, 288 (2): 256-263. 10.1006/viro.2001.1094.
49. Bryson Y, Dillon M, Bernstein DI, Radolf J, Zakowski P, Garratty E: Risk of acquisition of genital herpes simplex virus type 2 in sex partners of persons with genital herpes: a prospective couple study. J Infect Dis. 1993, 167 (4): 942-946.

50. Mertz GJ, Benedetti J, Ashley R, Selke SA, Corey L: Risk factors for the sexual transmission of genital herpes. Ann Intern Med. 1992, 116 (3): 197-202.
51. Looker KJ, Garnett GP: A systematic review of the epidemiology and interaction of herpes simplex virus types 1 and 2. Sex Transm Infect. 2005, 81 (2): 103-107. 10.1136/sti.2004.012039.
52. Schmidt OW, Fife KH, Corey L: Reinfection is an uncommon occurrence in patients with symptomatic recurrent genital herpes. J Infect Dis. 1984, 149 (4): 645-646.
53. Lakeman AD, Nahmias AJ, Whitley RJ: Analysis of DNA from recurrent genital herpes simplex virus isolates by restriction endonuclease digestion. Sex Transm Dis. 1986, 13 (2): 61-66. 10.1097/00007435-198604000-00001.

CHAPTER 9

In-vitro and in-vivo phenotype of type Asia 1 foot-and-mouth disease viruses utilizing two non-RGD receptor recognition sites

Pinghua Li[1], Zengjun Lu[1], Huifang Bao[1], Dong Li[1], Donald P King[2] Pu Sun[1], Xingwen Bai[1], Weijun Cao[1], Simon Gubbins[2], Yingli Chen[1], Baoxia Xie[1], Jianhong Guo[1], Hong Yin[1] and Zaixin Liu[1]

1State Key Laboratory of Veterinary Etiological Biology, National Foot and Mouth Disease Reference Laboratory, Key laboratory of Animal Virology of Ministry of Agriculture, Lanzhou Veterinary Research Institute, Chinese Academy of Agricultural Sciences, No. 1 Xujiaping, Yanchangbao, Lanzhou, Gansu 730046, PR China

2Institute for Animal Health, Pirbright Laboratory, Ash Road, Pirbright, Woking, Surrey, England GU24 0NF, UK

AUTHORS' CONTRIBUTIONS

PHL and ZJL conceived and designed the study. PHL and WJC constructed three FMDV full-length infectious cDNA clones. DL and XWB carried out the animal experiments. HFB and PS carried out the real-time quantitative RT-PCR assay. HY and ZXL supervised all aspects of the research. YLC, BXX and JHG passaged the three recombinant viruses respectively. PHL and DPK co-drafted the manuscript. SG aligned the data and conducted statistical analysis. All authors read and approved the final manuscript.

ABSTRACT

Background

Foot-and-mouth disease virus (FMDV) uses a highly conserved Arg-Gly-Asp (RGD) triplet for attachment to host cells and this motif is believed to be essential for virus viability. Previous sequence analyses of the 1D-encoding region of an FMDV field isolate (Asia1/JS/CHA/05) and its two derivatives indicated that two viruses, which contained an Arg-Asp-Asp (RDD) or an Arg-Ser-Asp (RSD) triplet instead of the RGD integrin recognition motif, were generated serendipitously upon short-term evolution of field isolate in different biological environments. To examine the influence of single amino acid substitutions in the receptor binding site of the RDD-containing FMD viral genome on virus viability and the ability of non-RGD FMDVs to cause disease in susceptible animals, we constructed an RDD-containing FMDV full-length cDNA clone and derived mutant molecules with RGD or RSD receptor recognition motifs. Following transfection of BSR cells with the full-length genome plasmids, the genetically engineered viruses were examined for their infectious potential in cell culture and susceptible animals.

Results

Amino acid sequence analysis of the 1D-coding region of different derivatives derived from the Asia1/JS/CHA/05 field isolate revealed that the RDD mutants became dominant or achieved population equilibrium with coexistence of the RGD and RSD subpopulations at an early phase of type Asia1 FMDV quasispecies evolution. Furthermore, the RDD and RSD sequences remained genetically stable for at least 20 passages. Using reverse genetics, the RDD-, RSD-, and RGD-containing FMD viruses were rescued from full-length cDNA clones, and single amino acid substitution in RDD-containing FMD viral genome did not affect virus viability. The genetically engineered viruses replicated stably in BHK-21 cells and had similar growth properties to the parental virus. The RDD parental virus and two non-RGD recombinant viruses were virulent to pigs and bovines that developed typical clinical disease and viremia.

Conclusions

FMDV quasispecies evolving in a different biological environment gained the capability of selecting different receptor recognition site. The RDD-containing FMD viral genome can accommodate substitutions in the receptor binding site without additional changes in the capsid. The viruses expressing non-RGD receptor binding sites can replicate stably in vitro and produce typical FMD clinical disease in susceptible animals.

1. BACKGROUND

Foot-and-mouth disease virus (FMDV) is an important animal pathogen that causes a severe vesicular disease in cattle, swine, sheep and other cloven-hoofed animals [1, 2]. The virus belongs to the Aphthovirus genus within the *Picornaviridae* family [3]. The genome is a positive-sense single-stranded RNA molecule that is encapsidated by 60 copies of each of the four structural polypeptides of which VP4 is internal and the others (VP1, VP2 and VP3) are exposed [4]. It has been shown that VP1 is the most variable among the capsid polypeptides, and it is widely used to characterize field strains of FMDV to provide data to support epidemiological investigations of disease outbreaks among livestock.

A major, highly variable antigenic site of FMDV is located at the exposed G-H loop comprising amino acids 134-160 of the capsid protein VP1 [4, 5, 6], which plays an important role in cell infection and is also a major target for protective host responses mediated via humoral immunity [5, 7, 8, 9]. This mobile loop contains a conserved Arg-Gly-Asp (RGD) motif that has been shown to be a major determinant in the interaction of the virus with cell surface receptors of the integrin superfamily [7, 10, 11]. Indeed, previous studies, using different approaches, have indicated that naturally occurring field isolates of FMDV bind to cells via these highly conserved surface-exposed RGD residues [11, 12]. In particular, it has been reported that FMD viruses utilize multiple RGD-dependent integrins of the αv subgroup to initiate infection, including αvβ3, αvβ6, αvβ1 and αvβ8 [13, 14, 15, 16, 17]. However, the RGD integrin recognition domain can become dispensable upon in-vitro passage of FMDV: multiple phenotypic changes that are associated with a limited number of amino acid substitutions at the capsid surface which may even include modifications within the RGD triplet [18, 19, 20, 21]. Tissue culture-adapted viruses can use heparan sulfate (HS) as a receptor to enter cells [18, 22], and can also dispense with their RGD integrin-binding motif [23]. These findings indicate the existence of alternative RGD-independent pathways for FMDV entry into cell.

In the present study we report that two viruses harboring alternative receptor binding sites (RDD or RSD) were generated after short-term passage of an FMDV field isolate (Asia1/JS/CHA/05) in different environments. The non-RGD receptor recognition motifs were stably maintained during subsequent passage in cell culture. To study the ability of an RDD-containing FMD viral genome to accommodate substitution in receptor binding site and non-RGD viruses to cause disease in susceptible animals, we assembled an RDD-containing FMDV (Asia1/JSp1c8) full-length cDNA clone and derived mutant clones harboring RGD or RSD motif with a single amino acid substitution (RDD→RGD, RDD→RSD) in the receptor binding site. Following transfection of BSR/T7 cell with three full-length plasmids, the resulting viruses were examined for their infectious potential in-vitro and in-vivo.

2. RESULTS

2.1 Sequence analysis of Asia1/JS/CHA/05 and its derivatives

Deduced amino acid sequence analysis of the 1D-encoding region showed that Asia1/JS/CHA/05 had a consensus RGD triplet at position 143-145 of VP1, while Asia1/JSp1c8 obtained an alternative RDD triplet at this position. However, careful inspection of the electropherograms from the Asia1/JSM4 VP1 gene sequencing reactions revealed the presence of two genetic subpopulations, one with RGD and the other with RSD at receptor binding site. To further investigate the genetic heterogeneity within these samples, 10 biological clones containing VP1 genes of each Asia1/JS/CHA/05, Asia1/JSp1c8 and Asia1/JSM4 were sequenced. The 10 clones obtained from each of the Asia1/JS/CHA/05 and Asia1/JSp1c8 viruses respectively encode RGD and RDD tripeptide at position 143-145 of VP1. For Asia1/JSM4, four clones encoded RSD and six clones maintain the RGD motif at the same position. These results were identical to the amino acid sequence analysis performed by direct sequencing of PCR amplicons. Additionally, amino acid sequence analysis of the capsid-coding regions of Asia1/JS/CHA/05, Asia1/JSp1c8, and Asia1/JSM4 showed that Asia1/JSp1c8 had seven amino acid substitutions in the capsid region (1 in 1A, 3 in 1B, 1 in 1C and 3 in 1D; Table 1) compared with Asia1/JS/CHA/05 and Asia1/JSM4.

Taken together, these results suggested that the RDD mutants became dominant or obtained equilibrium with the RGD and RSD subpopulations at an early phase of Asia1/JS/CHA/05 quasispecies evolution. Since RDD and RSD motifs are unusual in lacking the RGD integrin-recognition sequence, additional multiple passages were performed to determine its stability. Amino acid sequence of the VP1 gene of the viruses obtained from different passages of Asia1/JSp1c8 and Asia1/JSM4 revealed that the RDD and RSD sequence were genetically stable for at least 20 passages (Figure 1). The amino acid sequences of the G-H loop of viruses derived from different passages are summarized in table 2.

Table 1 Comparison of the P1 amino acid sequence of Asia1/JS/CHA/05, Asia1/JS/p1c8, and Asia1/JSM4

Capsid region	Amino acid residue position [a]	Asia1/JS/CHA/05	Asia1/JSM4	Asia1/JS/p1c8
1A	73	S	S	N
1B	107	I	I	V
	132	E	E	K
	134	D	D	G
1C	133	T	T	A
1D	144	G	G/S	D
	154	N	N	S
	202	K	K	E

[a] Amino acid residues are numbered from the amino terminus to the carboxyl terminus. Single letter amino acid code is used.

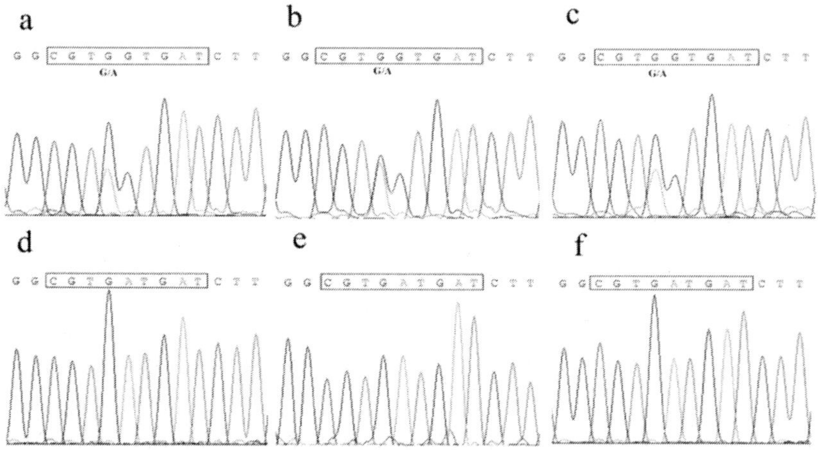

Figure 1 SEQUENCING ELECTROPHEROGRAMS OF THE VP1 PCR-AMPLICONS OF DERIVATIVES DERIVED FROM ASIA1/JSM4 AND ASIA1/JSP1C8. The nucleotides encoding receptor-binding tripeptide are boxed, (a, b, c) represent sequencing electropherograms of Asia1/JSM6c5, Asia1/JSM6c15, and Asia1/JSM6c20, respectively; (c, d, e) represent sequencing electropherograms of Asia1/JSp1c8, Asia1/JSp1c15, and Asia1/JSp1c20, respectively.

Table 2 Comparison of amino acid sequence at G-H loop of VP1 of the viruses derived from different origins and full-length plasmids

Virus/plasmid	**Encoded G-H loop amino acid sequence** [c]	**Additional amino acid changes in VP1**
Asia1/JS/CHA/05	TTYGEESSR**RGD**LAALARRVNNRLPTS	-
Asia1/JSp1	TTYGEESSR**RGD**LAALARRVNNRLPTS	-
Asia1/JSp1c4	TTYGEESSR**RDD**LAALARRVSNRLPTS	N154S
Asia1/JSp1c8	TTYGEESSR**RDD**LAALARRVSNRLPTS	N154S
Asia1/JSp1c20	TTYGEESSR**RDD**LAALARRVSNRLPTS	N154S
Asia1/JSM4	TTYGEESSR**RGD**LAALARRVNNRLPTS	-
	TTYGEESSR**RSD**LAALARRVNNRLPTS	-
Asia1/JSM6c20	TTYGEESSR**RGD**LAALARRVNNRLPTS	-
	TTYGEESSR**RSD**LAALARRVNNRLPTS	-
pRDD	TTYGEESSR**RDD**LAALARRVSNRLPTS	-
pRSD	TTYGEESSR**RSD**LAALARRVSNRLPTS	-
pRGD	TTYGEESSR**RGD**LAALARRVSNRLPTS	-
FMDV-RDD[a]	TTYGEESSR**RDD**FAALARRVSNRLPTS	L146F
FMDV-RSD[a]	TTYGEESSR**RSD**LAALARRVSNRLPTS	N154S
FMDV-RGD[a]	TTYGEESSR**RGD**FAALARRVSNRLPTS	L146F
FMDV-RDD/pig[b]	TTYGEESSR**RDD**LAALARRVSNRLPTS	-
FMFV-RDD/bovine[b]	TTYGEESSR**RDD**LAALARRVSNRLPTS	-
FMDV-RSD/pig[b]	TTYGEESSR**RSD**LAALARRVSNRLPTS	-
FMDV-RSD/bovine[b]	TTYGEESSR**RSD**LAALARRVSNRLPTS	-

[a] The rescued viruses were passaged 20 times in cell culture.

[b] Virus recovered from vesicular lesions, away from the inoculation site.

[c] Sequence data were obtained by RT-PCR of the VP1 capsid region. The dashes represent receptor binding triplet of the viruses derived from different origins and full-length plasmids.

Evidence that FMDVs can contain an RDD or RSD receptor-binding site increases the quasispecies complexity around the RGD-coding region.

2.2 Rescue of viable viruses from the full-length cDNA clones

To examine the influence of single amino acid substitutions in the receptor binding site of the RDD-containing FMD viral genome on virus viability and the ability of non-RGD viruses to cause disease in susceptible animals, we assembled a full-length cDNA clone of an RDD-containing FMDV and derived mutant clones containing RSD or RGD motifs with a single amino acid substitution in the receptor binding site (RDD→RGD, RDD→RSD). BSR-T7/5 cells were independently transfected with linearized-plasmids, pRDD, pRGD and pRSD. The typical FMDV cytopathic effect was clearly visible at 36 h post-transfection with pRDD and pRGD, and at 48 h post-transfection with pRSD. The three rescued viruses were named FMDV-RDD, FMDV-RGD, and FMDV-RSD, respectively.

To increase the virus titers, all rescued viruses were subjected to serial passage in BHK-21 cells, after which the VP1 sequence was analyzed to confirm that the recovered viruses had maintained the cDNA-encoded receptor binding motifs (Table 2). When the growth characteristics of the rescued viruses were compared with the parental virus Asia1/JSp1c8 by one-step growth kinetics assays, rescued viruses showed similar growth properties to the parental virus (Figure 2a). In addition, the plaque sizes of the parental virus and the rescued viruses were also similar (Figure 2b). These results suggest that single amino acid substitutions in the receptor binding site of Asia1/JSp1c8 virus do not affect virus viability.

2.3 The pathogenicity of the rescued viruses in cattle and swine

To investigate the pathogenicity of the non-RGD viruses in the natural host, we performed direct inoculation of parental virus Asia1/JSp1c8 and recombinant viruses (FMDV-RSD and FMDV-RDD) in cattle and pigs. After inoculation, a number of disease parameters were analyzed, including fever, clinical score, and viremia. The animals, except for the FMDV-RSD-inoculated animals, showed fever and extensive tissue damage at the inoculation sites by day 1 and achieved the maximal score of lesions on day 2-4. Some FMDV-RSD-inoculated animals developed fever and tissue damage by day 2 and achieved the maximal score of lesions on day 3-5. Two animals (infected with FMDV-RSD) had no evidence of tissue damage, except for occasional depression and anorexia when their body temperatures rose.

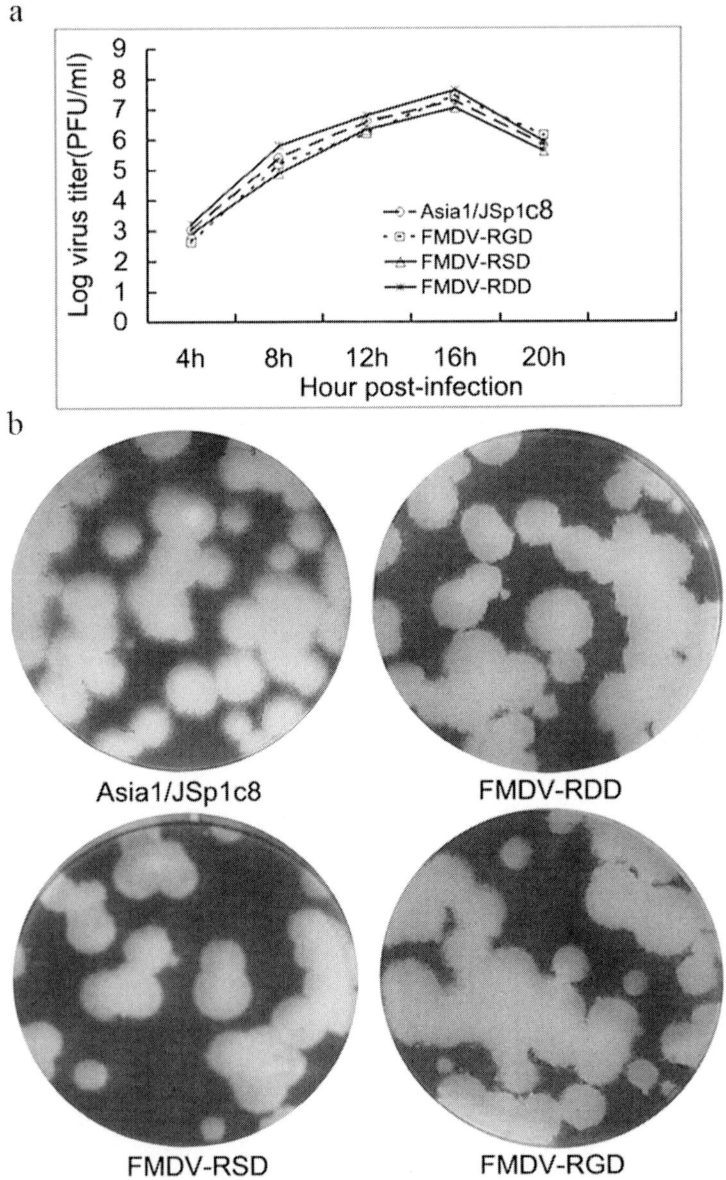

Figure 2 GROWTH CHARACTERISTICS OF THREE RESCUED VIRUSES IN CELL CULTURE COMPARED WITH PARENTAL VIRUS. (a), One-step growth curves of the parental and three cloned viruses. (b), Morphology of plaques formed in BHK-21 cell monolayers by the parental and three cloned viruses.

The Asia1/JSp1c8 and FMDV-RDD viruses produced more extensive tissue damage at the injected sites and induced fever and vesicles a day earlier than in the FMDV-RSD-inoculated animals. There were significant differences in lesion scores between RDD viruses (Asia1/JSp1c8 and FMDV-RDD) and RSD virus ($P < 0.05$, $P < 0.05$), however, no significant differences in lesion scores between cattle and pigs ($P > 0.05$). The lesion scores for the inoculated animals are summarized in table 3 and figure 3 shows the rectal temperature of all of the inoculated animals. The disease was characterized by viremia in all inoculated animals, including the animals that did not generate vesicular lesions. The level of viremia increased following inoculation, typically reaching a peak level after two or three days then decreasing to zero by day 8.

Table 3 Lesion scores of all animals on days-post inoculated

Virus	**Animal no.**	**Lesion scores of days-post inoculation** [a]							
		Day 1	**Day 2**	**Day 3**	**Day 4**	**Day 5**	**Day 6**	**Day 7**	**Day 8**
Asia1/JSp1c8	Bovine88	1	3	5	5	5	5	5	5
	Bovine91	1	3	3	4	4	4	4	4
	Pig451	1	4	5	5	5	5	5	5
	Pig453	1	1	2	4	4	4	4	4
	Pig454	1	3	5	5	5	5	5	5
FMDV-RDD	Bovine 96	1	3	4	5	5	5	5	5
	Bovine 99	1	3	5	5	5	5	5	5
	Pig 458	1	1	3	3	3	3	3	3
	Pig 459	1	2	4	5	5	5	5	5
	Pig 460	1	4	5	5	5	5	5	5
FMDV-RSD	Bovine 100	0	0	0	0	0	0	0	0
	Bovine 101	0	0	3	3	3	3	3	3
	Pig 461	0	1	3	3	4	4	4	4
	Pig 462	0	0	0	0	0	0	0	0
	Pig 465	0	2	3	3	3	3	3	3

[a] Lesion scores were calculated as described by Rieder *et al.* (2005).

In contrast to the lesion scores, there were no significant differences caused by viruses or species in terms of the maximum viremia (P = 0.56), time of maximum viremia (P = 0.75) or time of maximum rate of increase in viremia (P = 0.69). Virus RNA copies in the blood of the inoculated animals are summarized in table 4.

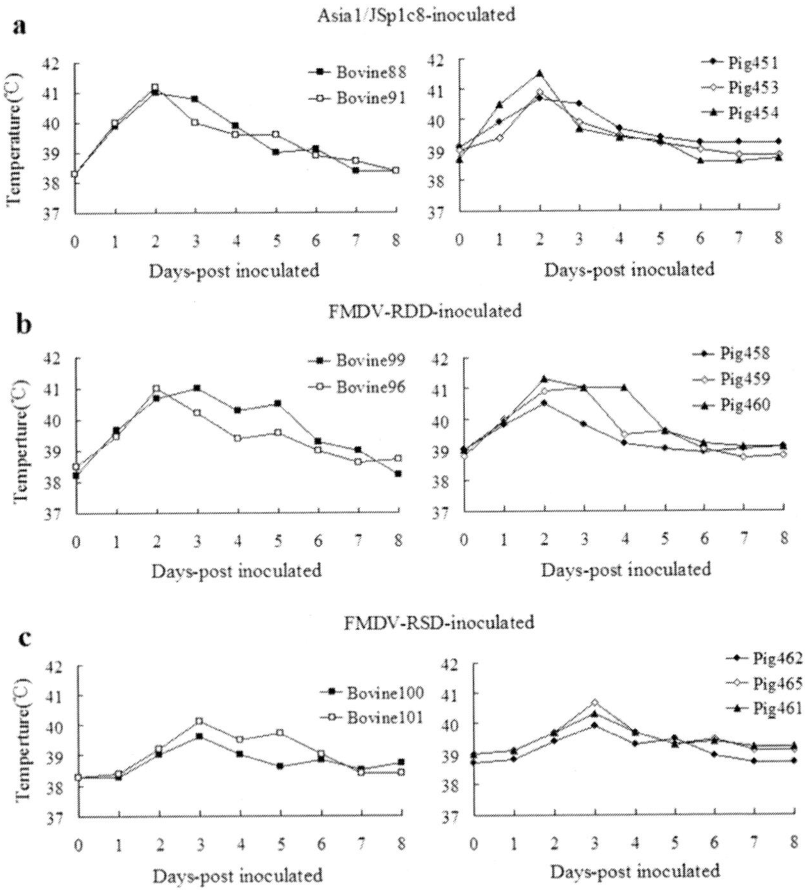

Figure 3 RECTAL TEMPERATURES OF ALL FMDV INOCULATED ANIMALS. (a), Temperatures in Asia1/JSp1c8-inoculated animals; (b), Temperatures in FMDV-RDD-inoculated animals; (c), Temperatures in FMDV-RSD-inoculated animals.

Table 4 Virus RNA copies detected in the blood of all animals on days-post inoculated

Virus	Animal no.	Virus RNA copies in the blood of days-post inoculation($\times 10^6$) [b]							
		Day 1	Day 2	Day 3	Day 4	Day 5	Day 6	Day 7	Day 8
Asia1 /JSp1 c8	Bovine88	0.1	14	4	0.9	2.6	1.1	0	0
	Bovine91	0.3	1.0	14.5	6	0.1	0	0	0
	Pig451	0.04	17	4.6	2.1	0.4	0	0	0
	Pig453	0.06	4	11.7	1	0.3	0	0	0
	Pig454	0.2	9	96.4	10	5	1.8	0.2	0
FMDV -RDD	Bovine 96	2	17.4	42.9	8.8	3.1	4.2	0	0
	Bovine 99	9	78.8	9.4	2.3	0.3	0	0	0
	Pig 458	0.03	0.6	22.5	5.5	3.9	1	0.2	0
	Pig 459	0.2	2.3	30.2	14.4	3.1	0.2	0	0
	Pig 460	0.3	2.8	36.9	15.1	2	0.3	0	0
FMDV -RSD	Bovine 100	0.02	0.2	7.8	3.8	2.1	0.2	0	0
	Bovine 101	0.1	3	12.6	16.2	9.8	6.2	2.3	0
	Pig 461	0.4	6.9	19.6	10.5	5.1	2.8	0.5	0
	Pig 462	0	0.1	14.6	7.1	1	0.9	0	0
	Pig 465	0.02	3.6	16.6	10.4	5.2	1.1	0.9	0

[b] The amount of virus in the blood was measured by real-time quantitative RT-PCR assay as described in materials and methods. Blood samples were collected at 1-8 dpi in inoculated animals.

Vesicular fluid was collected from inoculated animals, and each sample was separately processed for RT-PCR and nucleotide sequencing. The results revealed that the original receptor-binding motif did not change during growth in vivo.

3. DISCUSSION

The RGD integrin-binding motif within VP1 is highly conserved among FMDV field isolates, and is generally considered essential for virus viability via its interaction with cellular integrin heterodimers [24, 25, 26]. Biochemical evidence that small peptides containing the RGD sequence inhibited the adsorption of the virus to tissue culture cells [11], and genetically engineered virions containing either mutations or deletions of the RGD sequence were unable to bind to cells or cause disease in susceptible animals [12, 25, 27]. However, the RGD triplet may be dispensable upon short-term evolution of the virus harboring it in a constant environment [21, 28, 29]. In the present report, we have documented that non-RGD derivatives arose serendipitously during short-term passage of a FMDV field isolate in vitro and in vivo. One derivative containing an RDD triplet in the receptor-binding site was obtained from the serotype Asia 1 field isolate after a single cattle-to-pig transmission and subsequent BHK-21 in vitro passage. Sequence analysis of 10 biological clones of the VP1 encoding region of this population demonstrated that RDD viruses instead of the original RGD virus became predominant at an early phase of Asia1/JS/CHA/05 quasispecies evolution. Unexpectedly, however, both RGD and RSD viruses were obtained from the Asia1/JSM4 population that were generated after four serial passages of the Asia1/JS/CHA/05 field isolate in suckling mice, via intraperitoneal inoculation. The population equilibrium of RSD mutant and ancestor viruses was maintained after 20 passages of the Asia1/JSM6 population in BHK-21 cells. Although RDD- or RSD-containing FMDV are unusual, they were genetically stable upon extended replication in cell culture. Our results suggest that, in the context of the capsid proteins of Asia1/JS/CHA/05, a highly conserved RGD motif is not essential for replication in vitro and in vivo, suggesting functional flexibility of FMDV to enter cells in response to environmental modifications.

Like other RNA viruses, FMDV exists as closely related but non-identical genomes, termed viral quasispecies [30, 31]. Genetic diversity is an intrinsic property of the quasispecies, which arise due to the lack of proofreading activity during viral genome replication, a short replication cycle, and other environmental selective pressures [32, 33]. Our observations showed that evolution of FMDV population exhibited receptor binding motif diversity (genetic diversity) subjected to short-term passage of field isolate in different environments. From the standpoint of RNA virus population evolution, one possible scenario could explain this observation. The early interactions between viruses and host cells exert major selective force on virus populations, thus, the variants (RSD- and RDD-containing viruses) may already be present at low frequency in the natural population that are possibly more fit in new environments and become dominant strains. While this presumption is contrary to the view that the RGD triplet is highly conserved among natural isolates of FMDV, there is direct evidence that an RDD containing field virus was isolated from pigs during a type Asia 1 FMD outbreak in China. RDD-containing FMDV VP1 genes were amplified from sheep oesophageal-pharyngeal fluids (OP-fluids) collected during 2006 from a sheep herd in the region of China that had endemic Asia 1 serotype FMDV [34, 35]. The emergence of these non-RGD mutants in nature is likely to be influenced by specific epidemiological and immunological aspects of host-virus interaction as well as the quasispecies composition of the viral population [36, 37, 38, 39].

The molecular basis for this apparent relaxation of the requirement for the precise RGD sequence is not well understood, but functional replacements within the RGD triplet allow the use of an alternative receptor for entry of FMDV in BHK-21 cells. Since other FMDV lack the RGD motif, host cell recognition may be mediated through another integrin receptor or a non-integrin pathway, or use a third receptor (neither integrin-based nor HS) for entry into the host cell [18, 21, 40]. Further studies are required to analyze the interaction of these mutants with the major FMDV integrin receptors $\alpha v\beta 3$, $\alpha v\beta 6$, $\alpha v\beta 1$ and $\alpha v\beta 8$ identified to date, and to understand whether these viruses obtain alteration of cell tropism, antigenicity, and virulence.

To examine the influence of single amino acid substitutions in the receptor binding site of RDD-containing FMD viral genome on virus viability and the ability of non-RGD viruses to cause disease in susceptible animals, we constructed an FMDV Asia1/JS/p1c8 full-length clone and derived mutant molecules with RGD or RSD receptor recognition motifs. Following transfection of BSR cells with these clones, three recombinant viruses were rescued, in particular, six other amino acid differences in the P1 capsid region of Asia1/JS/CHA/05 and Asia1/JSM4 (compared with Asia1/JS/p1c8) did not affect rescue of viable RGD- and RSD-harboring viruses. Furthermore, in vitro growth properties of these viruses did not differ significantly. Our results showed that Asia1/JS/p1c8 viral genome can tolerate substitutions in the receptor binding site with no other changes in the capsid. The ability of the Asia1/JS/p1c8 viral genome to tolerate substitution of receptor binding sites may depend on the capsid sequence, because the Asp-143 Gly change of receptor recognition site was lethal in the context of the capsid proteins of FMDV C-S8c1. However, the same replacement yielded viable viruses in the context of the capsid protein of FMDV C-S8c1p100 and C-S8c1p213 [21, 41].

To assess the ability of non-RGD FMD viruses to cause disease in naturally susceptible animals, we performed experiment infections of cattle and pigs using the Asia1/JS/p1c8 and two non-RGD recombinant viruses.

Subsequent experiments showed that all viruses were able to cause disease in cattle and pigs and produce rapid onset of clinical signs, characteristic of infection with RGD field strains. The disease was characterized by viremia in all inoculated animals, including the individuals that did not generate vesicular lesions. Amongst these viruses, the RSD virus produced less tissue damage at the inoculation sites and induced fever and vesicles a day later than in the animals inoculated with RDD-containing viruses, which indicated a different degree of disease severity. The different virulence of these viruses was also supported by the maintenance of original receptor recognition sequence in vesicle samples obtained from infected animals. The reason for low virulence of RSD virus, which may bear on receptor usage, needs to be further explored. These observations are particularly interesting, since the presence of an RGD motif is believed to be the main determinant to direct FMDV to integrin-containing target tissues during infection in the natural host [42]. In addition, information currently available indicates that FMDV utilizes integrins for entry in the natural host, and there is no evidence of the use of alternative receptors in vivo [5, 14, 28]. Therefore, our results further support the possibility that a non-RGD-integrin interaction could be responsible for the generation of FMD in the natural host. Our study was the first to demonstrate the ability of an RDD containing natural isolate to cause disease in naturally susceptible animals, and will provide knowledge about the in vivo pathogenesis of non-RGD viruses.

4. CONCLUSION

FMDV quasispecies evolving in a different biological environment gained the capability of selecting different receptor recognition sites. Thus, the early interaction between the viruses and the host cells may exert major selective pressure on FMDV populations that contributes to the evolution and functional flexibility of FMDV to enter cells. Our studies using two non-RGD FMDVs not only show that there was an increase in the number of viable mutants with substitutions in the receptor-binding region, but also provide useful tools for studies of cell recognition by FMDV. Based on an RDD-containing full-length infectious cDNA clone, the RSD- and RGD-containing recombinant viruses were rescued, and single amino acid substitutions in the receptor-binding site did not affect virus viability. The viruses expressing non-RGD receptor binding sites can replicate stably in vitro and induce the disease in susceptible animals.

5. METHODS

5.1 Viruses and cells

FMDV Asia1/JS/CHA/05 utilized in this study was originally isolated from cattle in Wuxi, Jiangsu Province, China, in 2005. The complete genome sequence of this virus was published in GenBank (GenBank Accession: EF149009). FMDV Asia1/JSp1c8 is a viral population resulting from eight serial passages of Asia1/JSp1 virus in BHK-21 cells, as previously described [43], which was obtained from a pig infected by placing it in contact with an Asia1/JS/CHA/05 virus-inoculated cattle. FMDV Asia1/JSM4 is a viral population resulting from four serial passages of Asia1/JS/CHA/05 virus in suckling mice, via intraperitoneal inoculation. Figure 4 shows the passage history of Asia1/JS/CHA/05 field isolate in different environments.

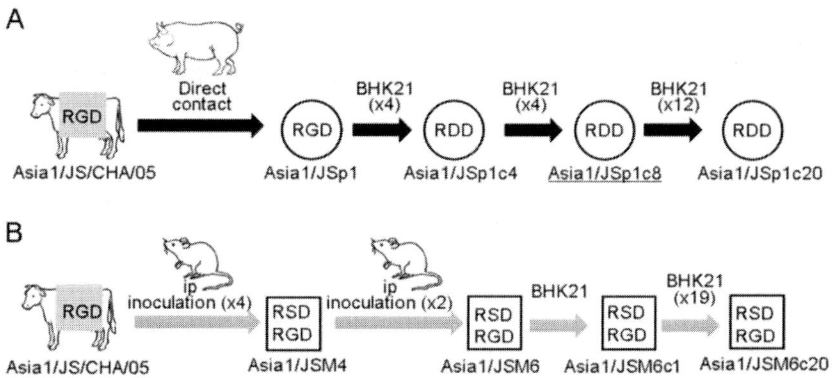

Figure 4 PASSAGE HISTORY AND ORIGIN OF FMDVS USED IN THIS STUDY DERIVED FROM A FIELD ISOLATE, ASIA1/JS/CHA/05. The Asia1/JSp1c8 and Asia1/JSM4 population with alternative RGD motifs were occasionally found by two different passage strategies (**A** and **B**). Nomenclature used for the passaged viruses is as follows: "p" denotes passage number in pig; "M" denotes passage number in suckling mice and "c" denotes passage number in BHK-21 cells. The virus used for construction of full-length infectious cDNA clone is underlined.

BSR-T7/5 cells (a cell line derived from BHK-21, which constitutively expresses T7 RNA polymerase [44]) were maintained in Glasgow minimal essential medium (GMEM) supplemented with 4% tryptose phosphate broth, 10% fetal bovine serum (FBS) and were additionally provided with G418 (1 mg mL^{-1}) on every second passage to ensure maintenance of the T7 polymerase gene. BHK-21 cells were grown in Eagle's minimal essential medium (EMEM) supplemented with 10% FBS.

5.2 RNA extraction, RT-PCR and nucleotide sequencing

RNA was extracted from virus stock of Asia1/JSp1c8, Asia1/JSM4, and Asia1/JS/China/2005 using RNeasy mini kit (Qiagen, Valencia, CA) according to the manufacturer's instructions. Viral cDNAs were synthesized from the viral RNAs, as previously described [45]. Briefly, viral cDNAs were synthesized using M-MLV reverse transcriptase (Invitrogen, Carlsbad, CA, USA) with NK61 primer (5'-GACATGTCCTCCTGCATCTG-3') and the VP1 coding regions were amplified by PCRs with the primer pair NK61/VP31 (5'-TAGTGCTGGYAARGACTTTG-3'). The PCRs were performed using PrimeSTAR HS DNA Polymerase (Takara, Dalian, China). PCR amplifications were carried out for 30 cycles of denaturation at 98°C for 20 s, annealing at 68°C for 1 min, and extension at 72°C for 8 min. Following amplification, the cDNA fragments were purified from agarose gels using a kit (Qiagen) and sequenced by Sunny Biotech (Shanghai, China). In order to detect heterogeneity of the VP1 gene, the amplicons were cloned into a pGEM-T vector (Promega, Madison, WI, USA) using standard molecular cloning techniques [46] and plasmids derived from 10 positive clones for each sample were sequenced. Additionally, the capsid-encoding regions of Asia1/JSp1c8, Asia1/JSM4, and Asia1/JS/CHA/05 were also amplified and sequenced.

5.3 Construction of genome-length cDNA clone of Asia1/JSp1c8 and derivation of G-H loop VP1 mutants

Recombinant DNA techniques were used according to standard procedures [46]. The viral RNA of Asia1/JSp1c8 was used as a template for first-strand cDNA synthesis with M-MLV reverse transcriptase by using specific oligonucleotide primers (E1', E2', E3', E4', and E5'). A total of five fragments (E1-E5; Figure 5), covering the complete virus genome, were subsequently amplified by PCR. Two fragments (E1 and E2 corresponding to nucleotide 1-390, 362-700) were amplified with the E1/E1' and E2/E2' primer pairs by PCR. T7 RNA polymerase promoter was introduced in the E1 primer. Cycling conditions for both PCRs were as follows: initial denaturation at 94°C for 1 min, 30 cycles of 98°C for 20 s, 68°C for 40 s, and then 72°C for 8 min. E12 fragments were generated by overlap PCR fusion E1 and E2 fragments with primer pair E1/E2'. PCR amplifications involved initial denaturation at 94°C for 1 min, followed by 30 cycles of 98°C for 20 s, 68°C for 1 min, then 72°C for 8 min. The amplicons were ligated into the pGEM-T vector, leading to the positive clone pGEME12. The other three fragments (E3, E4 and E5 corresponding to nucleotide 690-3101, 3090-5437 and 5425 to the 3'-end) were produced by PCR with primer pairs E3/E3', E4/E4', E5/E5'. Cycling parameters for three PCRs were as follows: initial denaturation at 94°C for 1 min, 30 cycles of 98°C for 20 s, 68°C for 3 min, and then 72°C for 10 min. The E3, E4 and E5 amplicons were cloned into the M-pSK vector with *XbaI/PstI*, *PstI/EcoRI*, and *EcoRI/NotI* sites, the resulting positive plasmids were designated pSKE3, pSKE4, and pSKE5, respectively. The M-pSK vector derived from pBluescriptSK (+) by removed T7 promoter and modified some restriction enzyme sites in the vector sequence, was synthesized by GenScript Biotech Company (Nanjing, China). To introduce the genetic tags into the genome of Asia1/JSp1c8, recombinant plasmid pSKE3Δ, which contained two synonymous mutations (1185A→G, 1185T→C) to eliminate the *Eco*RI site in the E3 fragment, were constructed by oligonucleotide-directed mutagenesis with PCR amplification of the parent plasmid pSKE3 using p1/p1'primer pair. PCR amplification was carried out for 18 cycles of denaturation at 95°C for 30 s, annealing at 55°C for 1 min, and extension at 68°C for 8 min. All recombinant plasmids were confirmed by complete DNA sequencing. Primers used to construct full-length cDNA clones of Asia1/JSp1c8 are listed in table 5.

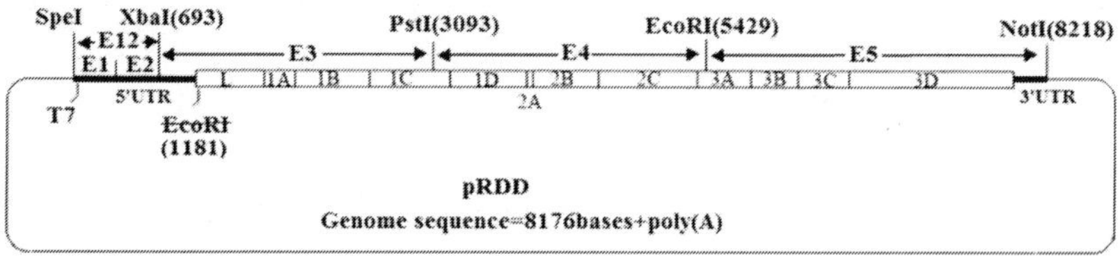

Figure 5 STRATEGY USED TO CONSTRUCT FMDV ASIA1/JSP1C8 FULL-LENGTH CDNA CLONE, PRDD. The location of restriction enzyme cleavage sites used to assemble the subcloned RT-PCR fragments (E1, E2, E3, E4, E5 and E12) are shown (numbered relative to nucleotide position in the virus genome). Thick lines and an open box represent the untranslated regions and the open-reading frame for the viral polyprotein, respectively. The thin line represents the vector sequence. FMDV cDNA is under the control of the T7 promoter.

The full-length virus genome was assembled by a series of ligation steps (Figure 5). First, a 2400-bp *XbaI-PstI* fragment was release from plasmid pSKE3Δ and cloned into plasmid pGEME12 digested with *PstI* and *XbaI*, leading to the construct pGEME123. A 3123-bp *SpeI-PstI* fragment of the pGEME123 was inserted into the pSKE4 plasmid digested with *SpeI* and *PstI*, the resulting plasmid pSKE1234. A 5429-bp *SpeI-EcoRI* fragment was release from plasmid pSKE1234 and ligated into plasmid pSKE5 digested with *EcoRI* and *SpeI*, the resulting plasmid named pRDD, which contained genome-length cDNA clone of Asia1/JSp1c8, was sequenced to confirm sequence fidelity.

Table 5 Sequences of the primers used for the construction of a full-length cDNA clone and mutants of FMDV Asia1/JSp1c8

Name	Nucleotide Sequence (5'→3')	Nucleotide Position (nt)
E1	CAGGATCC *TAATACGACTCACTATAGGG*TTGAAAAGG GGCGCTAGGGTC	1-21
E1'	TAAAACTTAGGGGGGGGGGGGGGGGGGGTGAAAG	361-390
E2	TTTCACCCCCCCCCCCCCCCCCCCCTAAGTTTTAC	362-391
E2'	CCTCTAGA CCTGGAAAGACCAGGC	677-700
E3	AGGTCTAGAGGGGTGACATTTTGT	690-713
E3'	GTCTGCAGCAGAAAGGTAAGGGAT	3078-3101
E4	CTGCTGCAGACTATGCTTACACTG	3090-3113
E4'	AAAGAATTC AATTGCTGCCTCATG	5414-5437
E5	AATTGAATTCTTTGAGGGAATGGTGCAC	5425-5452
E5'	TTGCGGCCGCTTT(38)	3'end
P1	ACAAGGAAAGATGGAGCTCACACTTCACAAC	1168-1198
P1'	GTTGTGAAGTGTGAGCTCCATCTTTCCTTGT	1168-1198
TR1	ACTGCATTCATTCTGAGTGGGA	2960-2984
TR1'	GGCAAGATCACCACGCCGCGAGGA	3679-3703(D→G)
TR2	TCCTCGCGGCGTGGTGATCTTGCC	3679-3703(D→G)
TR2'	5'-GAAGAAACTCGAGGCGACTTTGAC-3'	4342-4366
TR3	TCCTCGCGGCGTAGTGATCTTGCC	3679-3703(D→S)
TR3'	GGCAAGATCACTACGCCGCGAGGA	3679-3703(D→S)

Nucleotide positions of primers used for cloning are shown: numbering according to Asia1/JS/CHA/05 (Genbank Accession: EF149009). T7 promoter sequences are marked with italic type, the restriction sites are highlighted by underlining, silent mutations within the viral sequence are underlined and amino acid changes are shown in brackets.

Overlapping PCRs were used to introduce amino acid substitutions (144 D (gat) to G (ggt), 144 D (gat) to S (agt)) into the structural protein VP1 of Asia1/JSp1c8 virus. Individual parts were amplified with primer pairs TR1/TR1', TR2/TR2', TR1/TR3' and TR3/TR2' (Table 5), and then both overlapping PCR fusion reactions were performed by mixing PCR-amplified fragments with TR1/TR2' primer pair. The parameters of two PCRs as following: initial denaturation at 94°C for 1 min, 30 cycles of 98°C for 20 s, 68°C for 1 min, and then 72°C for 8 min. The two fused PCR fragments were digested with *Eco*RI and *Sac*II and cloned into the full-length plasmid pRDD. The mutated full-length cDNA clones named pRGD, and pRSD, respectively, were sequenced through the entire amplified regions to confirm the presence of the expected modifications.

5.4 Virus rescue

The plasmids pRDD, pRGD and pRSD were linearized with *Not*I and purified from agarose gels with columns (Qiagen). BSR-T7/5 cells (4-6 × 10^5 in a six-well plate) were transfected with mixtures containing 2 μg each of three linearized plasmids and 10 μL Lipofectamine 2000 (Invitrogen) according to the manufacturer's directions. As a negative control, Lipofectamine 2000 was also used to transfect BSR-T7/5 cells. After 6 h of incubation at 37°C, the cells were added to GMEM supplemented with 10% FBS and further incubated for 72 h at 37°C with 5% CO_2. The cell culture supernatants were harvested at 72 h post-transfection and were then serially passaged 10 times on BHK-21 cells to increase virus titers.

5.5 Replication kinetics of rescued FMDVs

Growth kinetics of the viruses was determined in BHK-21 cells. Confluent monolayers in 60 mm diameter plates were infected at a multiplicity of infection (MOI) of 10 PFU per cell with Asia1/JSp1c8 virus and the three genetically engineered viruses. After adsorption for 1 h, the monolayers were washed with 0.01 M phosphate-buffered saline (PBS; pH7.4), and maintained in DMEM supplemented with 2% FBS at 37°C with 5% CO_2. The virus-infected supernatants were collected at 4, 8, 12, 16 and 24 h after inoculation. Plaque assays were performed using gum tragacanth overlay and crystal violet staining, as previously described [47]. Virus titers (plaque-forming units (pfu) mL^{-1}) were determined on BHK-21, as described elsewhere [48].

5.6 Animal experiments

Nine 2-month-old pigs and six 1-year-old bovines were divided into three groups, each consisting of three pigs and two bovines. All animals were seronegative for FMDV non-structural protein (NSP) antibodies prior to experimental infection. Two non-RGD recombinant viruses and Asia1/JSp1c8 virus with a titer of 1.6×10^7pfu mL^{-1}, 1.3×10^7 pfu mL^{-1}, and 1.0×10^7 pfu mL^{-1}, respectively, were used to separately inoculate animals. Each pig was inoculated with 2 mL inoculum via the intramuscular route, each bovine received 1 mL intramuscularly and 1 mL via the tongue. Following inoculation, animals were carefully scored for appearance of lesions at inoculation sites and at other sites. Lesion scores were based on the number of sites affected that were distinct from actual injection sites. Scores were calculated as described by Rieder et al [28].

The viral load in the blood was assessed by real-time quantitative RT-PCR using the RNA Master SYBR green I kit (Roche), as specified by the manufacturer. Quantification was relative to a standard curve obtained with known amounts of FMDV O/CHA/99 RNA, using a procedure that has been described previously [49], except that the primers (patent pending) targeted the 3D non-structural protein were altered. The viral RNA was extracted from vesicular fluid (collected on selected days), reverse transcribed, and sequenced through the entire VP1 region as described above.

All animal studies were approved by the Review Board of Lanzhou Veterinary Research Institute, Chinese Academy of Agricultural Sciences (Permission number: SYXK-GAN-2004-0005). All animals used in this study were humanely bred during the experiment and euthanasia was carried out at the end of the experiment to reduce suffering.

5.7 Statistical analysis

Changes in viral titer over time for the in vitro passage experiments were modeled using linear models with virus and time since infection (treated as a factor) as fixed effects. Model selection proceeded by stepwise deletion of non-significant terms (as judged by *F*-tests) starting from a model including virus, time since infection and an interaction between these factors.

Lesion scores over time were modeled using linear mixed models with virus and species as fixed effects and animal identification number as a random effect. Model selection proceeded by stepwise deletion of non-significant terms (as judged by the Akaike information criterion; AIC) starting from a model including virus, species and an interaction between these factors.

The change in viremia (y, measured as viral RNA copy number) over time (t) was described using the following function,

$$\gamma(t) = \alpha\left(\frac{t}{\beta}\right)^{\beta^2/\gamma^2} \exp\left(-\frac{\beta}{\gamma^2}(t_j - \beta)\right),$$

where α is the maximum viremia, is the time at which maximum viremia occurs and is the time before (after) maximum viremia at which the maximum rate of increase (decrease) in viremia occurs. Parameters for each animal were estimated by fitting the curve to the data using the method of least-squares. Estimates for each animal were compared using Kruskal-Wallis tests to identify significant differences ($P < 0.05$) amongst animals infected with different viruses. This analysis was done for all animals and then repeated for cattle only and for swine only. Animal B99 was excluded from the analysis of viremia because robust estimates could not be obtained for the parameters.

ACKNOWLEDGEMENTS

We thank Karl-Klaus Conzelmann (Max von Pettenkofer Institute and Gene Center, Germany) for generously supplying the cells used in this study. This work was supported by National "863" project, 2011AA10A211.

REFERENCES

1. Bachrach HL: Foot-and-mouth disease virus. Annu Rev Microbiol. 1968, 22: 201-244. 10.1146/annurev.mi.22.100168.001221.
2. Thomson GR, Vosloo W, Bastos AD: Foot-and-mouth disease in wildlife. Virus Res. 2003, 91 (1): 145-161. 10.1016/S0168-1702(02)00263-0.
3. Belsham GJ: Distinctive features of foot-and-mouth disease virus, a member of the picornavirus family; aspects of virus protein synthesis, protein processing and structures. Prog Biophys Mol Biol. 1993, 60 (3): 241-260. 10.1016/0079-6107(93)90016-D.
4. Acharya R, Fry E, Stuart D, Fox G, Rowlands D, Brown F: The three-dimensional structure of foot-and-mouth disease virus at 2.9 A° resolution. Nature. 1989, 337 (6209): 709-716. 10.1038/337709a0.
5. Logan D, Abu-Ghazaleh R, Blakemore W, Curry S, Jackson T, King A, Lea S, Lewis R, Newman J, Parry N, Rowlands D, Stuart D, Fry E: Structure of a major immunogenic site on foot-and-mouth disease virus. Nature. 1993, 362 (6420): 566-568. 10.1038/362566a0.
6. Lea S, Hernández J, Blakemore W, Brocchi E, Curry S, Domingo E, Fry E, Abu Ghazaleh R, King A, Newman J, Stuart D, Mateu GM: The structure and antigenicity of a type C foot-and-mouth disease virus. Structure. 1994, 2 (2): 123-139. 10.1016/S0969-2126(00)00014-9.

7. Fox G, Parry N, Barnett PV, McGinn B, Rowlands DJ, Brown F: The cell attachment site on foot-and-mouth disease virus includes the amino acid sequence RGD (arginine-glycine-aspartic acid). J Gen Virol. 1989, 70 (Pt3): 625-637.
8. Strohmaier K, Franze R, Adam KH: Location and characterization of the antigenic portion of the FMDV immunizing protein. J Gen Virol. 1982, 59 (Pt2): 295-306.
9. Bittle JL, Houghten RA, Alexander H, Shinnick TM, Sutcliffe JG, Lerner RA, Rowlands DJ, Brown F: Protection against foot-and-mouth disease by immunization with a chemically synthesized peptide predicted from the viral nucleotide sequence. Nature. 1982, 298 (5869): 30-33. 10.1038/298030a0.
10. D'Souza S, Ginsberg M, Plow E: Arginyl-glycyl-aspartic acid (RGD): a cell adhesion motif. TiBS. 1991, 16 (7): 246-250.
11. Baxt B, Becker Y: The effect of peptides containing the arginine-glycine-aspartic acid sequence on the adsorption of foot-and-mouth disease virus to tissue culture cells. Virus Genes. 1990, 4 (1): 73-83. 10.1007/BF00308567.
12. Mason PW, Rieder E, Baxt B: RGD sequence of foot-and mouth disease virus is essential for infecting cells via the natural receptor but can be bypassed by an antibody-dependent enhancement pathway. Proc Natl Acad Sci USA. 1994, 91 (5): 1932-1936. 10. 1073/pnas.91.5.1932.
13. Berinstein A, Roivainen M, Hovi T, Mason PW, Baxt B: Antibodies to the vitronectin receptor (integrin αvβ3) inhibit binding and infection of foot-and-mouth disease virus to cultured cells. J Virol. 1995, 69 (4): 2664-2666.
14. Neff S, Sa-Carvalho D, Rieder E, Mason PW, Blystone SD, Brown EJ, Baxt B: Foot-and-mouth disease virus virulent for cattle utilizes the integrin αvβ3 as its receptor. J Virol. 1998, 72 (5): 3587-3594.
15. Jackson T, Sheppard D, Denyer M, Blakemore W, King AMQ: The epithelial integrin αvβ6 is a receptor for foot-and-mouth disease virus. J Virol. 2000, 74 (11): 4949-4956. 10.1128/JVI.74.11.4949-4956.2000.
16. Jackson T, Mould AP, Sheppard D, King AMQ: Integrin αvβ1 is a receptor for foot-and-mouth disease virus. J Virol. 2002, 76 (3): 935-941. 10.1128/JVI.76.3.935-941.2002.
17. Jackson T, Clark S, Berryman S, Burman A, Cambier S, Mu D, Nishimura S, King AMQ: Integrin αvβ8 functions as a receptor for foot-and-mouth disease virus: role of the β-chain cytodomain in integrin-mediated infection. J Virol. 2004, 78 (9): 4533-4540. 10.1128/JVI.78.9.4533-4540.2004.
18. Sa-Carvalho D, Rieder E, Baxt B, Rodarte R, Tanuri A, Mason PW: Tissue culture adaptation of foot-and-mouth disease virus selects viruses that bind to heparin and are attenuated in cattle. J Virol. 1997, 71 (7): 5115-5123.
19. Martínez MA, Verdaguer N, Mateu MG, Domingo E: Evolution subverting essentiality: dispensability of the cell attachment Arg-Gly-Asp motif in multiply passaged foot-and-mouth disease virus. Proc Natl Acad Sci USA. 1997, 94 (13): 6798-6802. 10.1073/pnas.94.13.6798.
20. Ruiz-Jarabo CM, Sevilla N, Da'vila M, Gomez-Mariano G, Baranowski E, Domingo E: Antigenic properties and population stability of a foot-and-mouth disease virus with altered Arg-Gly-Asp receptor-recognition motif. J Gen Virol. 1999, 80 (8): 1899-1909.
21. Baranowski E, Ruíz-Jarabo CM, Sevilla N, Andreu D, Beck E, Domingo E: Cell recognition by foot-and-mouth disease virus that lacks the RGD integrin-binding motif: flexibility in aphthovirus receptor usage. J Virol. 2000, 74 (4): 1641-1647. 10.1128/JVI.74.4.1641-1647.2000.
22. Jackson T, Ellard FM, Abu-Ghazaleh R, Brookes SM, Blakemore WE, Corteyn AH, Stuart DL, Newman JWI, King AMQ: Efficient infection of cells in culture by type O foot-and-mouth disease virus requires binding to cell surface heparan sulfate. J Virol. 1996, 70 (8): 5285-5287.
23. Fry EE, Newman JWI, Curry S, Najjam S, Jackson T, Blakemore W, Lea SM, Miller L, Burman A, King AMQ, Stuart DI: Structure of Foot-and-mouth disease virus serotype $A_{10}61$ alone and complexed with oligosaccharide receptor: receptor conservation in the face of antigenic variation. J Gen Virol. 2005, 86 (7): 1909-1920. 10.1099/vir.0.80730-0.

24. Mateu MG: Antibody recognition of picornaviruses and escape from neutralization: a structural view. Virus Res. 1995, 38 (1): 1-24. 10.1016/0168-1702(95)00048-U.
25. McKenna TS, Lubroth J, Rieder E, Baxt B, Mason PW: Receptor binding site-deleted foot-and-mouth disease (FMD) virus protects cattle from FMD. J Virol. 1995, 69 (9): 5787-5790.
26. Rieder E, Berinstein A, Baxt B, Kang A, Mason PW: Propagation of an attenuated virus by design: engineering a novel receptor for a noninfectious foot-and-mouth disease virus. Proc Natl Acad Sci USA. 1996, 93 (19): 10428-10433. 10.1073/pnas.93.19.10428.
27. Leippert M, Beck E, Weiland F, Pfaff E: Point mutations within the βG-βH loop of foot-and-mouth disease virus O1K affect virus attachment to target cells. J Virol. 1997, 71 (2): 1046-1051.
28. Rieder E, Henry T, Duque H, Baxt B: Analysis of a foot and- mouth disease virus type A24 isolate containing an SGD receptor recognition site in vitro and its pathogenesis in cattle. J Virol. 2005, 79 (20): 12989-12998. 10.1128/JVI.79.20.12989-12998.2005.
29. Baranowski E, Ruiz-Jarabo CM, Domingo E: Evolution of cell recognition by viruses. Science. 2001, 292 (5519): 1102-1105. 10.1126/science.1058613.
30. Domingo E, Martinez-Salas E, Sobrino F, de la Torre JC, Portela A, Ortin J, Lopez-Galindez C, Perez-Brena P, Villanueva R, Najera R, VandePol S, Steinhauer D, DePolo N, Holland J: The quasispecis (extreme heterogeneous) nature of viral RNA genome populations: biological relevance-a review. Gene. 1985, 40 (1): 1-8. 10.1016/0378-1119(85)90017-4.
31. Eigen M: On the nature of virus quasispecies. Trends Microbiol. 1996, 4 (6): 216-218. 10.1016/0966-842X(96)20011-3.
32. Holland JJ, Spindler K, Horodyski F, Grabau E, Nichol S, Vande Pol S: Rapid evolution of RNA genomes. Science. 1982, 215 (4540): 1577-1585. 10.1126/science.7041255.
33. Tosh C, Sanyal A, Hemadri D, Venkataramanan R: Phylogenetic analysis of serotype A foot-and-mouth disease virus isolated in India between 1977 and 2000. Arch Virol. 2002, 147 (3): 493-513. 10.1007/s007050200002.
34. Zheng HX, Jin Y, Shang YJ, Guo JH, Tian H, Yang YM, Liu XT, Cai XP: Comparative analysis of the genomes of foot-and-mouth disease virus isolates from porcine and cattle origin. Microbiology China. 2010, 37 (9): 1312-1319.
35. Wang JD, Lu ZJ, Bao HF, Cao YM, Guo JH, Liu XT, He SH, Yang CS, Liu ZX: Analysis of VP1 gene 3'terminal sequence of foot-and-mouth disease virus type Asia 1 derived from ovine O/P fluid. Chinese Veterinary Science. 2008, 38 (7): 559-562.
36. Domingo E, Holland JJ: RNA virus mutations and fitness for survival. Annu Rev Microbiol. 1997, 51 (1): 151-178. 10.1146/annurev.micro.51.1.151.
37. Domingo E, Gomez J: Quasispecies and its impact on viral hepatitis. Virus Res. 2007, 127 (2): 131-150. 10.1016/j.virusres.2007.02.001.
38. Perales C, Mateo R, Mateu MG, Domingo E: Insights into RNA virus mutant spectrum and lethal mutagenesis events: replicative interference and complementation by multiple point mutants. J Mol Biol. 2007, 369 (4): 985-1000. 10.1016/j.jmb.2007.03.074.
39. Coffin JM: HIV population dynamics in vivo: implications for genetic variation, pathogenesis and therapy. Science. 1995, 267 (5197): 483-489. 10.1126/science.7824947.
40. Zhao Q, Pacheco J, Mason PW: Evaluation of genetically engineered derivatives of a Chinese strain of foot-and-mouth disease reveals a novel cell-binding site which functions in cell culture and in animals. J Virol. 2003, 77 (5): 3269-3280. 10.1128/JVI.77.5.3269-3280.2003.
41. Gutierrez-Rivas M, Pulido MR, Baranowski E, Sobrino F, Saiz M: Tolerance to mutations in the foot-and-mouth disease virus integrin-binding RGD region is different in cultured cells and in vivo and depends on the capsid sequence context. J Gen Virol. 2008, 89 (Pt 10): 2531-2539.
42. Alexandersen S, Zhang Z, Donaldson AI, Garland AJ: The pathogenesis and diagnosis of foot-and-mouth disease. J Comp Pathol. 2003, 129 (1): 1-36. 10.1016/S0021-9975(03)00041-0.

43. Domingo E, Davila M, Ortin J: Nucleotide sequence heterogeneity of the RNA from a natural population of foot-and-mouth disease virus. Gene. 1980, 11 (3-4): 333-346. 10.1016/0378-1119(80)90073-6.
44. Buchholz UJ, Finke S, Conzelmann KK: Generation of bovine respiratory syncytial virus (BRSV) from cDNA: BRSV NS2 is not essential for virus replication in tissue culture, and the human RSV leader region acts as a functional BRSV genome promoter. J Virol. 1999, 73 (1): 251-259.
45. Mason PW, Bezborodova SV, Henry TM: Identification and characterization of a cis-acting replication element (cre) adjacent to the internal ribosome entry site of foot-and-mouth disease virus. J Virol. 2000, 76 (19): 9686-9694.
46. Sambrook J, Fitsch EF, Maniatis T: Molecular Cloning: A Laboratory Manual. 1989, Cold Spring Harbor, Cold Spring Harbor Press
47. Rieder E, Bunch T, Brown F, Mason PW: Genetically engineered foot-and-mouth disease viruses with poly(C) tracts of two nucleotides are virulent in mice. J Virol. 1993, 67 (9): 5139-5145.
48. Pacheco JM, Henry TM, O'Donnell VK, Gregory JB, Mason PW: Role of nonstructural proteins 3A and 3B in host range and pathogenicity of foot-and-mouth disease virus. J Virol. 2003, 77 (24): 13017-13027. 10.1128/JVI.77.24.13017-13027.2003.
49. Alexandersen S, Oleksiewicz MB, Donaldson AI: The early pathogenesis of foot-and-mouth disease in pigs infected by contact: a quantitative time-course study using TaqMan RT-PCR. J Gen Virol. 2001, 82 (Pt4): 747-755.

CHAPTER 10

Complete genome sequence of canine astrovirus with molecular and epidemiological characterisation of UK strains

Sarah L. Caddy[a, b,*] Ian Goodfellow[a, **]

[a] *Division of Virology, Department of Pathology, University of Cambridge, Addenbrookes Hospital, Cambridge CB2 2QQ, UK*

[b] *Section of Virology, Faculty of Medicine, Imperial College London, St. Mary's Campus, London W2 1PG, UK*

ABSTRACT

Astroviruses are a common cause of gastroenteritis in children worldwide. These viruses can also cause infection in a range of domestic and wild animal species. Canine astrovirus (CaAstV) was first identified in the USA, and has since been reported in dogs from Europe, the Far East and South America. We sought to determine whether CaAstV is circulating in the UK dog population, and to characterise any identified strains. Stool samples were collected from pet dogs in the UK with and without gastroenteritis, and samples were screened for CaAstV by qPCR. Four CaAstV positive samples were identified from dogs with gastroenteritis (4/67, 6.0%), whereas no samples from healthy dogs were positive ($p < 0.001$). Sequencing of the capsid sequences from the four CaAstV strains found significant genetic heterogeneity, with only 80% amino acid identity between strains. The full genome sequence of two UK CaAstV strains was then determined, confirming that CaAstV conforms to the classic genome organisation of other astroviruses with ORF1a and ORF1b separated by a frameshift and ORF2 encoding the capsid protein. This is the first report describing the circulation of CaAstV in UK dogs with clinical signs of gastroenteritis, and the first description of the full-length genomes of two CaAstV strains.

Keywords: Canine astrovirus, Gastroenteritis, Complete genome

1. INTRODUCTION

Astroviruses are small non-enveloped, positive sense RNA viruses with a wide host range. Astroviruses are classified into two genera; the *Mamastrovirus* genera includes astroviruses isolated from humans, pigs, cattle, cats and dogs, whereas astrovirus isolates from birds are categorised into the *Avastrovirus* genera (King et al., 2011). Astroviruses were first identified in 1975 in the stools of children with diarrhoea (Appleton and Higgins, 1975 ; Madeley and Cosgrove, 1975), and are now estimated to cause up to 10% of gastroenteritis cases in children worldwide (Moser and Schultz-Cherry, 2005).

Canine astrovirus (CaAstV) was first described in the USA, following the identification of star-shaped particles in diarrhoeic stools from a litter of beagles (Williams, 1980). Later studies have identified CaAstV in Italy, France, China, Korea and Brazil (Castro *et al.*, 2013; Choi *et al.*, 2014; Grellet *et al.*, 2012; Martella *et al.*, 2011 ; Zhu *et al.*, 2011).

The genome of astroviruses is typically 6.4–7.3 kb and divided into three open reading frames (ORFs), ORF1a, ORF1b and ORF2 with a 5′ untranslated region and a 3′ poly-A tail (King et al., 2011). ORF1 codes for the non-structural proteins and ORF2 encodes the capsid precursor protein. The complete coding sequence of ORF2, and a partial sequence of ORF1b has previously been determined for a number of CaAstV isolates (Toffan *et al.*, 2009 ; Zhu *et al.*, 2011), but to date, no full-length genome sequence of CaAstV has been reported.

The purpose of this study was to determine the prevalence of CaAstV in the UK dog population and to obtain where possible the complete genetic sequence of circulating strains. Four CaAstV strains were identified in dogs showing clinical signs of gastroenteritis and sequencing of ORF2 found significant genetic diversity between strains. The first full-length sequences of two CaAstV strains was subsequently determined.

2. METHODS

2.1. Sample collection

Stool samples were collected from dogs admitted to five participating veterinary clinics and a single animal shelter in counties in the south and east of England; Cambridgeshire, Kent, Lincolnshire, Middlesex and Suffolk. Ethical approval was not required for this study as all samples collected were animal waste products. With owner consent, dogs were recruited to the study if they passed stools whilst hospitalised. Stool samples were collected from dogs from the animal shelter if they passed diarrhoea. All stool samples were stored at −20 °C until and during transportation to the laboratory, where they were stored at −80 °C prior to nucleic acid extraction. As controls, stool samples were also collected from healthy dogs owned by veterinary staff at each clinic, as well as from dogs at participating boarding kennels. Basic case data was recorded for each dog from which a stool sample was collected, including age, breed, sex, reason for admission, and any recent history of enteric disease.

2.2. Detection of CaAstV in clinical samples

Stools were diluted to a final concentration of 10% w/v in phosphate-buffered saline, pH 7.2, and solids were removed by centrifugation at 8000 × *g* for 5 min. Viral nucleic acid was extracted from 140 μl of each clarified stool suspension using the GenElute™ Mammalian Total RNA Miniprep Kit (Sigma–Aldrich) according to the manufacturers' instructions. An internal extraction control (equine arteritis virus RNA) was added during the nucleic acid extraction process as previously described (Caddy et al., 2013). cDNA was generated by reverse transcription using MMLV reverse transcriptase enzyme (Life Technologies) and random hexamers (Life Technologies) with the reaction performed at 42 °C for 1 h, followed by an inactivation step at 70 °C for 10 min. qPCR was performed using primers targeting a conserved region of the viral RNA-dependent RNA polymerase as previously described (Martella et al., 2011). The sequences of primers used for PCR in this study are presented in Table 1. qPCR reactions were prepared using the MESA Blue qPCR MasterMix Plus for SYBR Assay (Eurogentech). Briefly, 2 μl cDNA was mixed with 2×MasterMix and 0.5 μM primers, then incubated at 95 °C for 10 min. The thermal cycle protocol used with a ViiA7 qPCR machine (AB Applied Biosystems) was as follows: 40 cycles of 94 °C, 15 s; 56 °C, 30 s; 72 °C, 30 s, followed by generation of a melt curve. Viral genome copy number was calculated by interpolation from a standard curve generated using serial dilutions of a standard DNA amplicon cloned from a positive control. The limit of detection was determined by the lowest dilution of control standard CNA reproducibly detected in the assay. All samples were additionally screened for the presence of canine parvovirus (CPV), canine enteric coronavirus (CECoV) and canine norovirus (CNV) using a 1-step qRT-PCR protocol (Caddy et al., 2013).

Table 1. CaAstV specific primers used in this study.

Primer target		Sequence	Reference
RdRp	625F-1	GTACTATACCRTCTGATTTAATT	Martella et al. (2011)
		AGACCAARGTGTCATAGTTCAG	
ORF1b–ORF2	AV12	TTT TTT TTT TTT TTT TTT GC	Toffan et al. (2009)
	s2m-rev	CCC TCG ATC CTA CTC GG	
ORF1a	683	TGAAGGACTGCTCAGAGTG	This paper
5′ RACE	682	ATGCAGCGACAAACACAACA	This paper
3′ RACE	894	CACAGCCCATTGAAGATG	This paper

2.3. Capsid sequencing

All samples positive to CaAstV by qPCR were subjected to conventional PCR to confirm the presence of CaAstV and to enable genome sequencing. The capsid of all positive samples was amplified from cDNA synthesised using SuperScript II Reverse Transcriptase (Invitrogen) according to the manufacturer's protocol with 0.5 μM AV12 primer (Table 1). The PCR reaction was performed using KOD hot start polymerase (EMD Millipore), with reverse primer s2m-rev, and the forward primer 625F-1 from the original qPCR assay. The amplification programme consisted of an initial 5 min step at 95 °C, followed by 35 cycles with 95 °C for 20 s, 58 °C for 30 s and 72 °C for 90 s. A final elongation step at 72 °C for 5 min was performed, followed by chilling to 4 °C.

PCR products were subsequently cloned into pCR-Blunt™ using the Zero Blunt PCR Cloning Kit (Life Technologies) according to the manufacturers protocol. Sequencing of the 5′ and 3′ regions of the plasmid insert was performed using pCR-Blunt™ specific primers by the University of Cambridge Biochemistry DNA Sequencing Facility. Sequencing primers for the central region of insert were then designed based on the primary sequence data to give a 200 nt overlap with each predecessor, and a second round of sequencing reactions was performed (sequencing primer details available on request). The complete capsid nucleotide sequence generated using this method was then confirmed by sequencing of PCR products generated from cDNA directly.

2.4. Full genome sequencing and phylogenetic analysis

PCR amplification of ORF1a and ORF1b was achieved using cDNA synthesised using SuperScript II Reverse Transcriptase (Invitrogen) according to the manufacturer's protocol with 626R-1 reverse primer (Table 1). A forward primer was designed approximately 300 bp from the 5′ end of ORF1a, based on a small genome fragment fortuitously amplified by mis-priming in a preliminary screen. The Zero Blunt PCR Cloning Kit was then used to clone the PCR product and sequencing performed using the protocol described above. Primers were designed sequentially to enable sequencing of the entire cloned product in both the 5′–3′ and 3′–5′ direction, with 200 nt overlaps between segments. Sequencing of PCR products directly was finally performed to ensure no mutations has arisen as a result of cloning. Each nucleotide was sequenced at least once in each direction, with an average coverage of 3.

Sequences at the extremity of the viral genome were determined using 5′ and 3′ RACE utilising a kit (Life Technologies) according to the manufacturers instructions, and gene specific primers as listed in Table 1. Direct sequencing was then performed as above.

Software used for sequence analysis included Vector NTI (Life Technologies), BLAST (http://blast.ncbi.nlm.nih.gov/Blast.cgi), ClustalW2 (http://www.ebi.ac.uk/Tools/msa/clustalw2) and Protein Variability Server software (Garcia-Boronat et al., 2008). Evolutionary analysis was conducted using MEGA version 6 (Tamura et al., 2013).

3. RESULTS

3.1. Sample collection

Stool samples and clinical data were collected from 67 dogs with severe gastroenteritis admitted to veterinary clinics or an animal shelter distributed across the UK between August 2012 and June 2014. Control samples were collected from 181 dogs without signs of gastroenteritis, from either veterinary inpatients with non-gastrointestinal illness, or dogs at boarding kennels or belonging to veterinary staff. In total 56 breeds of dog were represented. The mean age of dogs with gastroenteritis was 4.3 years (standard deviation 4.1 years) and the mean age of control animals was 6.1 years (standard deviation 3.9 years).

3.2. Identification of CaAstV isolates

Nucleic acid extraction and qPCR were performed on 248 stool samples. Samples were systematically tested for the presence of CaAstV using a SYBR-based qPCR screen. In addition, screening for CPV, CECoV, CNV and the internal extraction control (EAV) was also performed using a 1-step qPCR protocol. CaAstV was detected in a total of four samples. All four positive dogs were showing signs of gastroenteritis, whereas CaAstV was not detected in any dogs without gastroenteritis. The difference between the prevalence of CaAstV in dogs with gastroenteritis (6.0%) and prevalence in dogs without gastroenteritis was statistically significant (p = <0.001). CPV was detected in 10/67 dogs with gastroenteritis, and CECoV in 2/67 of the same group. Two of the four CaAstV positive dogs were also co-infected with CPV, but no co-infections with CECoV and CaAstV were identified. The age range of CaAstV positive dogs was from 7 weeks to 7 years, with a mean age 2.1 years (SD 3.3 years). Table 2 summarises the clinical information and viral screening results of the four CaAstV positive dogs.

3.3. Comparison of CaAstV capsid sequences

The complete coding sequence of the four CaAstV strains identified was determined using conventional PCR and cloning techniques. The ORF2 nucleotide and amino acid sequences of these strains were aligned using ClustalW2. The overall nucleotide identity between strains was 77.1–81.1%, whereas the amino acid identity was 79.3–86.3%. The four sequences were deposited in the GenBank database and assigned accession numbers KP404149–KP404152 (cases 1–4, respectively).

Table 2. Clinical and virological data for CaAstV positive cases.

Case	CaAstV strain	Age	Breed	Collection date	Additional viruses identified
1	Gillingham/2012/UK KP404149	7 weeks	Crossbreed	August 2012	CPV
2	Lincoln/2012/UK KP404150	7 years	Border Collie	August 2012	None
3	Huntingdon/2014/UK KP404151	1 year	Staffordshire Bull Terrier	February 2014	None
4	Braintree/2014/UK KP404152	10 weeks	Crossbreed	May 2014	CPV

A number of studies have reported the N-terminal and C-terminal regions of astrovirus capsids to be relatively conserved, whereas the central region is hypervariable. The human astrovirus capsid protein has previously been divided into three regions; the N terminus (amino acids 1–415), a variable central region (416–707), which includes a hypervariable section from 649 to 707, and a conserved C terminus (708–786) (Willcocks et al., 1995). An analogous approach for the CaAstV capsid was taken by Zhu et al. (2011), who divided the capsid into three regions for analysis: amino acids 1–446, 447–730, and 731–end. The four capsid sequences derived from this study have been analysed according to the latter scheme, and sequence identity compared (Fig. 1). Sequence analysis of the three regions clearly shows that the majority of sequence variation is concentrated in region II. A 24 nt deletion was identified in samples 2–4 in the 5′ end region II, which has previously been reported in Chinese CaAstV strains (Zhu et al., 2011). It is not possible to predict the location of this deletion on the capsid structure as it is beyond the region of the astrovirus capsid spike for which the crystal structure has been solved (Dong et al., 2011). However, sequence alignment with the human astrovirus type 8 capsid (GenBank AAF85964.1) shows this deletion to be located downstream of the caspase cleavage site required for virion maturation, which truncates the full length capsid protein (VP90) into the mature VP70 form (Banos-Lara and Méndez, 2010). Therefore it is predicted that the 24 nt deletion will not alter the mature virion.

Evolutionary analysis of the four CaAstV sequences from the study, alongside the seven previously reported full-length CaAstV capsid sequences are presented in Fig. 2. This analysis indicated that the UK strains do not cluster, contrary to a previous study, which analysed CaAstV strains from a single city (Zhu et al., 2011). Each UK strain is distinct from each other, and whereas one strain clusters most closely with the Chinese strains, the remainder group with strains identified in Italy over a number of years.

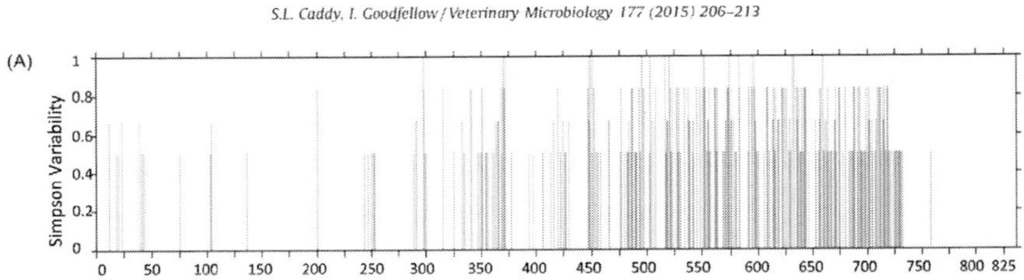

(B)

Region I

Strain	Gillingham/ 2012/UK	Lincoln/ 2012/UK	Huntingdon/ 2014/UK	Braintree/ 2014/UK
Gillingham/ 2012/UK	100.0	92.2	91.5	92.4
Lincoln/ 2012/UK	92.2	100.0	91.5	92.6
Huntingdon/ 2014/UK	91.5	91.5	100.0	94.0
Braintree/ 2014/UK	92.4	92.6	94.0	100.0

Region II

Strain	Gillingham/ 2012/UK	Lincoln/ 2012/UK	Huntingdon/ 2014/UK	Braintree/ 2014/UK
Gillingham/ 2012/UK	100.0	59.1	55.1	53.7
Lincoln/ 2012/UK	59.9	100.0	52.0	70.8
Huntingdon/ 2014/UK	55.1	52.0	100.0	49.1
Braintree/ 2014/UK	53.7	70.8	49.1	100.0

Region III

Strain	Gillingham/ 2012/UK	Lincoln/ 2012/UK	Huntingdon/ 2014/UK	Braintree/ 2014/UK
Gillingham/ 2012/UK	100.0	99.0	98.1	99.0
Lincoln/ 2012/UK	99.0	100.0	99.0	100.0
Huntingdon/ 2014/UK	98.1	99.0	100.0	99.0
Braintree/ 2014/UK	99.0	100.0	99.0	100.0

Figure. 1. Alignment and variability analysis of capsid amino acid sequences of the four CaAstV strains identified in the UK. (A) CaAstV capsid amino acid sequences were aligned using ClustalW2, then a variability scan was constructed using the Simpson diversity index and Protein Variability Server software (Garcia-Boronat et al., 2008). (B) Amino acid percentage identity between CaAstV strains in three different regions of the capsid protein.

3.4. Full-length CaAstV genome sequenced

The complete CaAstV genome was determined from two samples, isolated from a 7-week-old crossbreed dog (strain Gillingham/2012/UK, GenBank accession number KP404149) and a 7-year-old Border Collie (strain Lincoln/2012/UK, KP404150). The total length of CaAstV Gillingham/2012/UK is 6600 nt and CaAst/Lincoln/2012/UK is 6572 nt. Each genome encodes three ORFs; ORF1a, ORF1b, and ORF2 flanked by a 5′ UTR, and a 3′ UTR plus a poly-A tail (Fig. 3). In human astroviruses, the 5′ UTR is 85 nt in length, whereas data from both CaAstV strains would indicate that the CaAstV 5′ UTR is 45 nt. The 83 nt 3′ UTR of human astrovirus is identical to 3′ UTR in CaAstV Lincoln/2012/UK, whereas the 3′ UTR of CaAstV

Gillingham/2012/UK is 2 nt shorter. The nucleotide composition of both CaAstV strains is 29% A, 22% G, 26% T and 23% C. The G/C composition is 45%.

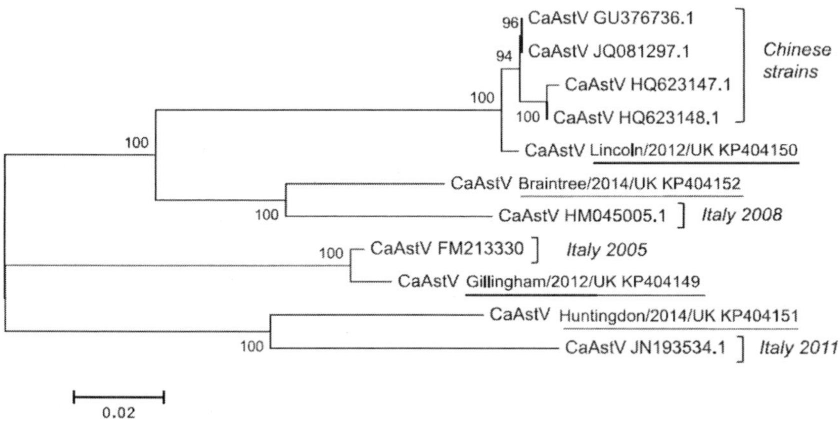

Figure. 2. Phylogenetic tree based on the full-length amino acid sequence of the capsid protein of CaAstVs. This includes the four UK strains identified in this study (underlined) and the seven strains previously sequenced and listed in GenBank. The tree was determined using the neighbour-joining method. The percentage of replicate trees in which the associated taxa clustered together in the bootstrap test (100 replicates) is shown next to the branches. Evolutionary analyses were conducted in MEGA6.

The ORF1a of non-canine astroviruses encodes a serine protease (Willcocks et al., 1994). Sequence alignment of CaAstV with astroviruses of other species shows a high degree of conservation in the predicted serine protease region (Fig. 3B). This is especially pronounced in the regions around the proposed catalytic triad of the serine protease.

ORF1a also encodes the viral genome-linked protein, VPg (Fuentes et al., 2012) (Fig. 4C). CaAstV VPg is predicted to start at aa 656, at a conserved QK cleavage site and is 90 aa in length. In other astroviruses, it has been proposed that the C-terminal VPg cleavage site is coded by Q(P/A/S/L) (Al-Mutairy et al., 2005), and the presence of a QS dipeptide at the same site in CaAstV is consistent with this prediction. The amino acid motif KGK(N/T)K is conserved at the N-terminal end of VPg sequences from both astroviruses and caliciviruses, and this is also identifiable in the CaAstV genome. Another conserved VPg motif is TEXEY, with mutagenesis studies indicating that the Y (tyrosine) residue covalently links VPg to viral RNA (Fuentes et al., 2012). Analysis of the CaAstV ORF1a sequence also identifies the conserved TEXEY motif at 684–688, thus this tyrosine is predicted to covalently link to the RNA genome. However, the CaAstV sequence diverges slightly from the other mamastroviruses studied, in that X of the motif corresponds to K, whereas this is E/Q in all other mamastroviruses.

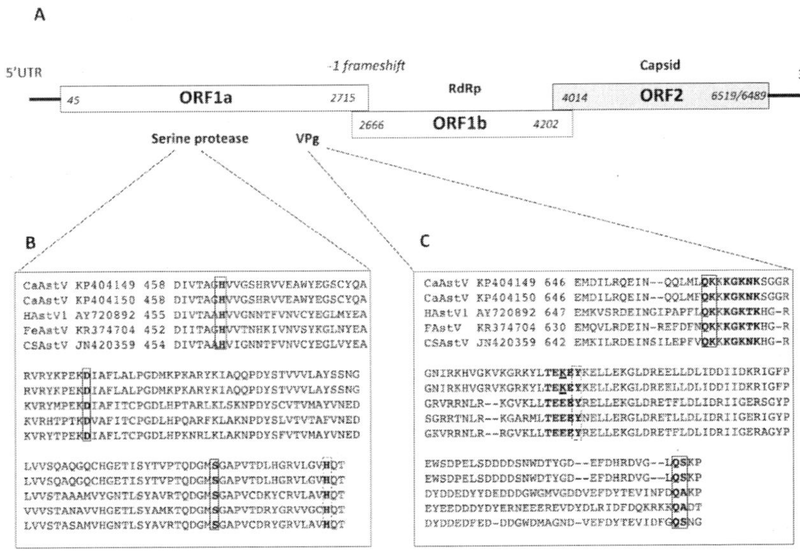

Figure. 3. Genome organisation of CaAstV. The genome is divided into three ORFs, with ORF1 divided by a frameshift site (A). The CaAstV of ORF1 is predicted to encode a serine protease (B) and a VPg protein (C). This is supported by alignment of the two CaAstV sequences with HAstV (human), FeAstV (feline) and CSLAstV (Californian sea lion) sequences. The suspected catalytic triad of the serine protease (B) is highlighted by the boxed residues, with the dotted box H residue (histidine) representing the substrate binding region. The boxed residues in (C) represent the C and N termini cleavage sites of VPg. The conserved KGK(N/T)K and TEXEY motifs of VPg proteins are highlighted in bold, with the TEKEY variation identified in CaAstV underlined. The tyrosine residues identified by a dashed box represent the site of covalent linkage of VPg to the RNA genome.

A −1 ribosomal frameshift site between ORF1a and ORF1b, present in human astroviruses (Marczinke et al., 1994), is also conserved in CaAstV. This translational frameshift is directed by the slippery heptamer sequence AAAAAAC at position 2666 in the CaAstV genome. A stem loop structure is predicted downstream of the slippery sequence, as shown in Fig. 4. The slippery sequence and downstream stem loop are highly conserved amongst mamastroviruses. The 3′ end of ORF1a overlaps with ORF1b by 49 nucleotides. This is shorter than the 71 nt overlap reported for human astroviruses.

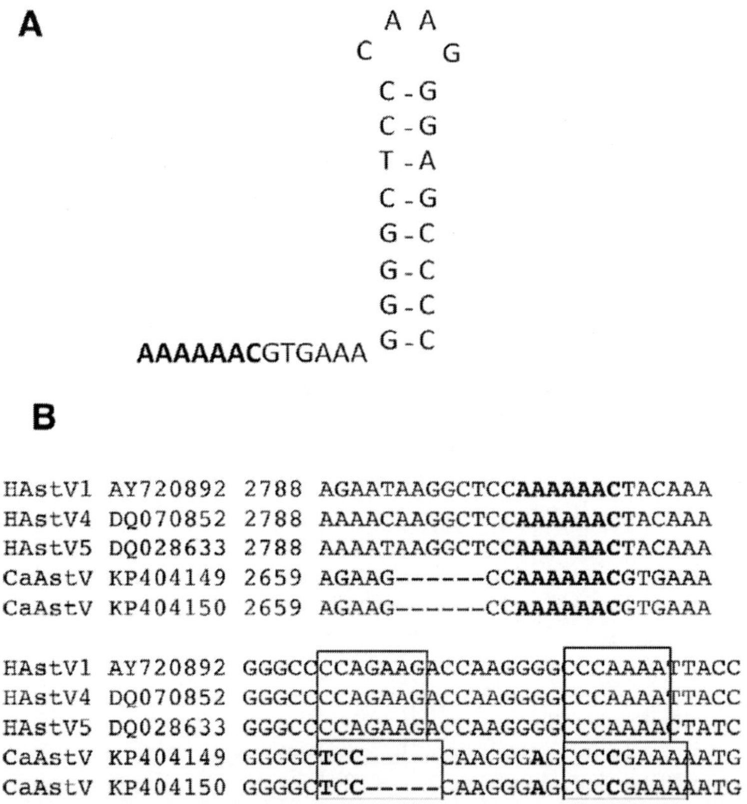

Figure. 4. Analysis of ORF1a/1b −1 frameshift site. (A) Predicted structure of the −1 frameshift site, directed by a slippery heptamer sequence and a downstream stem loop structure. (B) Sequence alignment of CaAstV and a range of human astrovirus (HuAstV) isolates shows this region to be highly conserved. The slippery heptamer sequence is highlighted in bold, and the stem loop structure is boxed, with nucleotides in bold representing differences.

ORF1b is predicted to code for an RNA dependent RNA polymerase (RdRp). The CaAstV sequence contains a YGDD motif at aa 1252, common to RdRps of a variety of RNA viruses, supporting this conclusion. ORF1b of CaAstV aligns with the RdRp of human astroviruses with 58–60% aa identity (HAV-1, 4, 5, and 8). There is a similar identity to feline astrovirus (56%) and porcine astrovirus (55%). CaAstV aligns most closely with the Californian sealion astroviruses (67–71%), though incomplete Californian sea lion astrovirus sequences were available.

As for all other astroviruses studied to date, an overlapping reading frame exists at the ORF1b–ORF2 junction of CaAstV. The CaAstV ORF1b–ORF2 overlap sequence is 8 nt, as reported for other mamastroviruses, hence ORF2 is in the same frame as ORF1a. As previously reported, the capsid sequence was found to have an in-frame start codon 180 nt upstream of the start codon homologous to other mamastrovirus genomes (Toffan et al., 2009).

The 6 aa C terminus of the VP1 (SRGHAE) is highly conserved with several mammalian AstVs. This motif is within a highly conserved nucleotide stretch, s2m, overlapping the termination codon of ORF2, and has been identified in both CaAstV strains sequenced.

The overall nucleotide identity between the two CaAstV strains sequenced was 88.5%. Sequence comparison of the individual ORFs is presented in Table 3. This clearly shows that ORF1b (RdRp) is most closely conserved, making this an ideal target for qPCR screens. Conversely, the capsid sequence is most diverse.

Table 3. Identity of the two full-length CaAstV sequences.

	Nucleotide (%)	Amino acid (%)
ORF1a	93.8	98.4
ORF1b	97.6	99.8
ORF2	77.5	80.8

A phylogenetic tree was constructed by multiple alignment of the full-length genome of the two CaAstV strains isolated in this study, and a number of astrovirus reference strains isolated from different mammalian species. This was achieved using MEGA6 software (Fig. 5).

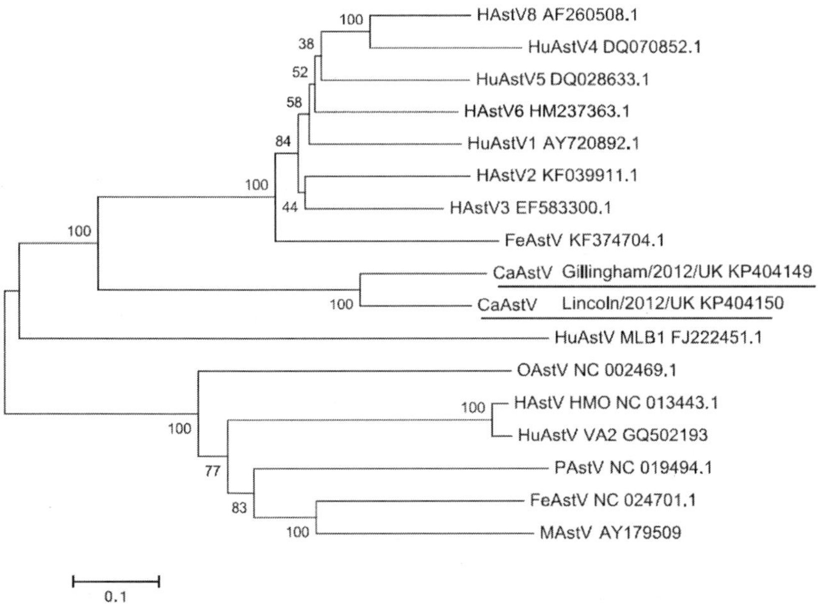

Figure. 5. Phylogenetic tree based complete genome nucleotide sequences of astroviruses from a range of mammalian species. GenBank accession numbers are listed for all sequences analysed, with the CaAstV sequences underlined. The tree was determined using the neighbour-joining method. The percentage of replicate trees in which the associated taxa clustered together in the bootstrap test (100 replicates) is shown next to the branches. Evolutionary analyses were conducted in MEGA6.

4. DISCUSSION

CaAstV has previously been detected sporadically in dogs across the world, but the association with disease, prevalence levels and genetic diversity is largely unknown. This study presents the first identification and molecular characterisation of CaAstV cases in dogs in the UK. Sequencing of the viral capsid for all four strains revealed extensive genetic diversity and the first full-length sequences for CaAstV were determined for two strains.

The prevalence of CaAstV in gastroenteritis cases in this study was shown to be 6.0%. This prevalence was unexpectedly high given that CaAstV has not previously been reported in the UK. It may be predicted that this is an underestimation of prevalence based on the population of dogs surveyed. The majority of previous CaAstV epidemiological studies have focused on dogs less than 6 months old, whereas this study included dogs of any age.

Serological studies have shown that exposure to CaAstV typically occurs in young animals, with dogs older than 3 months significantly more likely to be seropositive than younger dogs (Martella et al., 2011). This suggests that studies focusing only on young dogs will identify more positive CaAStV cases. However, the decision to survey dogs of any age in this study enabled detection of a CaAstV case in a 7-year-old dog. This is oldest case of CaAstV reported to date and highlights the need to have an index of suspicion for infectious causes of gastroenteritis in dogs of any age. Indeed although human astrovirus is more common in paediatric populations, infections in the elderly are reported (Fernández et al., 2011).

The pathology caused by CaAstV in dogs is uncertain as CaAstV has previously been detected in the stools of both healthy and diseased dogs (Grellet *et al.*, 2012 ; Martella *et al.*, 2011). However, our study shows a clear relationship between the presence of CaAstV RNA in stool samples, and the presence of clinical signs of gastroenteritis ($p < 0.001$). This finding is in agreement with two previous studies from Italy and China (Martella *et al.*, 2011 ; Zhu *et al.*, 2011), but is at odds with a French study which found no significant difference in CaAstV identification between diarrhoeic (27%) or healthy puppies (19%) (Grellet et al., 2012). Determination of the specific pathology induced by CaAstV infection in dogs is often confounded in a clinical setting by co-infections with other gastroenteric pathogens (e.g. CPV in dogs 1 and 4 in this study). Experimental studies will be required to confirm or refute the association of CaAstV with gastroenteritis, but this study does suggest CaAstV can cause disease.

Sequencing of the capsid region of each CaAstV strain identified in this study revealed significant sequence variation. This mirrors the variation previously identified within human astrovirus isolates (De Benedictis et al., 2011). At present, astroviruses are named according to the species in which they are isolated, and subsequent classification is based upon serotypes; these are defined if there is a 20-fold or greater two-way cross neutralisation titre (King et al., 2011). Sequence analysis has verified this classification, with human astroviruses 1–8 having 86–100% nucleotide identity within a serotype, based on capsid sequences (Noel et al., 1995). The nucleotide variation within the capsid region of the four CaAstV strains was shown to be 77.1–81.1%, which strongly suggests these strains are also different serotypes. Confirmation of this requires serological analysis, but unfortunately repeated attempts to grow the CaAstV isolates identified in this study in cell culture failed (data not shown).

Identification of four possible CaAstV serotypes circulating in the UK alone raises questions regarding the possible origins of these strains. Phylogenetic analysis of the UK capsid strains alongside the limited number of CaAstV sequences previously listed in GenBank was unexpected. There was no clustering of the UK strains, unlike the grouping of all Chinese strains. Instead UK strains each grouped with a different CaAstV isolate from either China or Italy. With such limited sequences available it is not possible to determine whether CaAstV strains have spread globally, or independent evolution has occurred. Clearly a high rate of evolution does occur within all astroviruses however, as their RNA genome facilitates introduction of point mutations and recombination events.

Given the strain diversity identified, it has been suggested that some CaAstV strains may be more pathogenic than others. This has previously been reported for astroviruses of mink, which show variation in their ability to invade the central nervous system in a strain related manner (Martella et al., 2012). Assessment of this risk will require wider epidemiological and clinical studies. Another concern raised by the existence of multiple circulating CaAstV strains is regarding future disease control. Management of viral causes of gastroenteritis in dogs is best achieved by widespread vaccination, exemplified by the widely used canine parvovirus vaccine. However, if CaAstV vaccine development is considered, the presence of multiple strains will make vaccine design challenging.

Full genome sequencing of two CaAstV isolates revealed them to be closely related and possess a typical astrovirus organisation. The first full length sequence of an astrovirus was for human astrovirus in 1994 (Willcocks et al., 1994), and relatively few full length sequences have since been determined. Sequence analysis of the CaAstV strains identifies the presence of a serine protease and VPg within ORF1a as for other astroviruses, which is separated from and conserved RdRp of ORF1b by a −1 frameshift.

In summary, this study has not only identified CaAstV circulating in the UK dog population, but also found significant genetic diversity within the CaAstV strains. Furthermore, full genome sequencing of two CaAstV isolates has enabled detailed molecular characterisation of this astrovirus species, and provides the astrovirus field with further examples of genome variation.

ACKNOWLEDGEMENTS

The authors would like to thank the veterinary teams at VetsNow Hospitals in Gillingham, Lincoln and Northolt, and Swayne and Partners, Bury St. Edmunds for assisting with stool sample collection. We would also like to thank Wood Green Animal Shelter, Abington Boarding Kennels and Greenlow Kennels in Cambridgeshire for samples. Further gratitude is expressed to James Warland at Cambridge Veterinary School for his work collecting samples and case management. This work was supported by a PhD studentship from the Medical Research Council to S.C. and a Wellcome

Trust Senior Fellowship to I.G. (Ref: WT097997MA). I.G. is a Wellcome Senior Fellow.

REFERENCES

1. Al-Mutairy et al., 2005 B. Al-Mutairy, J.E. Walter, A. Pothen, D.K. Mitchell Genome prediction of putative genome-linked viral protein (VPg) of astroviruses Virus Genes, 31 (2005), pp. 21–30 http://dx.doi.org/10.1007/s11262-004-2196-1
2. Appleton and Higgins, 1975 H. Appleton, P.G. Higgins Letter: viruses and gastroenteritis in infants Lancet, 1 (1975), p. 1297
3. Banos-Lara and Méndez, 2010 M.D.R. Banos-Lara, E. Méndez Role of individual caspases induced by astrovirus on the processing of its structural protein and its release from the cell through a non-lytic mechanism Virology, 401 (2010), pp. 322–332 http://dx.doi.org/10.1016/j.virol.2010.02.028
4. Caddy et al., 2013 S. Caddy, E. Emmott, L. El-Attar, J. Mitchell, A. de Rougemont, J. Brownlie, I. Goodfellow Serological evidence for multiple strains of canine norovirus in the UK dog population PLOS ONE, 8 (2013), p. e81596 http://dx.doi. org/10.1371/journal.pone.0081596
5. Castro et al., 2013 T.X. Castro, R.C.N. Cubel Garcia, E.M. Costa, R.M. Leal, M.D.P.T. Xavier, J.P.G. Leite Molecular characterisation of calicivirus and astrovirus in puppies with enteritis Vet. Rec., 172 (2013), p. 557 http://dx.doi.org/10.1136/vr.101566
6. Choi et al., 2014 S. Choi, S. Lim, Y. Kim, Y. Cho, J. Song, D. An Phylogenetic analysis of astrovirus and kobuvirus in Korean dogs J. Vet. Med. Sci., 78 (2014), pp. 1141–1145
7. De Benedictis et al., 2011 P. De Benedictis, S. Schultz-Cherry, A. Burnham, G. Cattoli Astrovirus infections in humans and animals – molecular biology, genetic diversity, and interspecies transmissions Infect. Genet. Evol., 11 (2011), pp. 1529–1544 http://dx.doi.org/10.1016/j.meegid.2011.07.024
8. Dong et al., 2011 J. Dong, L. Dong, E. Méndez, Y. Tao Crystal structure of the human astrovirus capsid spike Proc. Natl. Acad. Sci. U. S. A., 108 (2011), pp. 12681–12686 http://dx.doi.org/10.1073/pnas.1104834108
9. Fernández et al., 2011 J. Fernández, M. de Oña, S. Melón, M.E. Alvarez-Argüelles, J.A. Boga
10. Noroviruses as cause of gastroenteritis in elderly patients Aging Clin. Exp. Res., 23 (2011), pp. 145–147 http://dx.doi.org/10.3275/7040
11. Fuentes et al., 2012 C. Fuentes, A. Bosch, R.M. Pintó, S. Guix Identification of human astrovirus genome-linked protein (VPg) essential for virus infectivity J. Virol., 86 (2012), pp. 10070–10078 http://dx.doi.org/10.1128/JVI.00797-12
12. Garcia-Boronat et al., 2008 M. Garcia-Boronat, C.M. Diez-Rivero, E.L. Reinherz, P.A. Reche PVS: a web server for protein sequence variability analysis tuned to facilitate conserved epitope discovery Nucleic Acids Res., 36 (2008), pp. W35–W41 http://dx.doi. org/10.1093/nar/gkn211
13. Grellet et al., 2012 A. Grellet, C. De Battisti, A. Feugier, M. Pantile, S. Marciano, D. Grandjean, G. Cattoli Prevalence and risk factors of astrovirus infection in puppies from French breeding kennels Vet. Microbiol., 157 (2012), pp. 214–219 http://dx.doi. org/10.1016/j.vetmic.2011.11.012
14. King et al., 2011 A.M. King, M.J. Adams, E.B. Carstens, E.J. Lefkowitz Virus Taxonomy: Ninth Report of the International Committee on Taxonomy of Viruses Elsevier, London, UK (2011)
15. Madeley and Cosgrove, 1975 C.R. Madeley, B.P. Cosgrove Letter: 28 nm particles in faeces in infantile gastroenteritis Lancet, 2 (1975), pp. 451–452 Marczinke et al., 1994
16. B. Marczinke, A. Bloys, T. Brown, M. Willcocks, M.J. Carter, I. Brierley The human astrovirus RNA-dependent RNA polymerase coding region is expressed by ribosomal frameshifting J. Virol., 68 (1994), pp. 5588–5595

17. Martella et al., 2011 V. Martella, P. Moschidou, E. Lorusso, V. Mari, M. Camero, A. Bellacicco, M. Losurdo, P. Pinto, C. Desario, K. Bányai, G. Elia, N. Decaro, C. Buonavoglia Detection and characterization of canine astroviruses J. Gen. Virol., 92 (2011), pp. 1880–1887 http://dx.doi.org/10.1099/vir.0.029025-0
18. Martella et al., 2012 V. Martella, P. Moschidou, C. Catella, V. Larocca, P. Pinto, M. Losurdo, M. Corrente, E. Lorusso, K. Bànyai, N. Decaro, A. Lavazza, C. Buonavoglia Enteric disease in dogs naturally infected by a novel canine astrovirus J. Clin. Microbiol., 50 (2012), pp. 1066–1069 http://dx.doi.org/10.1128/JCM.05018-11
19. Moser and Schultz-Cherry, 2005 L.A. Moser, S. Schultz-Cherry Review pathogenesis of astrovirus infection Viral Immunol., 18 (2005), pp. 4–10 Noel et al., 1995
20. J.S. Noel, T.W. Lee, J.B. Kurtz, R.I. Glass, S.S. Monroe, J.S. Noel, T.W. Lee, J.B. Kurtz, R.I. Glass, S.S. Monroe Typing of human astroviruses from clinical isolates by enzyme immunoassay and nucleotide sequencing J. Clin. Microbiol., 33 (1995), pp. 797–801
21. Tamura et al., 2013 K. Tamura, G. Stecher, D. Peterson, A. Filipski, S. Kumar MEGA6: molecular evolutionary genetics analysis version 6.0 Mol. Biol. Evol., 30 (2013), pp. 2725–2729 http://dx.doi.org/10.1093/molbev/mst197
22. Toffan et al., 2009 A. Toffan, C.M. Jonassen, C. De Battisti, E. Schiavon, T. Kofstad, I. Capua, G. Cattoli Genetic characterization of a new astrovirus detected in dogs suffering from diarrhea Vet. Microbiol., 139 (2009), pp. 147–152 http://dx.doi.org/10. 1016/j.vetmic. 2009.04.031
23. Willcocks et al., 1994 M.M. Willcocks, T.D. Brown, C.R. Madeley, M.J. Carter The complete sequence of a human astrovirus J. Gen. Virol., 75 (1994), pp. 1785–1788
24. Willcocks et al., 1995 M.M. Willcocks, J.B. Kurtz, T.W. Lee, M.J. Carter Prevalence of human astrovirus serotype 4: capsid protein sequence and comparison with other strains Epidemiol. Infect., 114 (1995), pp. 385–391
25. Williams, 1980 F. Williams Astrovirus-like, coronavirus-like, and parvovirus-like particles detected in the diarrheal stools of beagle pup Arch. Virol., 66 (1980), pp. 215–226
26. Zhu et al., 2011 L.A. Zhu, W. Zhao, H. Yin, T.L. Shan, C.X. Zhu, X. Yang, X.G. Hua, L. Cui Isolation and characterization of canine astrovirus in China Arch. Virol., 156 (2011), pp. 1671–1675 http://dx.doi.org/10.1007/s00705-011-1022-z

CHAPTER 11

International *Clostridium difficile* animal strain collection and large diversity of animal associated strains

Sandra Janezic[1†], Valerija Zidaric[1†], Bart Pardon2, Alexander Indra[3], Branko Kokotovic4, Jose Luis Blanco[5], Christian Seyboldt[6], Cristina Rodriguez Diaz[7], Ian R Poxton[8], Vincent Perreten[9], Ilenia Drigo[10], Alena Jiraskova[11], Matjaz Ocepek[12], J Scott Weese[13], J Glenn Songer[14], Mark H Wilcox[15] and Maja Rupnik[1,16,17*]

[1]*National Laboratory for Health, Environment and Food, Maribor, Slovenia.*

[2]*Department of Large Animal Internal Medicine, Faculty of Veterinary Medicine, Ghent University, Merelbeke, Belgium.*

[3]*Austrian Agency for Health and Food Safety (AGES), Vienna, Austria.*

[4]*Technical University of Denmark, National Veterinary Institute, Copenhagen, Denmark.*

[5]*Complutense University, Madrid, Spain.*

[6]*Friedrich-Loeffler-Institut, Federal Research Institute for Animal Health, Institute of Bacterial Infections and Zoonoses, Jena, Germany.*

[7]*University of Liege, Faculty of Veterinary Medicine, Liege, Belgium.*

[8]*University of Edinburgh, Edinburgh, UK.*

[9]*University of Bern, Institute of Veterinary Bacteriology, Bern, Switzerland.*

[10]*IZSVe, Treviso, Italy.*

[11]*Charles University in Prague, 1st Faculty of Medicine, Prague, Czech Republic.*

[12]*University of Ljubljana, Veterinary Faculty, Ljubljana, Slovenia.*

[13]*University of Guelph, Ontario Veterinary College, Ontario, Canada.*

[14]*Department of Veterinary Science and Microbiology, University of Arizona, Tucson, USA.*

[15]*Department of Medical Microbiology, Leeds Teaching Hospitals, Leeds, UK.*

[16]*University of Maribor, Medical Faculty, Maribor, Slovenia.*

[17]*Centre of Excellence for Integrated Approaches in Chemistry and Biology of Proteins, Ljubljana, Slovenia.*

ABSTRACT

Background

Clostridium difficile is an important cause of intestinal infections in some animal species and animals might be a reservoir for community associated human infections. Here we describe a collection of animal associated *C. difficile* strains from 12 countries based on inclusion criteria of one strain (PCR ribotype) per animal species per laboratory.

Results

Altogether 112 isolates were collected and distributed into 38 PCR ribotypes with agarose based approach and 50 PCR ribotypes with sequencer based approach. Four PCR ribotypes were most prevalent in terms of number of isolates as well as in terms of number of different host species: 078 (14.3% of isolates; 4 hosts), 014/020 (11.6%; 8 hosts); 002 (5.4%; 4 hosts) and 012 (5.4%; 5 hosts). Two animal hosts were best represented; cattle with 31 isolates (20 PCR ribotypes; 7 countries) and pigs with 31 isolates (16 PCR ribotypes; 10 countries).

Conclusions

This results show that although PCR ribotype 078 is often reported as the major animal *C. difficile* type, especially in pigs, the variability of strains in pigs and other animal hosts is substantial. Most common human PCR ribotypes (014/020 and 002) are also among most prevalent animal associated *C. difficile* strains worldwide. The widespread dissemination of toxigenic *C. difficile* and the considerable overlap in strain distribution between species furthers concerns about interspecies, including zoonotic, transmission of this critically important pathogen.

Keywords: *Clostridium difficile*, Animals, Ribotyping, Geographic distribution, Strain collection

1. BACKGROUND

Clostridium difficile, an anaerobic sporogenic bacterium, is recognized as the major pathogen in healthcare associated intestinal infections in humans and also as an important animal pathogen. In addition to the potential for serious (including fatal) infections in animals, are companion and food animals considered as an important potential source for human community-acquired infections [1, 2, 3, 4]. This indicates the importance of preventive measures targeting animals and food [5]. Several studies have looked at similarity between strains isolated from humans and animals [1, 6, 7, 8, 9, 10], but typically focusing on limited number of different species and restricted to a narrow geographic region.

PCR ribotyping is currently the method of choice for differentiation of *C. difficile* strains. In humans more than 300 PCR ribotypes are recognized while the number of reported animal associated PCR ribotypes is much lower [1, 8, 11]. It could be expected that variety of animal associated *C. difficile* will increase with the increased number of typed animal isolates. To date piglets and pig farms are hosts and environments where *C. difficile* has been most extensively studied [3, 12, 13, 14, 15, 16, 17, 18]. For this reservoir the modes of transmission and potential association with human infections are also best understood [14, 16].

C. difficile strains can also be differentiated into toxinotypes according to the differences in the toxin A and toxin B encoding region (PaLoc) [19]. Toxinotypes V and XI are particularly often isolated from animals [19], but the reason for this association is not known. Many published studies report *C. difficile* PCR ribotypes, but not many do report the toxinotypes [9, 13].

The aim of this study was to collect *C. difficile* isolates from different countries and different animals and to compare them with classical agarose gel-based and sequencer-based PCR ribotyping and determine the toxinotypes.

2. RESULTS

Altogether 112 *C. difficile* isolates from 13 animal species were obtained from 14 laboratories from 12 different countries (Austria, Belgium, Canada, Czech Republic, Denmark, Germany, Italy, Scotland, Slovenia, Spain, Switzerland and USA) (Table 1, Additional file 1). Inclusion criteria were one strain (PCR ribotype) per animal species per laboratory. Each participating country contributed 1 to 24 isolates (Additional file 1). Strains were isolated between 1998 and 2012. Only five isolates were from 1998 to 2002, with the majority of isolates (n = 34) from 2011.

The majority of isolates were from pigs (n = 31; 27.7%) and cattle (n = 31; 27.7%), followed by poultry (n = 11; 9.8%), dogs (n = 10; 8.9%), horses, rabbits (7 isolates each; 6.3%), cats (n = 5; 4.5%), goats (n = 3; 2.7%), partridges and raccoons (2 isolates each; 1.8%), wild hare, crow and goose (1 isolate each; 0.9%).

Only 38 isolates were from animals with clinical signs, 56 were from clinically normal animals and for 18 isolates disease status of the animal was not known.

For 40 animals antibiotic use was unknown and further 52 animals had no history of antibiotic exposure. Prior use of antibiotics was reported for isolates from 20 animals. For 12 of them antibiotic was not specified and for the remaining eight antibiotics given to animals were amoxicillin/clavulanic acid, amoxicillin, colistin, enrofloxacin, cefovecin (6 months before sampling), gentamicin, oxytetracycline, and sulfonamide. Animals treated with antibiotics were cattle (8), cat (1), dog (1), horse (1), pig (2) and rabbit (7). Twenty isolates from animals with reported use of antibiotics belonged to 11 different PCR ribotypes (002, 012, 014/020, 027, 033, 045, 078, 127, (CE)013, (CE)032 and (CE)084). None of those PCR ribotypes was associated with a specific antibiotic.

2.1 Molecular characterization of strains with two PCR ribotyping approaches

Results of agarose gel-based and sequencer-based PCR ribotyping for 112 *C. difficile* isolates are presented in Table 1 and Additional file 1.

Using classical agarose-gel based PCR ribotyping all 112 isolates were distributed into 38 different PCR ribotypes (Table 1). With sequencer-based PCR ribotyping, 50 PCR ribotypes could be identified.

Table 1 PCR ribotypes and toxinotypes represented in the international collection of animal C. difficile strains

PCR ribotype	WEBRIBO type	Toxinotype*	Number of strains/different countries per animal species						In total	
			Cattle	Horse	Pig	Poultry	Cats and dogs	Others**	Nr. of strains (%)	Nr. of countries
078	078, 078/4	V/Btb+	5/4	2/2	8/7			1/1	16 (14,3)	9
014/020	014/0, 014/5, 020, 449,659	0	3/3	1/1	2/1	1/1	4/2	2/2	13 (11,6)	5
002	203, 209	0	2/2		2/2		1/1	1/1	6 (5,4)	5
012	012	0, XIX	2/2		1/1	1/1	1/1	1/1	6 (5,4)	4
010	010	tox-	1/1			1/1	2/2	1/1	5 (4,5)	4
033	033	XIa, XIb/Btb+	3/3	1/1					4 (3,6)	4
126	126,078ecdc	V/Btb+	2/2	1/1	1/1				4 (3,6)	4
150	AI-12	0			4/4				4 (3,6)	4
045	045, 598, PR4455, 413	V/Btb+	1/1		3/2	1/1		1/1	6 (5,4)	3
001	001, 001ecdc	0				2/2	2/1		4 (3,6)	3
005	005	0	1/1		1/1	1/1			3	3
(CE)013	AI-9-1	0	1/1					2/2	3	3
103	AI-82/1	0	1/1			1/1		1/1	3	3
(CE)288	660	XIb/Btb+	1/1	1/1				1/1	3	3
081	081	0	1/1		2/2				3	2
015	AI-8/0	0	1/1		1/1				2	2
027	027	IIIb/Btb+	1/1	1/1					2	2

029	029	0	1/1		1/1			2	2
(CE)050	050, AI-84	0	1/1	1/1				2	2
056	056	XII		1/1		1/1		2	2
SLO 024	652	V/Btb+				2/1		2	1
003	003	0					1/1	1	1
011/049	049/1	0		1/1				1	1
017	017	VIII					1/1	1	1
018	018	0			1/1			1	1
023	023	IV/Btb+			1/1			1	1
(CE)032	205	tox-					1/1	1	1
(CE)039	039/2	tox-				1/1		1	1
(CE)084	548	tox-					1/1	1	1
(CE)097	AI-60	0				1/1		1	1
127	651	VI/Btb+		1/1				1	1

258	446	XII						1/1	1	1
(CE)342	610	0	1/1						1	1
(CE)365	434	0	1/1						1	1
(CE)448	653	VI/Btb+			1/1				1	1
(CE)602	212	0	1/1						1	1
SLO 133	AI-15	XII			1/1				1	1
SLO 166	661	I						1/1	1	1
All		na	31/7	7/4	31/10	11/2	15/4	17/3	112	12***
Nr. of ribotypes per species			20	6	16	10	9	15		

*Btb + – presence of binary toxin CDT; **including racoons, wild hare, rabbits, goats, partridges, goose, and crow; ***12 countries participated; tox- -nontoxigenic strain (lacking the PaLoc and genes coding for binary toxin CDT); A "CE" prefix, eg. (CE)039, indicates that the PCR ribotype assigment was made in reference laboratory (CDRN Leeds) using the newer capillary electrophoresis-based approach. PCR-ribotypes in the table are ordered according to the number of countries in which the strains were found and then according to the number of strains belonging to that ribotype.

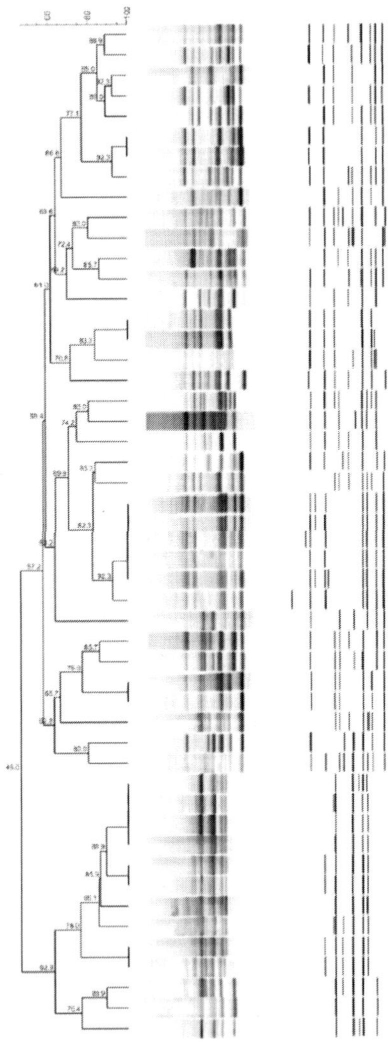

Strain	PCR-ribotype	CE ribotype	Toxinotype
ZZV11-3660	SLO 133	AI-15	XII
ZZV11-3915	011/049	049/1	0
ZZV11-3667	(CE)032	205	tox-
ZZV11-3865	150	AI-12	0
ZZV11-3952	103	AI-82/1	0
ZZV11-3920	001	001	0
ZZV12-4888	001	001ecdc	0
ZZV12-4721	056	056	XII
ZZV11-3654	(CE)084	548	tox-
ZZV12-4447	(CE)097	AI-60	0
ZZV08-903	003	003	0
ZZV08-630	023	023	IV
ZZV08-722	258	446	XII
ZZV11-3953	(CE)365	434	0
ZZV11-4078	002	203	0
ZZV07-487	002	209	0
CD 331 a	018	018	0
ZZV12-4723	010	010	tox-
ZZV11-3918	005	005	0
ZZV08-940	012	012	0
ZZV11-3909	029	029	0
ZZV11-3914	(CE)013	AI-9-1	0
ZZV12-4895	015	AI-8/0	0
ZZV11-3651	014/020	014/0	0
ZZV11-3653	014/020	014/5	0
ZZV11-4076	014/020	020	0
ZZV12-4452	014/020	449	0
ZZV08-1084	014/020	659	0
ZZV11-3908	(CE)602	212	0
ZZV11-3661	017	017	VIII
ZZV11-4111	081	081	0
ZZV12-4896	027	027	IIIb
ZZV11-3656	(CE)050	050	0
ZZV11-3910	(CE)050	AI-84	0
ZZV12-4885	SLO 166	661	I
ZZV11-3921	(CE)342	610	0
ZZV12-4453	(CE)039	039/2	tox-
ZZV11-3867	045	413	VI
ZZV11-3873	045	598	V
ZZV10-2995	045	045	V
ZZV10-2542	045	PR4455	V
ZZV11-3666	126	126	V
ZZV10-2553	126	078ecdc	V
ZZV11-3664	127	651	VI
ZZV12-4449	SLO 024	652	V
ZZV11-3665	078	078/4	V
ZZV12-4894	078	078	V
ZZV12-4892	(CE)288	660	XIb
ZZV08-919	033	033	XIa
ZZV11-3780	(CE)448	653	VI

Figure 1 Dendrogram showing similarities of banding patterns generated with classical agarose gel electrophoresis based PCR ribotyping for all 38 different PCR ribotypes included in the collection. For each "gel-based" PCR ribotype all profiles generated with capillary electrophoresis PCR ribotyping are added for comparison. A "CE" prefix, eg. (CE) 039, indicates that the ribotype assigment was made in reference laboratory (CDRN Leeds) using the newer capillary electrophoresis-based approach.

This is due to higher discriminatory power of capillary sequencer-based PCR ribotyping for some PCR ribotypes; 001 (001 and 001ecdc), 002 (203 and 209), 014/020 (014/0, 014/5, 020, 449 and 659), (CE)050 (050 and AI-84), 078 (078 and 078/4), 045 (045, 413, 598, PR4455) and 126 (126 and 078ecdc) (Table 1). Comparison of agarose gel PCR ribotypes with the corresponding band profiles generated by capillary electrophoresis-based PCR ribotyping is shown in Figure 1. The discrepancies in the nomenclature between both methods, observed for some of the strains [e.g. PCR ribotype 150 (gel-based) and AI-12 (sequencer-based)], are due to use of a different set of *C. difficile* reference strains and reflects the difficulties in attempts to unify PCR ribotyping nomenclature.

Two most common PCR ribotypes representing 25.9% of all strains were 078 and 014/020. PCR ribotypes with 5 or 6 representatives were 002, 012, 010 (non-toxigenic strain lacking the PaLoc and binary toxin CDT) and 045.

2.2 Molecular characterization of strains by toxinotyping

Within toxigenic isolates (n = 104; 92.9%) 11 different toxinotypes were identified (Table 1). More than a half of the isolates belonged to the nonvariant toxinotype 0 (n = 57; 54.9%). The variant toxinotypes were V (n = 27; 26.0%), XII (n = 4; 3.8%), VI (n = 3; 2.9%), XIa (n = 2; 1.9%), XIb (n = 5; 4.7%), III (n = 2; 1.9%), IV, I, VIII and XIX (n = 1; 1%). Eight strains (7.1%) from four different PCR ribotypes (010, (CE)032, (CE)039, (CE)084) were non-toxigenic.

2.3 Distribution of PCR ribotypes and toxinotypes in different animal hosts and in different countries

Table 1 shows distribution of *C. difficile* PCR ribotypes and toxinotypes from collected isolates throughout countries and animal species. Isolates from pigs and cattle were the most frequent and were received from 10 and 7 of the 12 participating countries, respectively. The variability of PCR ribotypes was accordingly also the largest in these two hosts; 20 PCR ribotypes came from cattle and 16 from pigs (Table 1). Distribution of the 15 most prevalent PCR ribotypes in countries is presented in Figure 2.PCR ribotype 014/020 was found in the majority of the animal species included in the collection (pig, cattle, horse, poultry, cat, dog, rabbit and a goat) across 5 different countries (Figure 2). PCR ribotypes that were found only in particular animal species, but in several different countries, were PCR ribotype 150, which was found only in pigs but in four different countries, and PCR ribotype 033, which was found in cattle in three different countries and only in a single case in horse.

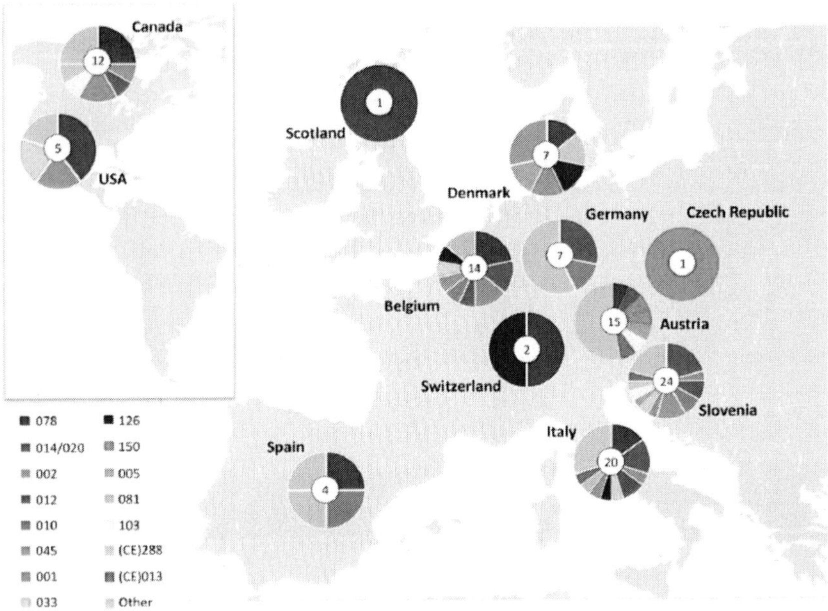

Figure 2 Geographical distribution of animal associated C. difficile PCR ribotypes from participating countries. Pie charts show proportion of 15 most prevalent PCR ribotypes in the collection for each participating country. The number in the center of pie chart represents the number of isolates from that country. The diversity of strains per country increases with the total number of strains contributed to the collection but also by strains from certain hosts (poultry in case of Slovenia and rabbits in case of Italy).

Toxinotype V strains of different PCR ribotypes were mainly associated with cattle, horses and pigs and were rarely found in other animals. Also toxinotype XI strains (a and b) were mainly associated with cattle and horses.

3. DISCUSSION

Several publications are available about isolation and characterization of *C. difficile* from animals, but they are usually limited to a specific geographical region and data of the different studies are sometimes difficult to compare due to nomenclature of PCR ribotypes that is not unified. The advantage of the international animal *C. difficile* strain collection is in performance of strain typing in the single setting, hereby minimizing the ambiguities in PCR ribotype designations. Additionally, two PCR ribotyping approaches have been used; the standard one using 'Cardiff' nomenclature and WEBRIBO based analysis (giving inter-laboratory comparable results independent of Cardiff/Leeds reference strains). This is not a prevalence study with collection of samples from defined number of farms or hosts as was done for hospitals and human isolates in Europe [20]. The isolates in this study were collected from different labs in different countries and the criterion was one PCR ribotype per animal species per lab. Therefore the collected strains are not reflecting the prevalence of PCR ribotypes but are giving a good basis to assess the diversity of animal associated *C. difficile*.

Our results show that variability of PCR ribotypes present in different animal species is large. Isolates from pigs and cattle were most common, perhaps reflecting the importance of *C. difficile* in pigs and concerns about zoonotic transmission from cattle (or more specifically, meat) (Table 1, Table 2). In contrast to cattle and pigs, are cats, dogs, rabbits and poultry clearly understudied and are probably associated with a broader variety of *C. difficile* PCR ribotypes than currently recognized. Only two countries have contributed poultry isolates to the collection, but they represent 10 different PCR ribotypes. Strains from captive rabbits were contributed only by one country, but represent 7 different PCR ribotypes (Additional file 1).

Some PCR ribotypes seem to be more often associated with a particular animal host. Many publications report PCR ribotype 078 in pigs [7, 10, 15] and type 033 in cattle [22, 24] (Table 2). Results presented here confirm this observation (Figure 2), but also suggest that the well-known animal-associated PCR ribotype 078 may not be currently present in animals in all countries. In addition to types 078 and 033 some other PCR ribotypes are typically associated with pigs, such as PCR ribotypes 150, 002, 045 and 081 (Table 1). Recent reports from Australia describe a new genotype in terms of PCR ribotype (237) and toxin genes (*tcd* B+, *tcd* A-, *cdt* A + and *cdt* B+) in pigs [18]. PCR ribotype 027 that was prevalent in humans in many countries within the past ten years was in this collection of animal strains found only in USA and Canada (Additional file 1).

Toxinotype V and XI and binary toxin positive strains were initially reported to be the prevalent population of strains isolated from animals (70-100%) [19, 29]. However, the results of this study show that nonvariant strains of toxinotype 0 are widespread in animals and that the proportion of binary toxin positive strains can be as low as 35.7%.

All isolates in the collection were distributed into 38 PCR ribotypes, but only a few of those contained five or more isolates, and the majority was represented only by a few or a single isolate (Table 1).

Table 2 Comparison of C. difficile PCR ribotypes detected in six animal species in different geographical regions

		This study	Netherlands	Germany	Switzerland	Australia	North America
Pigs	Number of ribotypes (reference)	16	3 [10], 1 [7]	20 [15]		1 [18]	7 [21]
	Most prevalent ribotypes	078, 150, 014/020, 045, 002, 081	078	078, 126, 002/2, 126, 413, 049, 598		237	078
Cattle	Number of ribotypes (reference)	20	2 [10]	17 [22]	5 [23]	21 [24]	7 [25] 3 [26]
	Most prevalent ribotypes	078, 014/020, 033, 002, 012, 126	012	033, 078, 045, 126	033, 003, 066, 070, 137	127, 033, 126, 056, 087	078, 017, 027, 014, 033
Cats and dogs	Number of ribotypes (reference)	9	9 [10]	5 [27]			
	Most prevalent ribotypes	014/020, 010, 001	010, 014, 039, 012	010, 014/020, 039, 045			
Goats and sheep	Number of ribotypes (reference)	3			2 [23]	7 [28]	
	Most prevalent ribotypes	010, 014/020, 045			001, 066	101, 137	

This resembles the situation with human strain collections with many different PCR ribotypes, but only a few of them having a large number of isolates [20]. However, as the inclusion criterion for the collection was one PCR ribotype per animal species per laboratory, the number of isolates of a given PCR ribotype does not reflect the actual prevalence of this PCR ribotype in animal host. But the high number of isolates of a given PCR ribotype in the collection does reflect its broader geographic presence and possibly broader spectrum of animal species from which it was isolated.

All animal-associated PCR ribotypes with four or more isolates reported here (Table 1) are among the most prevalent in diverse studies of human isolates [8, 11, 20, 30, 31]. In particular PCR ribotype 014/020 is currently the most prevalent type isolated in Europe and in some USA studies and is the only type that was present here in the majority of animal hosts (Table 1). Although type 014/020 is not recognized as hypervirulent and is not associated with outbreaks or severe disease in humans, its ability to colonize many diverse hosts and its ubiquitous presence indicates considerable endemic potential of this particular PCR ribotype [8].

4. CONCLUSIONS

PCR ribotype 078 is the most prevalent *C. difficile* type associated with this collection of animal isolates, but only with some hosts and in some countries. The second most prevalent PCR ribotype is 014/020 which, in contrast, has broader range of animal hosts. Variability of animal associated PCR ribotypes is substantial and is likely to increase with the number of typed animal isolates. Large overlap of animal associated *C. difficile* PCR ribotypes with human strains furthers concerns about interspecies, including zoonotic, transmission of this important pathogen. Moreover, strains that are prevalent in humans are also prevalent in different animals from different geographic areas, emphasizing that certain strains have a large potential for global dissemination.

5. METHODS

5.1 Inclusion criteria and requested isolate information

Laboratories from different countries with publications on *C. difficile* in animals were invited to contribute the isolates. Participating laboratories were asked to provide only a single representative PCR ribotype per animal species and laboratory/country. No formal structure was used to guide isolate submission.

Participating laboratories were asked to provide additional information about the individual isolate and the animal host: date and country/city of *C. difficile* isolation or specimen collection, molecular characterization (i.e. PCR ribotype, toxin genes), animal species, age, status (farm, domestic or wild animal), clinical signs and antimicrobial exposure history (when available).

5.2 Cultivation and storage of C. difficile isolates

Isolates were first inoculated onto selective medium (CLO, bioMerieux) and subcultured on blood agar plates (COH; bioMerieux). Species identification was confirmed by PCR amplification of *C. difficile* specific gene *cdd* using primers Tim 6 (5′-TCCAATATAATAAATTAGCATTCCA) and Struppi 6 (5′-GGCTATTACACGTAATCCAGATA) [32].

All isolates were stored in Microbank cryogenic vials (Pro-lab Diagnostics) at −80°C.

5.3 PCR ribotyping and toxinotyping

All isolates were characterized by toxinotyping, agarose gel-based PCR ribotyping and capillary gel electrophoresis-based PCR ribotyping.

Agarose gel-based PCR ribotyping was performed as described elsewhere [33]. PCR ribotypes were determined by comparison of banding patterns with the internal library using the BioNumerics software v7.1 (Applied Maths, Belgium). Banding patterns were compared with a reference library of 48 Cardiff type strains. Strains that were not consistent with those in the library were sent to reference laboratory (CDRN Leeds) for confirmation and are named with prefix (CE), indicating that the assignment was made with the newer capillary electrophoresis-based approach. Three strains generated new ribotyping profiles (not previously recognized in Leeds/Cardiff collection) and are designated with the internal nomenclature (SLO and 3-digit code).

Dendrograms were constructed using the Dice coefficient and the unweighted pair group method with arithmetic means (UPGMA). Position tolerance and optimization were set to 1.1%.

Capillary gel electrophoresis-based PCR ribotyping was performed as described previously [34]. Primers described by Bidet and colleagues [35] were used for amplification of intergenic spacer regions (ISRs), with forward primer labelled with a WellRED dye D4-PA (Sigma-Aldrich,

Germany). PCR products were analyzed with CEQ™ 8000 Genetic Analysis System (Beckman Coulter) using a 33 cm capillary and gel LPAI. CEQ 600-bp DNA size standard (Beckman-Coulter) was used to determine fragment lengths. Capillary separation conditions were as follows: samples injection voltage of 2 kV over 60 s, separation voltage of 4.8 kV with a total running time of 75 min and a capillary temperature of 50°C.

PCR ribotypes were determined with WEBRIBO database (https://webribo.ages.at/) [34]. Fragment sizes were also imported into BioNumerics software for comparison of banding patterns generated with classical agarose gel electrophoresis and sequencer-based capillary gel electrophoresis.

Toxinotyping was performed as described previously [36]; http://www.mf.uni-mb.si/mikro/tox). Non-toxigenic strains (without the PaLoc, region encoding toxins A and B) were confirmed by amplification of 115 bp long insert with primer pair Lok1/Lok3 [32]. Presence of binary toxin gene *cdt* B was detected as described previously [37].

NOTES

Sandra Janezic, Valerija Zidaric contributed equally to this work.

ACKNOWLEDGEMENTS

Part of this research has received funding from European Communities 7th Framework programme (FP7/2007-2011) under grant agreement No. 223585.

REFERENCES

1. Hensgens MP, Keessen EC, Squire MM, Riley TV, Koene MG, de Boer E, Lipman LJ, Kuijper EJ: *Clostridium difficile* infection in the community: a zoonotic disease?. Clin Microbiol Infect. 2012, 18 (7): 635-645. 10.1111/j.1469-0691.2012.03853.x.
2. Rupnik M, Songer JG: *Clostridium difficile* Its Potential as a Source of Foodborne Disease. Adv Food Nutr Res. 2010, 60: 53-66.
3. Squire MM, Riley TV: *Clostridium difficile* Infection in Humans and Piglets: A 'One Health' Opportunity. Curr Top Microbiol Immunol. 2012, 365: 299-314.
4. Weese JS: Clostridium difficile in food–innocent bystander or serious threat?. Clin Microbiol Infect. 2010, 16 (1): 3-10. 10.1111/j.1469-0691.2009.03108.x.
5. Rodriguez-Palacios A, Borgmann S, Kline TR, LeJeune JT: *Clostridium difficile* in foods and animals: history and measures to reduce exposure. Anim Health Res Rev. 2013, 14 (1): 11-29. 10.1017/S1466252312000229.

6. Bakker D, Corver J, Harmanus C, Goorhuis A, Keessen EC, Fawley WN, Wilcox MH, Kuijper EJ: Relatedness of human and animal *Clostridium difficile* PCR ribotype 078 isolates determined on the basis of multilocus variable-number tandem-repeat analysis and tetracycline resistance. J Clin Microbiol. 2010, 48 (10): 3744-3749. 10.1128/JCM.01171-10.
7. Debast SB, van Leengoed LA, Goorhuis A, Harmanus C, Kuijper EJ, Bergwerff AA: *Clostridium difficile* PCR ribotype 078 toxinotype V found in diarrhoeal pigs identical to isolates from affected humans. Environ Microbiol. 2009, 11 (2): 505-511. 10.1111/j.1462-2920.2008.01790.x.
8. Janezic S, Ocepek M, Zidaric V, Rupnik M: *Clostridium difficile* genotypes other than ribotype 078 that are prevalent among human, animal and environmental isolates. BMC Microbiol. 2012, 12 (1): 48-10.1186/1471-2180-12-48.
9. Jhung MA, Thompson AD, Killgore GE, Zukowski WE, Songer G, Warny M, Johnson S, Gerding DN, McDonald LC, Limbago BM: Toxinotype V *Clostridium difficile* in humans and food animals. Emerg Infect Dis. 2008, 14 (7): 1039-1045. 10.3201/eid1407.071641.
10. Koene MG, Mevius D, Wagenaar JA, Harmanus C, Hensgens MP, Meetsma AM, Putirulan FF, Van Bergen MA, Kuijper EJ: *Clostridium difficile* in Dutch animals: their presence, characteristics and similarities with human isolates. Clin Microbiol Infect. 2011, 18 (8): 778-784.
11. Keel K, Brazier JS, Post KW, Weese S, Songer JG: Prevalence of PCR ribotypes among *Clostridium difficile* isolates from pigs, calves, and other species. J Clin Microbiol. 2007, 45 (6): 1963-1964. 10.1128/JCM.00224-07.
12. Alvarez-Perez S, Blanco JL, Bouza E, Alba P, Gibert X, Maldonado J, Garcia ME: Prevalence of *Clostridium difficile* in diarrhoeic and non-diarrhoeic piglets. Vet Microbiol. 2009, 137 (3–4): 302-305.
13. Avbersek J, Janezic S, Pate M, Rupnik M, Zidaric V, Logar K, Vengust M, Zemljic M, Pirs T, Ocepek M: Diversity of *Clostridium difficile* in pigs and other animals in Slovenia. Anaerobe. 2009, 15 (6): 252-255. 10.1016/j.anaerobe.2009.07.004.
14. Keessen EC, Donswijk CJ, Hol SP, Hermanus C, Kuijper EJ, Lipman LJ: Aerial dissemination of *Clostridium difficile* on a pig farm and its environment. Environ Res. 2011, 111 (8): 1027-1032. 10.1016/j.envres.2011.09.014.
15. Schneeberg A, Neubauer H, Schmoock G, Baier S, Harlizius J, Nienhoff H, Brase K, Zimmermann S, Seyboldt C: *Clostridium difficile* Genotypes in Piglet Populations in Germany. J Clin Microbiol. 2013, 51 (11): 3796-3803. 10.1128/JCM.01440-13.
16. Keessen EC, Harmanus C, Dohmen W, Kuijper EJ, Lipman LJ: *Clostridium difficile* infection associated with pig farms. Emerg Infect Dis. 2013, 19 (6): 1032-1034. 10.3201/eid1906.121645.
17. Noren T, Johansson K, Unemo M: *Clostridium difficile* PCR ribotype 046 is common among neonatal pigs and humans in Sweden. Clin Microbiol Infect. 2014, 20 (1): 2-6. 10.1111/1469-0691.12296.
18. Squire MM, Carter GP, Mackin KE, Chakravorty A, Noren T, Elliott B, Lyras D, Riley TV: Novel molecular type of *Clostridium difficile* in neonatal pigs, Western Australia. Emerg Infect Dis. 2013, 19 (5): 790-792.
19. Rupnik M: Heterogeneity of large clostridial toxins: importance of *Clostridium difficile* toxinotypes. FEMS Microbiol Rev. 2008, 32 (3): 541-555. 10.1111/j.1574-6976.2008.00110.x.
20. Bauer MP, Notermans DW, van Benthem BH, Brazier JS, Wilcox MH, Rupnik M, Monnet DL, van Dissel JT, Kuijper EJ: *Clostridium difficile* infection in Europe: a hospital-based survey. Lancet. 2011, 377 (9759): 63-73. 10.1016/S0140-6736(10)61266-4.
21. Weese JS, Rousseau J, Deckert A, Gow S, Reid-Smith RJ: *Clostridium difficile* and methicillin-resistant Staphylococcus aureus shedding by slaughter-age pigs. BMC Vet Res. 2011, 7: 41-10.1186/1746-6148-7-41.
22. Schneeberg A, Neubauer H, Schmoock G, Grossmann E, Seyboldt C: Presence of *Clostridium difficile* PCR ribotype clusters related to 033, 078 and 045 in diarrhoeic calves in Germany. J Med Microbiol. 2013, 62 (Pt 8): 1190-1198.

23. Romano V, Albanese F, Dumontet S, Krovacek K, Petrini O, Pasquale V: Prevalence and genotypic characterization of *Clostridium difficile* from ruminants in Switzerland. Zoonoses Public Health. 2012, 59 (8): 545-548. 10.1111/j.1863-2378.2012.01540.x.
24. Knight DR, Thean S, Putsathit P, Fenwick S, Riley TV: Cross-sectional study reveals high prevalence of *Clostridium difficile* non-PCR ribotype 078 strains in Australian veal calves at slaughter. Appl Environ Microbiol. 2013, 79 (8): 2630-2635. 10.1128/AEM.03951-12.
25. Rodriguez-Palacios A, Stampfli HR, Duffield T, Peregrine AS, Trotz-Williams LA, Arroyo LG, Brazier JS, Weese JS: *Clostridium difficile* PCR ribotypes in calves, Canada. Emerg Infect Dis. 2006, 12 (11): 1730-1736. 10.3201/eid1211.051581.
26. Hammitt MC, Bueschel DM, Keel MK, Glock RD, Cuneo P, DeYoung DW, Reggiardo C, Trinh HT, Songer JG: A possible role for *Clostridium difficile* in the etiology of calf enteritis. Vet Microbiol. 2008, 127 (3–4): 343-352.
27. Schneeberg A, Rupnik M, Neubauer H, Seyboldt C: Prevalence and distribution of *Clostridium difficile* PCR ribotypes in cats and dogs from animal shelters in Thuringia, Germany. Anaerobe. 2012, 18 (5): 484-488. 10.1016/j.anaerobe.2012.08.002.
28. Knight DR, Riley TV: Prevalence of gastrointestinal *Clostridium difficile* carriage in Australian sheep and lambs. Appl Environ Microbiol. 2013, 79 (18): 5689-5692. 10.1128/AEM.01888-13.
29. Rupnik M: Is *Clostridium difficile*-associated infection a potentially zoonotic and foodborne disease?. Clin Microbiol Infect. 2007, 13 (5): 457-459. 10.1111/j.1469-0691.2007.01687.x.
30. Walk ST, Micic D, Jain R, Lo ES, Trivedi I, Liu EW, Almassalha LM, Ewing SA, Ring C, Galecki AT, Rogers MA, Washer L, Newton DW, Malani PN, Young VB, Aronoff DM: *Clostridium difficile* ribotype does not predict severe infection. Clin Infect Dis. 2012, 55 (12): 1661-1668. 10.1093/cid/cis786.
31. Waslawski S, Lo ES, Ewing SA, Young VB, Aronoff DM, Sharp SE, Novak-Weekley SM, Crist AE, Dunne WM, Hoppe-Bauer J, Johnson M, Brecher SM, Newton DW, Walk ST: *Clostridium difficile* ribotype diversity at six health care institutions in the United States. J Clin Microbiol. 2013, 51 (6): 1938-1941. 10.1128/JCM.00056-13.
32. Braun V, Hundsberger T, Leukel P, Sauerborn M, von Eichel-Streiber C: Definition of the single integration site of the pathogenicity locus in *Clostridium difficile*. Gene. 1996, 181 (1–2): 29-38.
33. Janezic S, Rupnik M: Molecular typing methods for *Clostridium difficile*: pulsed-field gel electrophoresis and PCR ribotyping. Methods Mol Biol. 2010, 646: 55-65. 10.1007/978-1-60327-365-7_4.
34. Indra A, Huhulescu S, Schneeweis M, Hasenberger P, Kernbichler S, Fiedler A, Wewalka G, Allerberger F, Kuijper EJ: Characterization of *Clostridium difficile* isolates using capillary gel electrophoresis-based PCR ribotyping. J Med Microbiol. 2008, 57 (Pt 11): 1377-1382.
35. Bidet P, Barbut F, Lalande V, Burghoffer B, Petit JC: Development of a new PCR-ribotyping method for *Clostridium difficile* based on ribosomal RNA gene sequencing. FEMS Microbiol Lett. 1999, 175 (2): 261-266. 10.1111/j.1574-6968.1999.tb13629.x.
36. Rupnik M, Avesani V, Janc M, von Eichel-Streiber C, Delmee M: A novel toxinotyping scheme and correlation of toxinotypes with serogroups of *Clostridium difficile* isolates. J Clin Microbiol. 1998, 36 (8): 2240-2247.
37. Stubbs S, Rupnik M, Gibert M, Brazier J, Duerden B, Popoff M: Production of actin-specific ADP-ribosyltransferase (binary toxin) by strains of *Clostridium difficile*. FEMS Microbiol Lett. 2000, 186 (2): 307-312. 10.1111/j.1574-6968.2000.tb09122.x.

CHAPTER 12

Exposure to environmental stressors result in increased viral load and further reduction of production parameters in pigs experimentally infected with PCV2b

Robert Patterson[a], Amanda Nevel[a], Adriana V. Diaz[a], Henny M. Martineau[a], Theo Demmers[b], Christopher Browne[a], Bettina Mavrommatis[a,1], Dirk Werling[a,*]

[a]*Royal Veterinary College, Department of Pathology and Pathogen Biology, Hawkshead Lane, AL9 7TA, UK*

[b]*Royal Veterinary College, Department of Clinical Sciences, Hawkshead Lane, AL9 7TA, UK*

ABSTRACT

Porcine circovirus type 2 (PCV2) has been identified as the essential, but not sole, underlying infectious component for PCV-associated diseases (PCVAD). Several co-factors have been suggested to convert an infection with PCV2 into the clinical signs of PCVAD, including co-infection with a secondary pathogen and the genetic background of the pig. In the present study, we investigated the role of environmental stressors in the form of changes in environmental temperature and increased stocking-density on viral load in serum and tissue, average daily weight gain (ADG) and food conversion rate (FCR) of pigs experimentally infected with a defined PCV2b strain over an eight week period. These stressors were identified recently as risk factors leading to the occurrence of severe PCVAD on a farm level. In the current study, PCV2-free pigs were housed in separate, environmentally controlled rooms, and the experiment was performed in a 2 × 2 factorial design. In general, PCV2b infection reduced ADG and increased FCR, and these were further impacted on by the environmental stressors. Furthermore, all stressors led to an increased viral load in serum and tissue as assessed by qPCR, although levels did not reach statistical significance. Our data suggest that there is no need for an additional pathogen to develop PCVAD in conventional status pigs, and growth retardation and clinical signs can be induced in PCV2 infected pigs that are exposed to environmental stressors alone.

Keywords: PCV2; PCVAD; Environment; Housing; Temperature; qPCR; ADG; FCR

1. INTRODUCTION

Porcine circovirus 2 (PCV2) is a non-enveloped single-stranded DNA virus with a diameter of less than 20 nm that was isolated from diseased pigs in the late 1990s (Allan et al., 1998 ; Mankertz, 2012). The virus has a small genome of less than 2000 nucleotides compromising three open reading frames. Thus far, there are three reported genotypes of PCV2. These are PCV2a, PCV2b and PCV2c, each of which have several subtypes. In the UK the dominant strain circulating on severely infected farms belongs to the PCV2b genotype as determined by a recent study (Wieland et al., 2012). In the early days of PCV2 occurrence, the virus was associated with post-weaning multi-systemic wasting syndrome (PMWS) in pigs, which had a major impact on the swine industry primarily due to high post-weaning mortality rates on affected units. Pigs that develop PMWS usually present with a range of clinical signs including diarrhoea, jaundice, respiratory distress, wasting and enlarged lymph nodes. Pigs most commonly at risk are between eight and fifteen weeks of age, which is due to a rapid decline of maternally derived antibodies in the serum (Lukert, 1999). As the disease progresses, mortality declines and subsequent production losses are due to reduced growth rates, increased feed conversion ratios and deaths in post-weaned pigs through subsequent secondary infections (Alarcon et al., 2011b). For these reasons, the nomenclature has moved away from PMWS and it is now more often termed PCV-associated disease (PCVAD) (Opriessnig and Langohr, 2013). Control of PCVAD has presented many challenges due to its nature as multi-factorial disease. Current control strategies include implementation of an "all in-all out" replacement system, as well as high standards of hygiene and biosecurity on the pig farm. However, due to the lack of a viral envelope, PCV2 is notoriously resistant to inactivation by common detergents and disinfectants (Ellis, 2000 ; Ghebremariam and Gruys, 2005). Vaccination is an additional control strategy and there are currently five commercially available PCV2 vaccines. All of these are against the PCV2a genotype and are either an inactivated whole virus vaccine or subunit vaccine based on the Cap protein, which is the major immunogenic protein of PCV2 (Beach and Meng, 2012). None of the vaccines on the market have been tested in single pathogen controlled challenge experiments due to the lack of an available challenge model. Registration and licensing requirements have been based on either field studies or from co-infected pigs (where at least one other pathogen is used in an artificial challenge model) (Beach and Meng, 2012).

Although PCV2 is the necessary causal agent, many studies have demonstrated that infection with the virus alone is not sufficient to induce clinical signs of PCVAD in pigs; however, there is increasing evidence of subclinical infection (Alarcon et al., 2011b; Allan et al., 1998; Ellis et al., 1998 ; Nayar et al., 1997). Young et al. (2011) demonstrated that in pig farms with no visible PCVAD, vaccination significantly improved production parameters (Young et al., 2011). PCV2 is considered to be widespread in swine, whereas PCVAD is usually fairly sporadic (Ghebremariam and Gruys, 2005 ; Larochelle et al., 2003), therefore subclinical PCVAD is likely to be underestimated. Indeed a number of studies have demonstrated that, while PCV2 is detected in both PCVAD-affected and apparently healthy pigs, viral load is higher in PCVAD sufferers (Alarcon et al., 2011b; Brunborg et al., 2004; Grau-Roma et al., 2009; Ladekjaer-Mikkelsen et al., 2002 ; Sibila et al., 2004) and vaccination of such herds results in a benefit to production indicating that disease is present, but not detected. There has been much speculation and research into the co-factors contributing to the development of PCVAD in PCV2 infected pigs. These include the genotype of the infecting virus with PCV2b having a higher association with disease than PCV2a (Rose et al., 2012) as well as the immune status of the sow and the age at which the piglet is exposed to PCV2 (Grau-Roma et al., 2009). A number of studies have also demonstrated that co-infection of pigs with porcine reproductive and respiratory syndrome virus (PRRSV), porcine parvovirus (PPV) and Mycoplasma hyopneumoniae can exacerbate PCV2 infection and can lead to PCVAD. This co-infection has been shown to increase PCV2 replication in the host and also to modify cytokine production and profile (Opriessnig and Halbur, 2012 ; Segales et al., 2013). Recently, the role of environmental factors and their contribution to the onset of PCVAD has been explored. Housing conditions, hygiene, biosecurity and husbandry have all been linked to PCVAD development (Segales et al.,

2013). A recent study demonstrated that reduced pen size and cross-fostering in farrowing crates alter the course of PCV2 infection, favouring earlier infections and therefore possibly exacerbating disease (Andraud et al., 2009). It seems that PCVAD is a truly multifactorial disease and disease progression may not only be dependent on PCV2 infection and one other contributing factor but could depend on a multitude of factors. Currently, little is known about the impact of these different co-factors in the outcome and severity of disease. During their life, pigs are exposed to many environmental stressors in addition to weaning; these include changes in temperature, mixing, noise and shipping. Many of these have been shown to suppress the immune system and therefore increase susceptibility to disease (Kelley, 1980 ; McGlone et al., 1993). However, whether these environmental stressors affect progression of PCV2 infection is currently poorly understood.

Using recently identified environmental risk-factors for occurrence of PCVAD in a herd (Alarcon et al., 2011a ; Alarcon et al., 2011b), we investigated the role of these potential co-factors on PCV2 infection with the aim of developing a disease model for PCVAD which does not rely on gnotobiotic pigs.

2. MATERIAL AND METHODS

2.1. Ethics statement

All animal studies were performed according to the regulations and guidance provided under the UK Home Office Animals (Scientific Procedures) Act 1986. Experimental protocols were approved under project licence number PPL 70/7219, as well as the RVC Ethics and Welfare Committee.

2.2. Animals

In a cross-sectional study of 114 farms in England in 2008, antibodies against PCV2 were detected in 99.1% of herds and PCV2 was detected by PCR on 90.4% of farms, indicating a nearly endemic infection (Wieland et al., 2010). However, we were able to purchase a total of 54 large white × landrace pigs of a similar age from a commercial farm that tested PCV2 free as by PCR and antibody ELISA before study recruitment. Pigs were randomly allocated to nine groups (n = 6). Eight of the nine groups were replicated in a 2 × 2 factorial design to maximise experimental achievement by reducing animal use (Shaw et al., 2002), meaning that a total of 12 animals were tested for each condition, with an additional six pigs in the 9th group which served as a functional control and were not exposed to PCV2 or any environmental stress.

2.3. Virus preparation for infection

To avoid potential false-positive results due to LPS contaminations of reagents/media, all substances used were tested for their LPS content using the Endosafe-PTS system (Charles River, Charleston, USA). Samples with endotoxin content below 0.01 EU ml^{-1} were considered as LPS-free.

A previously characterized PCV2b isolate from the UK (GenBank accession number JX193799; (Wieland et al., 2012) was used to generate the virus stock for the experimental infections. The virus was propagated in type-I IFNKO PK15-ALR-NPro cells, free of PCV1 and PCV2. Cells were cultured in MEM containing Earle's salts supplemented with 10% (v/v) tetracycline-free foetal bovine serum (FBS) (Clontech, Saint-Germain-en-Laye, France). To induce the tetracycline-regulated expression of the IFN-KO, tetracycline was added 2 h pre-inoculation. Cell monolayers were inoculated at 50% confluence. After 18 h incubation at 37 °C, the inoculum was removed and retained, and the monolayer treated with 300 mM D-glucosamine in Hanks balanced salt solution (HBSS) for 30 min at 37 °C. After removal of the glucosamine and subsequent washes with HBSS, cultures were overlaid with retained inoculum. Cultures were overlaid with media 24 h post-inoculation. After further three days incubation, media was removed and retained and virus harvested from trypsinised cells by freeze-thawing. The cell lysate was clarified by centrifugation at 1000 × g for 5 min at 4 °C and the supernatant added to the retained media. This virus suspension was then concentrated approximately 10-fold using dialysis tubing (Spectra/Por, Biotech Cellulose Ester membrane; Spectrum Europe B.V.) in polyethylene glycol (PEG 12000 flake; Whyte Chemicals Ltd.) at 4 °C. Concentrated virus suspension was subsequently dialysed in MEM overnight and aliquoted. PCV2 stocks were titrated on PK15-ALR-NPro cells as described elsewhere. The titre of the virus stock was determined by qPCR.

2.4. Experimental design

In the study design, four treatment groups, one challenged only control group (each with n = 6 animals, repeated in a 2 × 2 study design for a total of 12 animals per treatment in two separate rooms), and one unchallenged control group were allocated to nine identical rooms, all of which had an isolated environmental system, allowing for control of airflow, humidity and temperature. At four weeks of age (Day 0) treatment groups and challenged control groups were inoculated intra-nasally with 1 × 10^{10} PCV2 particles in 5 ml media. Non challenged controls (C) were inoculated with 5 ml of virus free media. One treatment group was inoculated with virus (V) but not subjected to other environmental stressors. The remaining groups, all inoculated with virus, were subjected to either high stocking density (V SD), high environmental temperature (V T) or both, high stocking density and high temperature combined (V SD T). High stocking density was calculated as DEFRA guidelines minus 25% for the average weight of pig in that group. Areas were altered weekly, after weekly weights were recorded. The three control groups were housed at an optimal stocking rate based on DEFRA guidelines. In high environmental temperature rooms, the air temperature was maintained at 5 °C above upper critical temperature for pigs and was between 30 °C and 35 °C throughout the study. Air temperature was altered weekly according to the size of the pigs. Air temperature for control pigs was maintained at the thermo-neutral zone and was between 20 °C and 25 °C during the study. Throughout this study, two husbandry teams were employed to avoid transfer of virus between treatment and the unchallenged control room. All pigs were observed daily for signs of disease. Feed intake for each pen was recorded daily and pigs were weighed weekly. Blood samples were collected weekly and serum was stored at −80 °C until analysis. On day 56 pigs were humanely killed by captive bolt stunning, pithing and bleeding. Necropsies were performed and samples of mesenteric lymph node, inguinal lymph node, lung and bone marrow were collected and stored at −20 °C for later analysis of PCV2 load by qPCR.

2.5. Measurement of weight and calculation of feed conversion ratios

Each pig was weighed at weekly intervals. Additionally, mass of feed consumed in kg was determined weekly for each room. Food conversion ratios (FCR) were calculated by dividing the mass of feed consumed per week, per room, by the sum of weight gains per room for the corresponding week interval.

2.6. Isolation of total DNA from tissue and serum samples

Total DNA was isolated from between 25 and 50 mg of tissue sample collected from all pigs at post-mortem using a DNeasy blood and tissue kit (Qiagen) according to the manufacturer's instructions. Isolated DNA was eluted in sterile and nucleic acid-free water (Sigma-Aldrich) and the DNA concentration was determined using an Infinite 200 Nanoquant spectrophotometer (Tecan). Eluted DNA was stored at −20 °C. In addition,

DNA was isolated from 200 μl cell-free serum collected from pigs weekly in one room of each treatment during the course of the study. The QiAMPMinElute Virus spin kit (Qiagen) was used and the manufacturer's instructions were followed. Isolated DNA was eluted in sterile and nucleic acid-free water (Sigma-Aldrich) and the DNA concentration was determined using an Infinite 200 Nanoquant spectrophotometer (Tecan). Eluted DNA was stored at −20 °C.

2.7. Quantification of PCV2 in tissue or serum samples

The number of PCV2 copies per ng of DNA isolated from each tissue/serum sample was determined by comparison to known standards using qPCR. Each sample was measured in triplicate and in a final volume of 20 μl per well in Microamp Fast Optical 48-microtiter well plates (Applied Biosystems). Each well contained, 2 μl of DNA (standard or test sample), 10 μl of 2 × TaqMan Universal Master Mix II (Applied Biosystems), 50 pmol of each primer (Forward: 5′-GCTCTYTATCGGAGGATTAC-3′, Reverse: 5′-ATAAAAACCATTACGAWGTGATA-3′) (MWG) and 2.5 μM of TaqMan probe (5′FAM-CCATGCCCTGAATTTCCATATGAAAT-3′TAMRA) (Applied Biosystems). The volume of each well was adjusted to 20 μl by addition of nuclease-free water (Sigma-Aldrich). The TaqMan probe and the primers were designed to target a partial (137 bp) sequence of the ORF1 of PCV2 (Grierson et al., 2004). Standard measurements were performed in a 10-fold serial dilution from 10^9 PCV2 copies to 0 copies. After plate set up, the qPCR was performed in a StepOne Real-time PCR machine (Applied Biosystems) and StepOne software version 2.2.2 (Applied Biosystems). The cycling conditions were 95 °C for 10 min for polymerase activation followed by 40 cycles of denaturation at 95 °C for 15 s and annealing/extension at 55 °C for 1 min. Data were analysed using either StepOne software version 2.2.2 (Applied Biosystems) or Excel 2010 (Microsoft). Ct values of triplicate sample and standard measurements were averaged and this average Ct value was used to calculate the PCV2 ORF1 copy number in samples by comparison to known standards. The copy number in each sample was then divided by the nanogram of DNA in that sample to give the PCV2 copy number per ng of DNA.

2.8. Statistical analysis

Average daily weight gain (ADG) values were calculated for each pig by dividing the difference in weights from one week to the previous week by the number of days between both measurements. ADG (gram per day) were analysed through a One Way Analysis of Variance (ANOVA) selecting cases by study week with a Least Square Difference (LSD) Post-hoc, using IBM

SPSS statistics 21 software package. A P-value <0.05 was considered significant.

The mean copy number of viral DNA found in tissue and serum samples was calculated for each room using GraphPad Prism (Version 6 for Windows). Differences between group means were statistically verified by ANOVA after assessment for normal distribution, followed by a modified t-test (Bonferroni) for selected time-points. A P-value < 0.05 was considered significant.

3. RESULTS

Four pigs either died or were euthanized before the end of the study. Of these four pigs, one pig from the V SD group was euthanized at week 5 post inoculation, after veterinary inspection concluded it was showing signs of PCVAD (pyrexia, dehydration, emaciation, lethargy, increased respiratory effort). In addition, one pig from the V T group was found dead the day after weighing and blood sampling at week 4. The pig had evidence of diarrhoea ante-mortem. Post-mortem examination revealed small intestine torsion, possibly secondary to gas build up from the diarrhoea and precipitated by handling for the blood sampling. The other two pigs from the V T group were euthanized for lameness at week 7 post inoculation considering their welfare.

3.1. The combination of PCV2b infection and environmental stressors further impacts on growth parameters

Each pig in the study was weighed weekly and the mass of feed consumed by pigs in each room throughout each week was recorded. The mean weekly mass (Fig. 1) and the mean average daily weight gain (ADG, Fig. 2) and feed conversion ratios (FCR, Table 1) were recorded for each room.

At week 8, the mass (kg) of pigs in the C group (43 kg ± 0.75 kg) compared with pigs in the V SD T group showed a significant difference of 8 kg (35 kg ± 1.25 kg, $p < 0.05$; Fig. 1). Differences between control animals and animals in other groups were just below the level of significance.

There was no statistically significant difference in mean ADG (Fig. 2) between groups as determined by one-way ANOVA at week 1 ($F(4,49) = 0.929$, $p = 0.455$) and week 2 ($F(4,49) = 0.571$, $p = 0.685$) post inoculation.

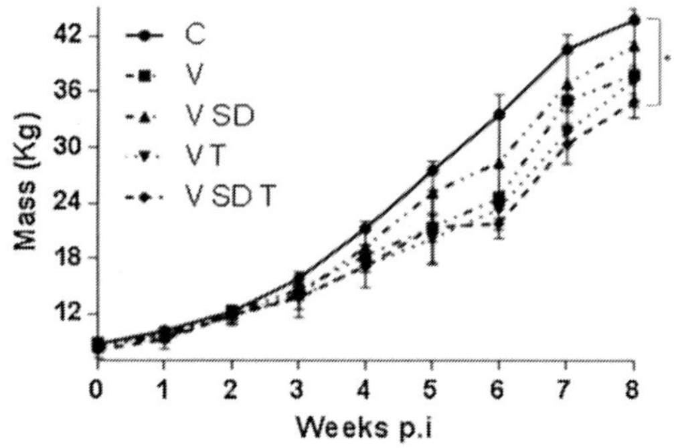

Figure. 1. Mean weekly mass of pigs infected with *PCV2b* for 8 weeks and subjected to different environmental stresses. Pigs were infected with PCV2b and subjected to high stocking density (SD), high temperatures (T) or both (SD T). Pigs were weighed weekly and the average weight for the pigs subjected to each condition was recorded. Significant different values were analysed by one-way ANOVA, with $^{*}p < 0.05$.

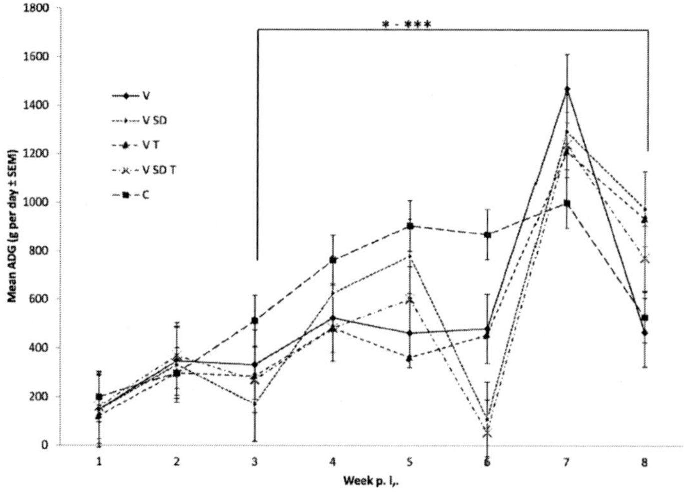

Figure. 2. Mean Average daily weight gain (ADG) of treatment groups per study week. With the exception of control pigs (C), all pigs were infected with PCV2b (V) on week 0 and subjected to high stocking density (SD), high temperatures (T) or both (SD T). Weights were recorded weekly up to 8 weeks post inoculation. Average daily weight gain per week was calculated for each pig, and a mean ADG per treatment group estimated. Stars shown above the graph indicate significant ANOVA p-values between the C and treatment groups, with: $^{*}p < 0.05$; $^{**}p < 0.01$; $^{***}p < 0.001$. P-values for LSD post hoc test are presented within the text.

Table 1. Mean weekly feed conversion ratio for groups per room.

Week p.i	R1 (V)	R2 (V SD)	R3 (V T)	R4 (V SD T)	R5 (C)	R7 (V SD)	R8 (V SD T)	R9 (V T)	R10 (V)
1	1.13	1.04	1.54	1.08	1.04	1.41	0.83	1.13	1.30
2	1.12	1.44	0.99	1.02	1.32	0.95	0.86	1.51	0.84
3	1.97	3.74	1.78	2.30	1.35	3.02	1.79	1.92	1.77
4	1.59	1.29	1.66	1.89	1.32	1.60	1.21	1.59	1.78
5	1.38	1.69	1.35	1.77	1.44	1.76	1.66	19.19	6.14
6	7.73	23.25	13.43	25.00	1.72	8.98	19.16	1.81	1.85
7	1.18	1.19	1.04	1.43	1.93	1.45	1.00	1.42	1.20
8	2.18	2.02	1.65	1.74	2.93	1.38	1.69	1.25	7.67

From week 3 to week 8 post inoculation, the ADG means differed significantly between the groups, as analysed by one-way ANOVA, with p-values ranging from $p < 0.05$ to $p < 0.001$. At week 3, the LSD post-hoc test revealed that the ADG was statistically significantly lower in the V group (332.14 ± 194.27 g/day, $p < 0.05$), V SD group (170.23 ± 108.70 g/day, $p < 0.001$), V T group (285.71 ± 95.54 d/day, $p < 0.01$) and the V SD T group (270.24 ± 102.01 g/day, $p < 0.01$) compared with the C group (511.90 ± 29.16 g/day).

At week 4, the LSD post-hoc test revealed that the ADG was statistically significantly lower in the V group (523.81 ± 226.56 g/day, p, 0.05), the V T group (482.14 ± 180.51 g/day, $p < 0.01$) and the V SD T group (482.14 ± 175.29 g/day, $p < 0.01$) compared with the C group (761.90 ± 147.54 g/day). As before, there was no significance between the values obtained for the other groups.

At week 5, the LSD post-hoc test revealed that the ADG was statistically significantly lower in the V group (461.58 ± 303.33 g/day, $p < 0.05$), the V T group (389.61 ± 344.58 g/day, $p < 0.001$) and the V SD T group (601.19 ± 159.60 g/day, $p < 0.05$) compared with the C group (904.76 ± 173.01 g/day).

At week 6, the LSD post-hoc test revealed that the ADG was statistically significantly lower in the V group (480.16 ± 372.72 g/day, $p < 0.05$), the V SD group (103.90 ± 187.26 g/day, $p < 0.001$), the V T group (422.08 ± 386.58 g/day, $p < 0.01$) and the V SD T group (53.57 ± 101.59 g/day, $p = 0.0001$) compared with the C group (869.05 ± 70.23 g/day).

At week 7, the LSD post-hoc test revealed that the ADG was statistically significantly higher in V group (1470.24 ± 123.56 g/day, $p < 0.001$), V SD group (1305.19 ± 209.69 g/day, $p < 0.01$), V T group (1198.41 ± 150.58 g/day, $p < 0.05$) and V SD T group (1238.09 ± 186.10 g/day, $p < 0.05$) compared with the C group (1000.00 ± 135.53 g/day). Until this time-point, no statistically significant differences were seen between V SD and V T group, V SD and V SD T as well as V T and V SD T.

At week 8, the LSD post-hoc test revealed that the ADG was statistically significantly higher in V SD group (962.12 ± 239.69 g/day, $p < 0.05$) and V T group (953.70 ± 273.58 g/day, $p < 0.05$) compared with the C group (527.78 ± 208.61 g/day). Statistically significant differences ($p < 0.05$) were seen between V group (465.28 ± 308.68 g/day) and V SD T group (770.83 ± 261.72 g/day). There were no statistically significant differences between C and V group ($p = 0.641$), C and V SD T group ($p = 0.075$), V SD and V T ($p = 0.944$), V SD and V SD T ($p = 0.092$) and V T and V SD T ($p = 0.126$).

Over the entire study period, C group had the highest ADG (636.06 g/day ± 13.79), followed by the V group (528.79 g/day ± 29.19), the V SD group (518.45 g/day ± 40.12), the V T group (457.08 g/day ± 41.00). Lowest ADG was recorded for the V SD T group (488.03 g/day ± 23.38). Throughout the study, the greatest variation in ADG was observed within the V SD and V T groups, which also had the highest standard error among groups. Moreover, both these groups had pigs, which presented with clinical signs of PCV-systemic disease, which would have affected their ADG and FCR.

Expected FCR for pigs in the UK (Garth Stockmanship Standards) varies from 1.3 at weaning to 2.3 at 12 weeks of age, which coincides with the length of the study. FCR of pigs in the C group was within these ranges, ending the study with a FCR of 2.93, meaning they gained 1 kg of live weight for every 2.93 kg of feed consumed (**Table 1** and **Table 2**). In comparison, inoculated pigs presented a wider variation in FCR during the study, with high to very high FCR at week 5 in a V group room and a V T group room, respectively, and very high FCR for the rest of treatment rooms at week 6 (**Table 2**). This was mainly due to low weight gain although feed consumed was lower for all treatment rooms compared with control group at week 6 post inoculation. Interestingly at week 5, room 9 (V T) consumed less feed than the control in contrast to room 10 (V) were feed consumption was higher than in the control group (**Table 2**).

Table 2. Average weekly feed conversion ratio for each pig infected with PCV2b and subjected to environmental stress.

Week	Control	Virus only	Virus and SD	Virus and T	Virus, T and SD
1	0.92	1.01	1.06	1.11	0.81
2	1.32	0.98	1.20	1.25	0.94
3	1.35	1.86	3.38	1.85	2.04
4	1.31	1.68	1.44	1.63	1.55
5	1.44	5.51	1.66	4.31	1.72
6	1.72	7.06	16.02	7.69	22.36
7	71.93	1.16	1.32	1.53	1.22
8	3.08	5.83	1.97	1.71	2.04

3.2. PCV2b viral load in the serum and tissue samples is increased by environmental stressors

To assess whether exposure of pigs to environmental stress would impact on the viral load detected, in addition to the weight differences seen, serum (Fig. 3) and tissues samples (Fig. 4) were analysed by qPCR for PCV2 DNA. All pigs were serum PCV2b negative at week 1 post-infection and uninfected controls remained serum PCV2b negative throughout the eight-week study (Fig. 3). Viraemia in PCV2b-infected pigs always peaked at three weeks post infection and then dropped rapidly up to eight weeks post infection regardless of which environmental stressors were also applied. Interestingly, a much higher peak in serum viraemia was detected in V SD T pigs compared with V SD or V T pigs (Fig. 3), but the differences did not reach statistical significance due to large variation between pigs in the same group. These results indicate that exposure to environmental stresses potentially increases the viral load in the serum of PCV2 infected pigs.

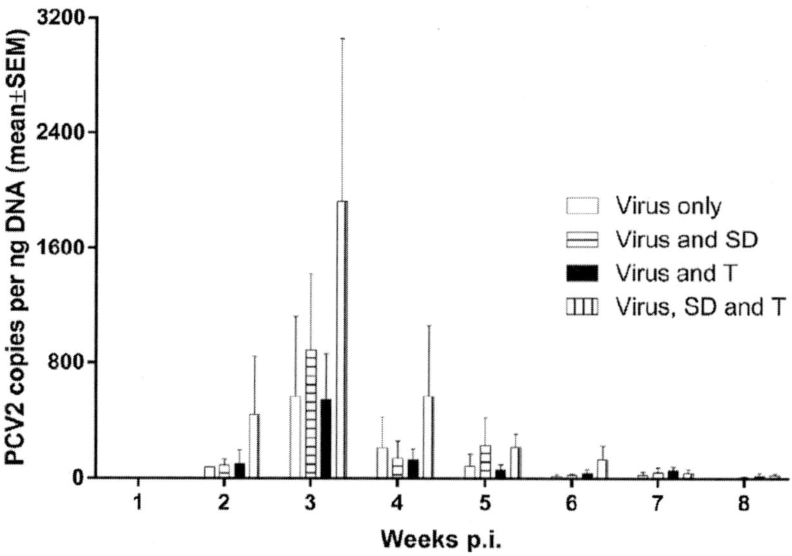

Figure. 3. Viral copies in serum samples taken from pigs infected with PCV2b and subjected to different environmental stresses. Pigs were infected with PCV2b (V) and subjected to high stocking density (V SD), high temperatures (V T) or both (V SD T). Serum samples were taken weekly up to 8-weeks post-infection and tested for viral load by qPCR. Values are expressed as mean ± standard error of the mean (SEM) PCV2 copies numbers per ng DNA.

In addition to serum samples, the amount of PCV2 DNA in samples obtained from mesenteric lymph nodes, inguinal lymph nodes, lung and bone marrow were analysed. All samples were collected post-mortem at eight weeks post infection, and viral DNA was analysed as described. A comparison between the different tissue samples showed that the highest viral load was observed in the mesenteric lymph node (**Fig. 4**A), followed by the inguinal lymph node (**Fig. 4**B), lung (**Fig. 4**C) and bone marrow (**Fig. 4**D). Independent of the tissue tested, there was a tendency that viral load was highest in samples obtained from pigs exposed to all conditions (V SD T), with the exception of the samples taken from the inguinal lymph node. Furthermore, there was a tendency that samples taken from infected pigs exposed to a temperature stress (V T) showed elevated levels of PCV2 DNA in all tissues. In contrast, the amount of PCV2 detected in non-infected controls was negligible for all tissues tested. Additionally, PCV2 DNA copy numbers detected in pigs exposed to PCV2 alone (V) remained low for every tissue tested. Similar to the results obtained with serum samples, none of the differences reached the level of significance due to a large variation in PCV2 copy number between pigs in the same group.

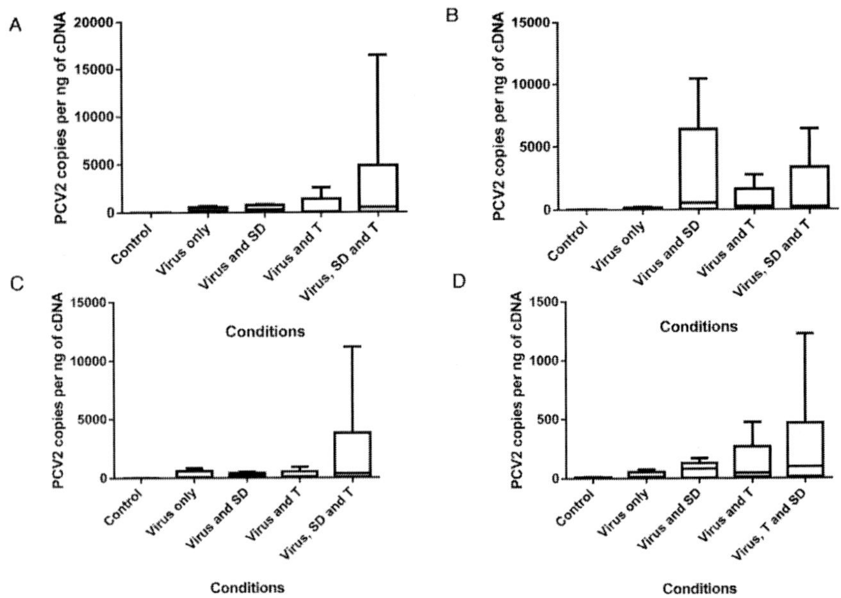

Figure. 4. Viral copies in mesenteric lymph node (A), inguinal lymph node (B), lung (C) and bone marrow (D) taken from pigs infected with PCV2b for 8 weeks and subjected to different environmental stresses. Pigs were infected with PCV2b (V) and subjected to high stocking density (V SD), high temperatures (V T) or both (V SD T). Tissue samples were collected post-mortem at 8-weeks post-infection and tested for viral load by qPCR. Values are expressed as mean ± standard error of the mean (SEM) PCV2 copies numbers per ng DNA.

4. DISCUSSION

In a recent study performed by our group, we identified stocking density and temperature as farm level risk factors linked with the severity of PMWS occurrence (Alarcon et al., 2011a). Therefore, the present study assessed whether these "environmental stressors" directly impact on PCV2 viral load, and thus could be used to develop a PMWS model which does not rely on the use of gnotobiotic pigs. To our knowledge, this study evaluates for the first time, the roles of these specific environmental stressors likely to be encountered in the field, on PCV2 replication under experimental conditions. We demonstrated that subjecting PCV2 infected pigs to environmental stress led to decreased body weight (**Fig. 1**) and ADG (**Fig. 2**) which was associated with an increased feed conversion ratio (Table 1). Furthermore, the negative effects on production parameters observed by subjecting pigs to environmental stressors were associated with increased levels of PCV2 DNA detected in the serum (**Fig. 3**), as well as the mesenteric lymph node, inguinal lymph node, lung and bone marrow (**Fig. 4**A–D).

There is overwhelming evidence that infection with PCV2 does not necessarily lead to PCVAD but can persist as sub-clinical infection. Recent research suggested a variety of co-factors, including other pathogens that are necessary to induce clinical signs of PCVAD (Opriessnig and Halbur, 2012). This phenomenon would mirror other veterinary diseases such as Staphylococcus aureus infection in pigs where it has been demonstrated that infection of pigs with Influenza A virus greatly increases the pathology to secondary S. aureus colonization in the respiratory tract (Smith et al., 2011). However, in the case of PCV2, co-infection with a secondary pathogen does not necessarily trigger the onset of PCVAD (Alarcon et al., 2011b). More recently, research has sought to determine the involvement of environmental co-factors which can potentially lead to signs of disease in PCV2 infected pigs (Alarcon et al., 2011b). Indeed, it is well established that sub-optimal husbandry conditions negatively impact on growth performance, food conversation rates, and general thriving of pigs (Prunier et al., 2010). This effect seems to be mediated through a constant elevated production of stress hormones, such as cortisol (Marco-Ramell et al., 2011 ; O'Connor et al., 2010). High cortisol levels have been shown to induce immune-suppression, leading to increased susceptibility to infection in different vertebrate species including animals and humans (Kick et al., 2012 ; Merlot et al., 2013). Furthermore, high growth performances of current commercial pig lines are suspected to favour the development of mild chronic stress, increasing disease susceptibility (Prunier et al., 2010 ; Rutherford et al., 2006). Overall, the changes observed in our treatment groups are in-line with earlier reported impacts of temperature and stocking density on ADG (Huynh et al., 2005 ; Kerr et al., 2005).

Pigs in the C group had better growth rates (ADG) for the duration of the study compared with all other treatment groups with mean ADG followed in order by V, V SD, V T and V SD T group with the lowest. Significant differences were seen in ADG from week 3 until week 8 post inoculation which coincides with detectable viral loads, possibly reflecting the impact of viraemia which peaked at day 21. At week 3 and 6 post inoculation, mean ADG was statistically significantly lower in all groups that received PCV2 compared with the C group. Also, FCR during week 6 of the study were higher in all groups compared with the C group, attributed mainly to low weight gain. The highest FCR in week 6 was seen in the V SD T group which was over 4 times the value of the V pigs. Groups V T and V SD T had similar mean ADG values at week 3, 4 and 7, which were lower than the V and C groups at week 3, 4 and 6 and only lower than the V group at week 7. A lower ADG value in V SD T than V T was seen in week 8 although this was not statistically significant. Significant differences between V T and V SD T were seen in weeks 5 and 6. A recent meta-analysis of the impact of high environmental temperature on growth rates concluded that although there was inconsistency between studies, overall there was a negative impact on growth rates (Renaudeau et al., 2011). Interestingly however, the ADG over the time-span of the experiment was further affected by the presence of PCV2, and in pigs exposed to all three stressors, virus load was highest. These data indicate a potential interaction of environmental stress and PCV2 replication. Similar to our findings, sub-optimal husbandry stress has been shown to increase susceptibility to Mycobacteria infection in zebrafish which was shown to be associated with elevated body levels of cortisol (Ramsay et al., 2009).

The increased replication of PCV2 observed in this study due to environmental stress being placed on the pigs can be attributed to a few factors. Firstly, pigs kept in high stocking density are more likely to come into contact with any infected pigs already in the pen and therefore are more regularly exposed to virus. Secondly, increased stress on the pig as an individual can lead to increase in serum levels of the stress hormone cortisol. This increase in serum cortisol levels has been demonstrated in pigs in response to food and water deprivation (Parrott et al., 1989), heat stress (Becker et al., 1985), social stress (Parrott and Misson, 1989a) and shipping of animals (McGlone et al., 1993; Nyberg et al., 1988 ; Parrott et al., 1989b). Higher levels of serum cortisol associated with a higher level of stress have been demonstrated to be associated with a reduced natural killer (NK) cell cytotoxicity as well as a reduced number of circulating lymphocytes in the blood (McGlone et al., 1993). As NK cells are an essential component of the innate immune system and are heavily involved in the early response to viral infection, the increased PCV2 replication observed in stressed pigs in this study may be explained by a similar inhibition in NK cell function. Further study into the cortisol levels and NK cell function in the system described in this study would indeed be interesting and could potentially further develop the PCV2 infection model.

Results from the present study showed that on average, PCV2 copies in the mesenteric lymph node, lung and bone marrow are higher in pigs from the V SD T group compared with those subjected to only one environmental stress or C pigs. In the inguinal lymph node however, it was observed that the average PCV2 copy number was highest in the V SD group. This high average was a result of one animal having a very high PCV2 copy number in this tissue. Although the daily behaviour of the pigs was not monitored, such high PCV2 copy number in the inguinal lymph node of a single pig may be a result of that pig being submissive. McGlone et al. found increased levels of serum cortisol and an associated decrease in NK activity in pen mates that were determined submissive compared with those that were determined dominant (McGlone et al., 1993).

In conclusion, inoculation with virus reduced ADG in all treatment groups at the time of peak viraemia, and in combination with high stocking density and high temperatures exacerbated the effect. Additionally, environmental stress on the animal seems to cumulatively increase viral DNA replication, which may lead to more viral particles in the tissues tested here. In essence, it seems that we have reproduced PCVAD in pigs without the need for a secondary pathogen. That being said, the large variation in viral copy number found in tissues of pigs from the same treatment group indicates that the immune response of each individual animal may play a role in the outcome of PCV2 infection. The results presented here further strengthen the case for housing pigs in favourable conditions in order to increase production parameters and reduce susceptibility to disease.

ACKNOWLEDGMENT

This work was funded by a grant (BB/FO18394/1) from the BBSRC 'Combating Endemic Diseases of Farmed Animals for Sustainability' (CEDFAS) initiative, with contributions from BPEX, Biobest Laboratories and Zoetis Animal Health. We thank the staff of the Biological Service Unit of the RVC, especially A. Wallis and C. Davies for their technical help. This manuscript represents publication number PPB_00864 of the RVC.

REFERENCES

1. Alarcon, P., Velasova, M., Mastin, A., Nevel, A., Stark, K.D., Wieland, B., 2011a. Farm level risk factors associated with severity of post-wean-ing multi-systemic wasting syndrome. Prev. Vet. Med. 101, 182–191.

2. Alarcon, P., Velasova, M., Werling, D., Stark, K.D., Chang, Y.M., Nevel, A., Pfeiffer, D.U., Wieland, B., 2011b. Assessment and quantification of post-weaning multi-systemic wasting syndrome severity at farm level. Prev. Vet. Med. 98, 19–28.
3. Allan, G., Meehan, B., Todd, D., Kennedy, S., McNeilly, F., Ellis, J., Clark, E.G., Harding, J., Espuna, E., Botner, A., Charreyre, C., 1998. Novel porcine circoviruses from pigs with wasting disease syndromes. Vet. Rec. 142, 467–468.
4. Andraud, M., Rose, N., Grasland, B., Pierre, J.S., Jestin, A., Madec, F., 2009. Influence of husbandry and control measures on porcine circovirus type 2 (PCV-2) dynamics within a farrow-to-finish pig farm: a model-ling approach. Prev. Vet. Med. 92, 38–51.
5. Beach, N.M., Meng, X.J., 2012. Efficacy and future prospects of commer-cially available and experimental vaccines against porcine circovirus type 2 (PCV2). Virus Res. 164, 33–42.
6. Becker, B.A., Nienaber, J.A., Christenson, R.K., Manak, R.C., DeShazer, J.A., Hahn, G.L., 1985. Peripheral concentrations of cortisol as an indicator of stress in the pig. Am. J. Vet. Res. 46, 1034–1038.
7. Brunborg, I.M., Moldal, T., Jonassen, C.M., 2004. Quantitation of porcine circovirus type 2 isolated from serum/plasma and tissue samples of healthy pigs and pigs with postweaning multisystemic wasting syn-drome using a TaqMan-based real-time PCR. J. Virol. Methods 122, 171–178.
8. Ellis, J., Hassard, L., Clark, E., Harding, J., Allan, G., Willson, P., Strokappe, J., Martin, K., McNeilly, F., Meehan, B., Todd, D., Haines, D., 1998. Isola-tion of circovirus from lesions of pigs with postweaning multisys-temic wasting syndrome. Can. Vet. J. La revue veterinaire canadienne 39, 44–51.
9. Ellis, J.a.A.G., 2000. The natural history of porcine circoviruses. In: 16th International Pig Veterinarians Society Congress (Melbourne, Australia). pp. 3–15
10. Ghebremariam, M.K., Gruys, E., 2005. Postweaning multisystemic wast-ing syndrome (PMWS) in pigs with particular emphasis on the causative agent, the mode of transmission, the diagnostic tools and the control measures. A review. Vet. Quart. 27, 105–116.
11. Grau-Roma, L., Hjulsager, C.K., Sibila, M., Kristensen, C.S., Lopez-Soria, S., Enoe, C., Casal, J., Botner, A., Nofrarias, M., Bille-Hansen, V., Fraile, L., Baekbo, P., Segales, J., Larsen, L.E., 2009. Infection, excretion and seroconversion dynamics of porcine circovirus type 2 (PCV2) in pigs from post-weaning multisystemic wasting syndrome (PMWS) affect-ed farms in Spain and Denmark. Vet. Microbiol. 135, 272–282.
12. Grierson, S.S., King, D.P., Sandvik, T., Hicks, D., Spencer, Y., Drew, T.W., Banks, M., 2004. Detection and genetic typing of type 2 porcine circoviruses in archived pig tissues from the UK. Arch. Virol. 149, 1171–1183.
13. Huynh, T.T., Aarnink, A.J., Verstegen, M.W., Gerrits, W.J., Heetkamp, M.J., Kemp, B., Canh, T.T., 2005. Effects of increasing temperatures on physiological changes in pigs at different relative humidities. J. Anim. Sci. 83, 1385–1396.
14. Kelley, K.W., 1980. Stress and immune function: a bibliographic review. Annales de recherches Vet. Ann. Vet. Res. 11, 445–478. Kerr, C.A., Giles, L.R., Jones, M.R., Reverter, A., 2005. Effects of grouping unfamiliar cohorts, high ambient temperature and stocking density on live performance of growing pigs. J. Anim. Sci. 83, 908–915.
15. Kick, A.R., Tompkins, M.B., Flowers, W.L., Whisnant, C.S., Almond, G.W., 2012. Effects of stress associated with weaning on the adaptive immune system in pigs. J. Anim. Sci. 90, 649–656.
16. Ladekjaer-Mikkelsen, A.S., Nielsen, J., Stadejek, T., Storgaard, T., Krakowka, S., Ellis, J., McNeilly, F., Allan, G., Botner, A., 2002. Reproduction of postweaning multisystemic wasting syndrome (PMWS) in immunos-timulated and non-immunostimulated 3-week-old piglets experi-mentally infected with porcine circovirus type 2 (PCV2). Vet. Microbiol. 89, 97–114.
17. Larochelle, R., Magar, R., D'Allaire, S., 2003. Comparative serologic and virologic study of commercial swine herds with and without postweaning multisystemic wasting syndrome. Can. J. Vet. Res. 67, 114–120.
18. Lukert, P., 1999. Porcine cicovirus. In: Diseases of Swine8th ed. Iowa State University Press, Ames, USA.

19. Mankertz, A., 2012. Molecular interactions of porcine circoviruses type 1 and type 2 with its host. Virus Res. 164, 54–60.
20. Marco-Ramell, A., Pato, R., Pena, R., Saco, Y., Manteca, X., Ruiz de la Torre, J.L., Bassols, A., 2011. Identification of serum stress biomarkers in pigs housed at different stocking densities. Vet. J. 190, e66–e71.
21. McGlone, J.J., Salak, J.L., Lumpkin, E.A., Nicholson, R.I., Gibson, M., Norman, R.L., 1993. Shipping stress and social status effects on pig perfor-mance, plasma cortisol, natural killer cell activity, and leukocyte numbers. J. Anim. Sci. 71, 888–896.
22. Merlot, E., Quesnel, H., Prunier, A., 2013. Prenatal stress, immunity and neonatal health in farm animal species. Animal: Int. J. Anim. Biosci. 7, 2016–2025.
23. Nayar, G.P., Hamel, A., Lin, L., 1997. Detection and characterization of porcine circovirus associated with postweaning multisystemic wast-ing syndrome in pigs. Can. Vet. J. La revue veterinaire canadienne 38, 385–386.
24. Nyberg, L., Lundstrom, K., Edfors-Lilja, I., Rundgren, M., 1988. Effects of transport stress on concentrations of cortisol, corticosteroid-binding globulin and glucocorticoid receptors in pigs with different halothane genotypes. J. Anim. Sci. 66, 1201–1211.
25. O'Connor, E.A., Parker, M.O., McLeman, M.A., Demmers, T.G., Lowe, J.C., Cui, L., Davey, E.L., Owen, R.C., Wathes, C.M., Abeyesinghe, S.M., 2010. The impact of chronic environmental stressors on growing pigs Sus scrofa (Part 1): stress physiology, production and play behaviour. Animal: Int. J. Anim. Biosci. 4, 1899–1909.
26. Opriessnig, T., Halbur, P.G., 2012. Concurrent infections are important for expression of porcine circovirus associated disease. Virus Res. 164, 20–32.
27. Opriessnig, T., Langohr, I., 2013. Current state of knowledge on porcine circovirus type 2-associated lesions. Vet. Pathol. 50, 23–38.
28. Parrott, R.F., Misson, B.H., 1989a. Changes in pig salivary cortisol in response to transport simulation, food and water deprivation, and mixing. Brit. Vet. J. 145, 501–505.
29. Parrott, R.F., Misson, B.H., Baldwin, B.A., 1989b. Salivary cortisol in pigs following adrenocorticotrophic hormone stimulation: comparison with plasma levels. Brit. Vet. J. 145, 362–366.
30. Prunier, A., Heinonen, M., Quesnel, H., 2010. High physiological demands in intensively raised pigs: impact on health and welfare. Animal: Int. J. Anim. Biosci. 4, 886–898.
31. Ramsay, J.M., Watral, V., Schreck, C.B., Kent, M.L., 2009. Husbandry stress exacerbates mycobacterial infections in adult zebrafish, Danio rerio (Hamilton). J. Fish Dis. 32, 931–941.
32. Renaudeau, D., Gourdine, J.L., St-Pierre, N.R., 2011. A meta-analysis of the effects of high ambient temperature on growth performance of growing-finishing pigs. J. Anim. Sci. 89, 2220–2230.
33. Rose, N., Opriessnig, T., Grasland, B., Jestin, A., 2012. Epidemiology and transmission of porcine circovirus type 2 (PCV2). Virus Res. 164, 78–89.
34. Rutherford, K.M., Haskell, M.J., Glasbey, C., Lawrence, A.B., 2006. The responses of growing pigs to a chronic-intermittent stress treatment. Physiol. Behav. 89, 670–680.
35. Segales, J., Kekarainen, T., Cortey, M., 2013. The natural history of porcine circovirus type 2: from an inoffensive virus to a devastating swine disease? Vet. Microbiol.
36. Shaw, R., Festing, M.F., Peers, I., Furlong, L., 2002. Use of factorial designs to optimize animal experiments and reduce animal use. ILAR J./Natl. Res. Council Instit. Lab. Anim. Resour. 43, 223–232.
37. Sibila, M., Calsamiglia, M., Segales, J., Blanchard, P., Badiella, L., Le Dimna, M., Jestin, A., Domingo, M., 2004. Use of a polymerase chain reaction assay and an ELISA to monitor porcine circovirus type 2 infection in pigs from farms with and without postweaning multisystemic wast-ing syndrome. Am. J. Vet. Res. 65, 88–92.
38. Smith, E.A., Kumar, S.R., Deventhiran, J., Cecere, T.E., Leroith, T., McGilliard, M., Elankumaran, S., Mullarky, I.K., 2011. A time course for suscepti-bility to staphylococcus aureus respiratory infection during influenza in a Swine Model. Influenza Res. Treatment 2011, 846910.
39. Wieland, B., Alarcon, P., Velasova, M., Nevel, A., Towrie, H., Pfeiffer, D., Wathes, C., Werling, D., 2010. Prevalence of endemic pig diseases in England: an overview six months into a

large scale cross sectional study on post-weaning multisystemic wasting syndrome (PMWS). Pig J. 63, 20–24.
40. Wieland, B., Werling, D., Nevel, A., Rycroft, A., Demmers, T.G., Wathes, C.M., Grierson, S., Cook, A.J., Done, S.H., Armstrong, D., Wathes, C.M., 2012. Porcine circovirus type 2 infection before and during an out-break of postweaning multisystemic wasting syndrome on a pig farm in the UK. Vet. Rec. 170, 596.
41. Young, M.G., Cunningham, G.L., Sanford, S.E., 2011. Circovirus vaccination in pigs with subclinical porcine circovirus type 2 infection compli-cated by ileitis. J. Swine Health Prod. 19 (3), 175–180. R. Patterson et al. / Veterinary Microbiology 177 (2015) 261–269 269

CHAPTER 13

Fecal prevalence, serotype distribution and antimicrobial resistance of *Salmonellae* in dairy cattle in central Ethiopia

Tadesse Eguale[1], Ephrem Engidawork[2], Wondwossen A. Gebreyes[3], Daniel Asrat[4], Haile Alemayehu[1], Girmay Medhin[1], Roger P. Johnson[5] and John S. Gunn[6]

[1]*Aklilu Lemma Institute of Pathobiology, Addis Ababa University, P.O. Box 1176, Addis Ababa, Ethiopia.*

[2]*Department of Pharmacology and Clinical Pharmacy, School of Pharmacy, College of Health Sciences, Addis Ababa University, Churchill Avenue, P.O. Box 1176, Addis Ababa, Ethiopia.*

[3]*Department of Veterinary Preventive Medicine, The Ohio State University, 1920 Coffey Rd., Columbus, OH 43210, USA.*

[4]*Department of Microbiology Immunology & Parasitology, School of Medicine, College of Health Sciences, Addis Ababa University, Churchill Avenue, P.O. Box 9086 Addis Ababa, Ethiopia.*

[5]*Laboratory for Foodborne Zoonoses, 110 Stone Road West, Guelph, ON N1G 3W4, Canada.*

[6]*Department of Microbial Infection and Immunity, Center for Microbial Interface Biology, The Ohio State University,Biomedical Research Tower, 460 West 12th, Columbus, OH 43210-1214, USA.*

AUTHORS' CONTRIBUTIONS

TE, EE, WG, JSG and DA, participated in conception of the study and review of the draft manuscript. TE was involved in sample collection laboratory investigation and preparation of the draft manuscript. HA participated in laboratory work. RPJ was involved in serotyping, and phage typing of the isolates as well as preparation of the manuscript. GM participated in data analysis. All authors read and approved the final manuscript.

ABSTRACT

Background

Salmonellae are major worldwide zoonotic pathogens infecting a wide range of vertebrate species including humans. Consumption of contaminated dairy products and contact with dairy cattle represent a common source of non-typhoidal *Salmonella* infection in humans. Despite a large number of small-scale dairy farms in Addis Ababa and its surrounding districts, little is known about the status of *Salmonella* in these farms.

Results

Salmonella was recovered from the feces of at least one animal in 7.6 % (10/132) of the dairy farms. Out of 1203 fecal samples examined, 30 were positive for *Salmonella* resulting in a weighted animal level prevalence of 2.3 %. Detection of diarrhea in an animal and in a farm was significantly associated with animal level ($p = 0.012$) and herd level ($p < 0.001$) prevalence of *Salmonella*. Animal level prevalence of *Salmonella* was significantly associated with age ($p = 0.023$) and study location; it was highest among those under 6 months of age and in farms from Adaa district and Addis Ababa ($p < 0.001$). Nine different serotypes were identified using standard serological agglutination tests. The most frequently recovered serotypes were *Salmonella* Typhimurium (23.3 %), *S.* Saintpaul (20 %), *S.* Kentucky (16.7 %) and *S.* Virchow (16.7 %). All isolates were resistant or intermediately resistant to at least one of the 18 drugs tested. Twenty-six (86.7 %), 19 (63.3 %), 18 (60 %), 16 (53.3 %) of the isolates were resistant to streptomycin, nitrofurantoin, sulfisoxazole and tetracycline , respectively. Resistance to 2 drugs was detected in 27 (90 %) of the isolates. Resistance to 3 or more drugs was detected in 21 (70 %) of the isolates, while resistance to 7 or more drugs was detected in 11 (36.7 %) of the isolates. The rate of occurrence of multi-drug resistance (MDR) in

Salmonella strains isolated from dairy farms in Addis Ababa was significantly higher than those isolated from farms outside of Addis Ababa ($p = 0.009$). MDR was more common in *S.* Kentucky, *S.* Virchow and *S.* Saintpaul.

Conclusion

Isolation of *Salmonella* serotypes commonly known for causing human salmonellosis that are associated with an MDR phenotype in dairy farms in close proximity with human population is a major public health concern. These findings imply the need for a strict pathogen reduction strategy.

Keywords: Dairy farm, Feces, *Salmonella enterica,* Multi-drug resistance

1. BACKGROUND

Salmonella is a diverse bacterial species comprising over 2600 serotypes [1]. *Salmonella* commonly colonizes a range of animal hosts such as mammals, amphibians, reptiles, birds and insects [2]. There are 2 species of *Salmonella*: *Salmonella enterica* and *Salmonella bongori. Salmonella enterica* is further classified into 6 subspecies (*Salmonella enterica* subspecies *enterica, S. enterica* Subspecies *salmae, S. enterica* Subspecies *arizonae, S. enterica* Subspecies *diarizonae, S. enterica* Subspecies *hautenae* and *S. enterica* Subspecies *indica*). Most of the *Salmonella* serotypes are part of *S. enterica* subspecies *enterica*, and over 99 % of human and animal infections are caused by serotypes under this subspecies [3].

Diseases caused by *Salmonella* represent an important public health problem among the common bacterial foodborne pathogens worldwide. It is estimated that globally 93.8 million cases and 155,000 deaths are associated with gastroenteritis due to *Salmonella* species annually. Of these cases, 85.6 % were estimated to be foodborne [4]. Human salmonellosis has been associated with contaminated food products, mainly those of animal origin such as poultry, beef, pork and dairy products, as well as direct contact with infected animals [5, 6, 7].

Various serotypes of *Salmonella* have been isolated from the feces of apparently healthy dairy cattle. *Salmonella* in dairy animals may exist as a normal microbiota of the gastrointestinal population, or as a transient member of the gastrointestinal microbial population [8].

All sick, recovered and asymptomatic cattle can shed *Salmonella* through feces and the organism can survive for a long time in favorable environments outside the host [9]. Fecal shedding of *Salmonella* can increase intra-herd transmission, accidental spread to other herds, environmental contamination and risk of human infection [10]. Consumption of raw milk, inadequately pasteurized milk, improperly cooked beef from culled dairy cattle, contaminated water and direct animal contact are the major routes of acquiring dairy associated salmonellosis in humans [6].

In Ethiopia, there are large numbers of small-scale peri-urban dairy farms mainly situated close to areas of public residence. Most of these farms are located very close to Addis Ababa, capital city of the country, or reside within the city in a very close proximity with human populations. The consumption of raw milk and its derivatives is common in Ethiopia, posing high risk of infection with dairy-associated foodborne pathogens. Such pathogens include *Salmonella spp., Klebsiella* spp., *Enterobacter* spp., and *Escherichia coli*, which have been identified in milk products in Ethiopia [11]. Gram positive pathogens such as *Staphylococcus aureus, Bacillus cereus, Listeria monocytogenes* and *Enterococcus* spp. have also been frequently isolated from milk [12, 13].

Occurrence of non-typhoidal *Salmonella* serotypes commonly infecting humans in dairy cattle, particularly, those stains resistant to antimicrobial agents commonly used in human medicine, are a serious threat to human health. Some multi-drug resistant (MDR) *Salmonella* outbreaks in humans have been linked to exposure to dairy farms or contaminated dairy products [6, 14]. Information on the prevalence, serotype distribution and antimicrobial susceptibility of *Salmonella* in dairy farms is vital to implementation of appropriate strategies to prevent the introduction and spread of the pathogen in the farm as well as to reduce the risk of human salmonellosis. Knowledge on the serotypes circulating in dairy farms would inform scientists/clinicians on the role of dairy cattle as a source of human *Salmonella* infections. A previous study conducted in Addis Ababa has shown farm level prevalence of 47.8 % and animal level prevalence of 7.7 % [15]. However, this study involved small sample size and the isolates were not serotyped. Given the relative lack of information concerning the prevalence and serotype distribution of *Salmonella* spp. in dairy farms in Ethiopia, the present study was designed to investigate animal level and herd level fecal prevalence of *Salmonella*, serotype distribution and antimicrobial resistance profiles of *Salmonella* in dairy farms in and around Addis Ababa, Ethiopia. It also attempted to investigate the association of farm size, occurrence of diarrhea in the farm and age of animals with prevalence of *Salmonella* in these dairy farms.

2. METHODS

2.1 Study design, study area and sampling of study animals

A cross-sectional study was conducted in Addis Ababa and in five districts of the Oromia region located at the outskirt of Addis Ababa, namely: Sebeta, Barake, Welmera, Sululta and Adaa (Fig. 1). In these areas, the interaction between animal and human population is very high due to high density of the human populations and the large number of peri-urban dairy farming facilities. These areas are the major sources of dairy milk supply to Addis Ababa. Sampling of study herds and animals was conducted from June to December 2013. Study animals were selected from 132 dairy herds (Addis Ababa; $n = 38$; Adaa, $n = 12$; Sebeta, $n = 21$; Sululta, $n = 24$; Welmera; $n = 18$). Inclusion of herd in the sampling was based on representation of the area under study, willingness of the owners, geographical accessibility, and the herd having a minimum of 5 cattle. The largest herd size contained 398 head of cattle. Farms were categorized into small (5–20 animals in a herd), medium (21–50 animals in a herd) and large (more than 50 animals). Mean herd size of small, medium and large farms was 12.6, 31.7 and 100.4, respectively. In total 1203 fecal samples were collected from healthy as well as diarrheic cattle during the study period. The study design was cross-sectional implying a one point fecal sample collection from a given herd and hence there was no repeated fecal sample collection.

2.2 Sample collection, *Salmonella* isolation and identification

Fecal samples were collected directly from the rectum using disposable gloves into sterile zippered plastic bags and transported to the Microbiology Laboratory, Aklilu Lemma Institute of Pathobiology, in an ice box within 3–4 h of collection. Isolation and identification of *Salmonella* was conducted using conventional methods [16, 17]. Briefly, 10 g of feces was pre-enriched in 90 ml of buffered peptone water (BPW) (Becton Dickinson, Sparks, MD) and incubated overnight at 37 °C. A 100 µl pre-enriched suspension was added into 9.9 ml of Rappaport-Vassiliadis enrichment Broth (RVB) (Oxoid, USA) and incubated at 42 °C for 24 h. At the same time, 1 ml of suspension was also transferred to 10 ml of Tetrathionate broth (TTB) (Oxoid, USA) and incubated for 24 h at 37 °C. It was then streaked from both RVB and TTB to Xylose Lysine Tergitol 4 (XLT-4) (Oxoid, USA) selective media and the plates were incubated at 37 °C for 24 to 48 h. Presumptive *Salmonella* colonies were further investigated biochemically using Triple Sugar Iron agar, Urea, Citrate and Lysine Iron Agar slants. Those colonies with typical *Salmonella* biochemical properties were then further confirmed by genus specific PCR [18]. A reference strain of *S.* Typhimurium (ATCC 14028) was used as a positive control during biochemical analysis and PCR. One confirmed *Salmonella* isolate from each positive sample was stored at −80 °C in 20 % glycerol until further testing.

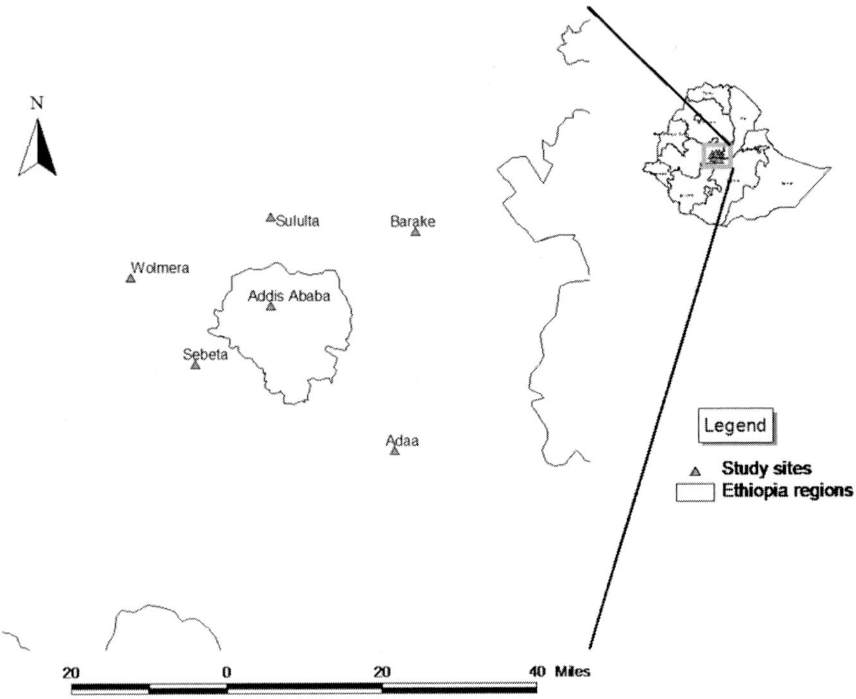

Figure. 1 Locations of Addis Ababa and surrounding districts where fecal samples from dairy cattle were collected (source: Original Ethiopian shape file was obtained from Ethiopian Mapping Agency)

2.3 Data collection

Information such as herd size, housing condition, types of antimicrobials commonly used in the farm, age, sex of animals and presence of diarrhea in a farm was recorded using a purposively designed questionnaire. A farm was categorized as a diarrheic farm if one or more animals in the herd were diarrheic at the time of sample collection. Collection of data was performed at the time of fecal sample collection from each farm.

2.4 *Salmonella* serotyping and phage typing

Salmonella isolates were serotyped and phage typed at the World Organization for Animal Health (OIÉ) Reference Laboratory for Salmonellosis of the Public Health Agency of Canada's National Microbiology at Guelph. Serovars were determined by serum agglutination according to the White-Kauffmann-Le Minor scheme [19, 20], with identification of somatic (O) antigens by slide agglutination tests [21] and flagellar (H) antigens by a microplate agglutination technique [22]. *S.* Typhimurium isolates were phage typed by the methods developed initially by Callow [23] and extended by Anderson et al. [24] and Rabsch [25] with 30 reference phages obtained directly from the WHO Reference Laboratory for phage typing of *Salmonella* species at Public Health England or from the same source via Canada's National Microbiology Laboratory at Winnipeg. These typing phages were number 1–35 with discontinued use of phages 9, 30, 31, 33 and 34. Internal reference strains of phage type 1 (fully susceptible) and phage type 124 (susceptible to only one phage) were included as controls. Isolates that reacted with the phages but did not conform to any recognized phage type were designated atypical (AT), while those that did not react with any of the typing phages were designated untypeable (UT).

2.5 Antimicrobial susceptibility testing

Susceptibility of the isolates to 18 antimicrobials was determined using the Kirby-Bauer disk diffusion method according to the guidelines of the Clinical and Laboratory Standards Institute [26]. The following antimicrobials (Sensi-Discs, Becton, Dickinson and Company, Loveton, USA) and disc potencies (μg) were used: amikacin (30), amoxicillin + clavulanic acid (20/10), ampicillin (10), cefoxitin (30), ceftriaxone (30), cephalothin (30), chloramphenicol (30), ciprofloxacin (5), gentamicin (10), kanamycin (30), nalidixic acid (30), neomycin (30), nitrofurantoin (100), streptomycin (10), sulfisoxazole (1000), sulfamethoxazole + trimethoprim (23.75/1.25), trimethoprim (5) and tetracycline (30). The interpretation of the categories of susceptible, intermediate or resistant was based on the CLSI guidelines [26]. For the purpose of analysis, all readings classified as intermediate were considered as resistant unless indicated.

E. coli ATCC 25922 was used as a quality control organism.

2.6 Statistical analysis

The data analysis method that fits to survey data as it is implemented in STATA version 12 was used to estimate prevalence of *Salmonella* and to investigate its association with pre-specified background characteristics. In animal level analysis, the probability of selecting a given animal from a given herd was considered as a weighting variable. Animal level prevalence of *Salmonella* was calculated as the weighted percentage of *Salmonella* culture-positive fecal samples among the total number of animals examined. Herd level prevalence of *Salmonella* was calculated as the percentage of herds with one or more *Salmonella* culture-positive fecal samples among the total number of herds sampled. Association of weighted animal level prevalence and selected background characteristics was assessed using pearson chi-square within survey command of STATA software. Association of herd level *Salmonella* positivity and pre-specified characteristics was assessed using pearson chi-square. The difference between mean numbers of antimicrobials to which isolates were resistant was compared using a student *t*-test. Results were reported as being statistically significant whenever the *p*-value was less than 0.05.

2.7 Ethics statement

Ethical clearance for the study was obtained from the National Research Ethics Review Committee, Ethiopia. Informed oral consent was obtained from the farm owners at the time of sample collection.

3. RESULTS

3.1 Prevalence and risk factors

Weighted animal level *Salmonella* prevalence was 2.3 % and at least one *Salmonella* positive animal was detected in 7.6 % (10/132) of the herds examined. There was no significant difference in prevalence of *Salmonella* between male and female animals. Significant difference in the prevalence of *Salmonella* was observed across different age groups ($p = 0.023$) and the largest was observed in cattle less than 6 months old. Similarly, animal level prevalence of *Salmonella* among study sites was significantly different ($p < 0.001$): highest prevalence was 5 % in Adaa followed by 4.2 % in Addis Ababa and 2.0 % in Sebeta (Table 1). However, herd level prevalence of *Salmonella* was not significantly different among study sites (Table 2). Diarrhea was detected in 34 of 1203 animals. Three of the diarrheic animals were positive for *Salmonella* whereas 27 out of 1169 non-diarrheic animals were positive for *Salmonella.* These 3 *Salmonella* positive diarrheic animals were a 2 week old calf infected with *S.*Typhimurium var. Copenhagen on a farm in Adaa district, a 3 month old calf infected with *S.* kentucky in Addis Ababa and a 6 year old cow infected with *S.* Dublin in the Sebeta district.

Detection of diarrhea in an animal was significantly associated with animal level *Salmonella* carriage ($p = 0.012$) (Table 1). Detection of diarrhea in one or more animals in a farm was also significantly associated with herd level prevalence of *Salmonella* ($p < 0.001$) (Table 2). Out of 255 fecal samples collected from 24 diarrheic herds, 22 were positive for *Salmonella* whereas, only 8 of the 948 fecal samples collected from 108 non-diarrheic herds were positive for *Salmonella.* Six of 24 (25 %) of diarrheic herds were positive for *Salmonella* while only 4 of 108 (3.7 %) of non-diarrheic herds were positive for *Salmonella* ($p < 0.001$).

Table 1 Animal level prevalence of *Salmonella* and its unadjusted association with selected characteristics

Characteristics	Categories	Number	Weighted [a]percent positive for *Salmonella*	*p*-value
Sex	Male	101	1.4	0.483
	Female	1102	2.4	
Age	<6 month	280	4.5	0.023
	6 months–2 years	162	0.0	
	2 years–5 years	143	2.9	
	5 years–8 years	496	1.6	
	≥8 years	122	2.5	
Study site	Sebeta	141	2.0	<0.001
	Addis Ababa	319	4.2	
	Adaa	184	5.0	
	Barake	151	0.4	
	Welmera	195	0.0	
	Sululta	213	0.2	
Herd Size	Small [5–20)	480	1.8	0.117
	Medium [20–50)	369	2.1	
	Large [50+	354	4.3	
Have diarrhoea	No	1169	2.1	0.012
	Yes	34	9.4	
Overall		1203	2.3	

[a]The result was weighted by the probability of selecting animals from its respective farm

Table 2 Herd level prevalence of *Salmonella* and its unadjusted association with selected farm level characteristics

Characteristics	Categories	Number of farms studied	Percent positive for *Salmonella*	*p*-value
Study site	Sebeta	20	5.0	0.372
	Addis Ababa	38	13.2	
	Adaa	12	16.7	
	Barake	19	5.3	
	Welmera	18	0.0	
	Sululta	25	4.0	
Herd Size	Small [5–20)	79	3.8	0.047
	Medium [20–50)	33	9.1	
	Large [50+	20	20.0	
Farm diarrhoea status	Diarrheic	24	25.0	<0.001
	Non-diarrheic	108	3.7	
Overall		132	7.6	

Table 3 *Salmonella* serotype distribution and number and percent of intermediate and resistant isolates to antimicrobial agents

Serotype	Number	No. of intermediately resistant and resistant isolates																									
		Amp		Amc		Cf		Cip		Gm		K		Tmp		Te		Su		S		Nitro		Na		N	
		I	R	I	R	I	R	I	R	I	R	I	R	I	R	I	R	I	R	I	R	I	R	I	R	I	R
Aberdeen	1	–	–	–	–	–	–	1	–	–	–	1	–	–	–	1	–	1	–	1	–	–	1	1	–	1	–
Dublin	3	–	–	–		2	–	–	–	–	–	–	–	–	–	–	–	1	–	3	–	–	–	–	–	–	–
I:6,7,14:–l,w	1	–	1	–	1	–	1	1	–	–	–	–	–	–	–	1	–	–	–	1	–	–	1	1	–	–	–
Kentucky	5	–	5	2	3	–	5	–	5	–	5	4	–	–	1	–	5	–	5	–	5	4	1	–	5	–	2
LivingstoneVar.14+	1	–	–	–	–	–	–	1	–	–	–	–	–	–	–	–	–	–	–	–	–	–	–	–	1	–	–
Mikawasima	1	–	–	–	–	–	–	–	–	–	–	–	–	–	–	–	–	1	–	–		1	–	–	–	–	–
Saintpaul	6	1	–	–	–	1	–	–	–	–	–	4	–	–	–	4	–	3	2	5	1	2	3	–	–	3	–
Typhimurium	7	–	–	–		2	–	–	–	–	–	3	–	–	–	4	–	3	–	7	–	1	1	1	–	1	–
Virchow	5	–	3	3	–	–	3	1	–	1	1	1	–	–	–	–	1	1	1	1	2	1	3	–	2	–	–
Total	30	1	9	5	4	5	9	4	5	1	6	13	–	–	–	10	6	10	8	18	8	9	10	3	8	5	2
% R		3.3	30	16.7	13.3	16.7	30	13.3	16.7	3.3	20	43.3	–	–	3.3	33.3	20	33.3	26.7	60	26.7	30	33.3	10	26.7	16.7	6.7
% (I + R)		33.3		30		46.7		30		23.3		43.3		3.3		53.3		60		86.7		63.3		36.7		23.3	

Since all isolates were susceptible to Amikacin, Chloramphenicol, Cefoxitin, Ceftriaxone and Sulfamethoxazole + Trimethoprim, they were not included in the table

Amp Ampicillin, *Amc* Amoxicillin and clavulanic acid, Cf Cephalothin, *Cip* Ciprofloxacin, *Gm* Gentamicin, *K* Kanamycin, *Tmp* Trimethoprim, *Te* Tetracycline, *Su* Sulfisoxazole, *S* Streptomycin, *Nitro* Nitrofurantoin, *Na* Nalidixic acid, *N* Neomycin, *I* Intermediate, *R* Resistant

Amp Ampicillin, *Amc* Amoxicillin and clavulanic acid, Cf Cephalothin, *Cip* Ciprofloxacin, *Gm* Gentamicin, *K* Kanamycin, *Tmp* Trimethoprim, *Te* Tetracycline, *Su* Sulfisoxazole, *S* Streptomycin, *Nitro* Nitrofurantoin, *Na* Nalidixic acid, *N* Neomycin, *I* Intermediate, *R* Resistant

There was no significant difference in animal level prevalence of *Salmonella* among animals from farms of different herd size ($p = 0.117$) (Table 1). However, herd level prevalence of *Salmonella* was significantly higher in farms with large herd size ($p = 0.047$) (Table 2). All 30 *Salmonella* isolates were obtained from herds that were kept completely indoors, while none was recovered from farms that allowed their animals to graze outside occasionally or those where cattle were totally outdoors.

3.2 *Salmonella* serotype distribution

Nine different serotypes were identified (Table 3). *S.* Typhimurium, grouped with its variant *S.* Typhimurium var. Copenhagen, was the most common (7/30, 23.3 %) and was isolated from seven animals on three farms in three study sites (Adaa, Sululta and Barake districts). *S.* Saintpaul (6, 20 %) was isolated only from a single farm in the Adaa district, *S.* Kentucky (5, 16.7 %), and *S.* Virchow (5, 16.7 %) were isolated from five animals each in two different farms in Addis Ababa, while *S.* Dublin (3, 10 %) and one isolate (1, 3.3 %) of *S.* Livingstone var.14+, *S.* I: 6, 7, 14:-: I,w, *S.* Mikawasima and *S.* Aberdeen were isolated from one animal on five different farms. Two different serotypes were isolated from two farms in the present study (Table 4).

3.3 Antimicrobial resistance

The common antimicrobials used in the farms were oxytetracycline, penicillin + streptomycin, and sulfonamide in 94.6, 81.8 and 13.6 % of the farms, respectively. Resistance patterns of the isolates are shown in Table 4. All isolates were resistant to at least one of the 18 antimicrobials tested. Twenty-six (86.7 %), 19 (63.3 %), 18 (60 %), 16 (53.3 %) of the isolates were resistant to streptomycin, nitrofurantoin, sulfisoxazole and tetracycline, respectively. Resistance to two or more antimicrobials was recorded in 90 % of the isolates, while resistance to 3 or more antimicrobials was detected in 21 (70 %) of the isolates. MDR to 7 or more antimicrobials were detected in 11 (36.7 %) of the isolates. The five *S.* Kentucky isolates were resistant to 10–13 antimicrobials (Table 4). One isolate (*S.* Kentucky) from a farm in Addis Ababa was resistant to 13 out of 18 antimicrobials tested. All isolates were susceptible to amikacin, chloramphenicol, cefoxitin, ceftriaxone and sulfamethoxazole + trimethoprim.

Table 4 *Salmonella* serotypes isolated from dairy cattle in various study sites and their antimicrobial resistance pattern

Number	Study site	Farm Code	Serotype	R-pattern	
				Intermediate	Resistant
1	Adaa	DZC --03	Aberdeen	CipKTeSuSNaN	NitroS
2	Adaa	DZC --03	Saintpaul	NitroSuS	-
3	Adaa	DZC --03	Saintpaul	TeSuS	Nitro
4	Adaa	DZC --03	Saintpaul	AmpCfKTeSNaN	SuNitro
5	Adaa	DZC --03	Saintpaul	KTeN	SuSNitro
6	Adaa	DZC --03	Saintpaul	KS	-
7	Adaa	DZC --03	Saintpaul	NitroKTeSuSN	-
8	Adaa	DZC-06	Typhimurium var. copehagen PT 193	CfKS	-
9	Adaa	DZC-06	Typhimurium var. copehagen PT 193	CfTeSuS	-
10	Adaa	DZC-06	Typhimurium var. copehagen PT U285	KTeSuSNitroNaN	-
11	Adaa	DZC-06	Typhimurium var. copehagen PT193	TeS	-
12	Addis Ababa	AAC-25	I:6,7,14:-:l,w	CipTeSNa	AmpAmcCfNitro
13	Addis Ababa	AAC-38	Kentucky	KNitro	AmpAmcCfCipGmTeSuSNa
14	Addis Ababa	AAC-25	Kentucky	Nitro	AmpAmcCfCipGmTeSuSNa
15	Addis Ababa	AAC-38	Kentucky	KNitro	AmpAmcCfCipGmTeSuSNa

16	Addis Ababa	AAC-38	Kentucky	AmcKNitro	AmpCfCipGmTeSuSNaN
17	Addis Ababa	AAC-38	Kentucky	AmcK	AmpCfCipGmTmpTeSuSNitroNaN
18	Addis Ababa	AAC-25	Livingstone Var.14+	Cip	Na
19	Addis Ababa	AAC-09	Mikawasima	SuNitro	...
20	Addis Ababa	AAC-23	Virchow	Amc	AmpCf
21	Addis Ababa	AAC-23	Virchow	AmcK	AmpCfCipGmTeSuSNitroNa
22	Addis Ababa	AAC-23	Virchow	AmcSu	AmpCfSNitroNa
23	Addis Ababa	AAC-23	Virchow	GmSNitro	...
24	Addis Ababa	AAC-24	Virchow	...	Nitro
25	Barake	BAR- 18	Typhimurium PT Atypical	KSuS	Nitro
26	Barake	BAR- 18	Typhimurium PT 67	S	...
27	Sebeta	SC-04	Dublin	Cf,S	...
28	Sebeta	SC-04	Dublin	CfSuS	...
29	Sebeta	SC-04	Dublin	S	...
30	Sululta	Suc-07	Typhimurium var. copehagen PT Atypical	TeS	...

PT Phagetype, *Amp* Ampicillin, *Amc* Amoxicillin and clavulanic acid, Cf Cephalothin, *Cip* Ciprofloxacin, *Gm* Gentamicin, *K* Kanamycin, *Tmp* Trimethoprim, *Te* Tetracycline, *Su* Sulfisoxazole, *S* Streptomycin, *Nitro* Nitrofurantoin, *Na* Nalidixic acid, *N* Neomycin

There was a statistically significant difference in the rate of occurrence of MDR between isolates obtained from dairy farms in Addis Ababa and outside of the city limits of Addis Ababa. The mean ± standard error of mean (SEM) number of antimicrobials to which isolates obtained from Addis Ababa were resistant was 7.23 ± 1.32, while isolates obtained outside of Addis Ababa were resistant to 4 ± 0.62 antimicrobials ($p = 0.01$). Resistance to first line antimicrobial agents in human medicine for treatment of *Salmonella* like beta-lactam and quinolones was also more common in isolates obtained from Addis Ababa. The extent of MDR varied with the serotype, as the overall MDR was more common in *S.* Kentucky, *S.* Virchow and *S.* Saintpaul compared to strains from other serotypes. Interestingly, all of the 5 Kentucky strains were resistant to nalidixic acid and ciprofloxacin (Table 4).

4. DISCUSSION

Food animals are the primary sources for transmitting non-typhoidal *Salmonella* to humans [27]. Outbreaks of salmonellosis in humans has been linked to improperly pasteurized dairy products, undercooked beef, water runoff from farms, and direct animal or fecal contact [28]. In the current study, farm level prevalence of *Salmonella* was 7.6 % and individual animal level prevalence was 2.3 %, which is much lower than the previous studies in the USA, where 31 % of dairy farms had at least one cow shedding *Salmonella* in feces and 7.3 % of individual animals were shedding [29]. It is also much lower than a previous study conducted in Addis Ababa [15] that reported farm level fecal prevalence of 47.8 % and individual animal level prevalence of 7.7 %. A study on slaughtered cattle in Addis Ababa recovered *Salmonella* from 7.1 % of apparently healthy animals [30]. Recent study in Jordan showed 23 % and 4 % of herd level and individual animal level prevalence of *Salmonella* in dairy farms, respectively, which is also higher than our finding [31]. This difference could be due to differences in the *Salmonella* isolation protocol employed in each study, seasonal variation in *Salmonella* shedding of animals as well as other factors such as herd size and age composition [28, 29]. Most of the farms in the current study had small herd size. Moreover, animals of all age groups in the farm were sampled in the current study unlike the other two studies [15, 29] which involved only lactating cows.

This study also showed that *Salmonella* shedding was common in farms that keep animals completely indoors while none was detected in those that occasionally graze outside or are totally outdoors. Similarly, higher prevalence of *Salmonella* was reported in swine kept indoors than those kept outdoors [32]. This probably is due to free cycling of *Salmonella* between animals in a limited host environment once the pathogen gets access to the farm in animals kept indoors. The fact that the use of processed feed is more common in animals kept indoors than those kept outdoors might also suggest the possibility of indoor kept animals being infected with *Salmonella* from contaminated animal feed. A previous study has also shown livestock waste generated by animals consuming a diet principally composed of grass were less likely to harbor *Salmonella* spp. [33].

In this study, the larger the herd size, the higher the probability of having *Salmonella* positive animals in the farm, which is in agreement with previous reports [34, 35, 36, 37]. This could be due to overcrowding of animals in the larger herds, especially those housed indoors, increasing animal to animal contact which enhances transmission of pathogens within the herd. Moreover, the larger the number of animals in the herd, the higher the probability of having a few weak and stressed animals, which increases the likelihood of continuous shedding of *Salmonella* from these cattle. Asymptomatic carrier cattle have been reported to shed *Salmonella* for up to 18 months [38]. Additionally, in the absence of mechanized feeding and milking systems in Ethiopia, several animal attendants are involved in daily activities of the large farms with the possibility of serving as a source of dissemination among individual animals in a farm. Contrary to the above findings, another study has reported that there is no association of herd size and *Salmonella* shedding [29].

The strong association of individual animal and herd level prevalence of *Salmonella* with detection of diarrhea in one or more animals suggests that *Salmonella* is one of the causes of diarrhea in dairy cattle in the study population. Detection of more *Salmonella* from diarrheic as well as non-diarrheic cattle in farms with one or more diarrheic animal in the herd might be due to the presence of carrier animals shedding *Salmonella* to other animals without showing clinical manifestations post infection or after recovery from clinical salmonellosis [39].

The higher *Salmonella* recovery rate in young animals in the current study is presumably due to the lack of an adequate adaptive immune response in the young calves compared to adult animals. Also, co-infection with multiple enteric pathogens is common in calves and may compromise their immune system. In addition, relative lack of protective microflora in calves may also predispose them to pathogenic organisms [37]. A previous study has similarly reported an inverse relationship of calf age and the prevalence of *Salmonella* [40].

The dominant serotypes isolated from dairy cattle in the current study, *S.* Typhimurium, *S.* Saintpaul, *S.* Virchow and *S.* Kentucky, are among the common causes of non-typhoidal salmonellosis in humans [41, 42, 43]. There is no previous report in Ethiopia showing serotype distribution of

Salmonella in dairy cattle. The top three serotypes in slaughtered cattle in Addis Ababa were *S.* Mishmarhaemek, *S.* Typhimurium and *S.* Enteritidis [30]. In another study conducted in north Ethiopia in slaughtered cattle, *S.* Typhimurium and *S.* Newport were the two dominant serotypes recovered [44].

The observed high resistance to streptomycin and tetracycline is not surprising since these antimicrobials are commonly used in most of the farms for management of bacterial infections. Similar high resistance rates were reported to streptomycin (77 %) and tetracycline (65.5 %) in *Salmonella* isolates obtained from different food animals from Ethiopia [45]. Another study also reported 75 and 46.9 % resistance to streptomycin and tetracycline in *Salmonella* isolated from different food items and personnel in Addis Ababa [46]. Though nitrofurantoin and sulfonamide were less commonly used in the farms during the study period, large proportion of isolates were resistant to these agents. This is probably due to the fact that these antimicrobials had been used in the animal health sector for a long time in the country and *Salmonella* had already developed resistance. A previous study conducted in Addis Ababa [15] showed 83 % of *Salmonella* isolates from dairy farms to be resistant to 2 or more antimicrobials out of 10 antimicrobials tested. Unlike the previous study [15] that reported 100 % resistance to ampicillin, in the current study, only 33.3 % of the isolates exhibited resistance to ampicillin. However, most of the isolates obtained from farms in Addis Ababa were resistant to ampicillin.

Resistance to ciprofloxacin was not reported in the previous study [15], but in the current study, 30 % of the isolates were resistant to ciprofloxacin. Despite detection of resistance to ciprofloxacin and nalidixic acid in the current study, from our interview with farm owners during sampling, none of the farms was using ciprofloxacin or other quinolone antimicrobials to treat their dairy cattle, and use of quinolones is not a regular practice in veterinary medicine in Ethiopia. However, a similar high percentage of resistance to ciprofloxacin was observed in *S.* Kentucky isolates carrying *Salmonella* Genomic Island K (SGI1-K) collected during 2000–2008 from meat of swine, cattle and poultry [45]. Also, recently isolated *S.* Kentucky strains from diarrheic patients in Addis Ababa were resistant to several antimicrobials including ciprofloxacin [47]. MDR *S.* Kentucky belonging to a single clone resistant to quinolones and carrying SGI1-K has been reported from European travelers returning from different African and Asian countries [48]. This occurrence of MDR *S.* Kentucky in both humans and animals in the region might be due to this specific clone widely circulating in Africa.

The high MDR in *Salmonella* from dairy cattle is alarming. The incidence of MDR in *Salmonella* has increased in the last few decades globally [49].

Infection of humans with MDR strains of *Salmonella* has been reported to be associated with increased burden of morbidity, extended hospitalization, increased risk of invasive illness and increased mortality, compared to those infected with susceptible strains [50, 51, 52]. In fact, the increase in MDR observed in *Salmonella* isolates from dairy farms in Addis Ababa compared to those out of Addis Ababa could be due to greater availability of antimicrobial agents and extensive use of antimicrobials in both animals and humans, fostered by a highly populated city where animals and humans live in close proximity.

5. CONCLUSION

The occurrence of MDR *Salmonella* serotypes that commonly cause human salmonellosis in dairy herds residing in close proximity with human populations warrants the need for strict biosecurity and intervention strategies to control these *Salmonella* isolates in dairy farms and to protect human and animal health. Paramount is the resistance to ciprofloxacin, which is of great concern as this antimicrobial is among the last options for treatment of complicated non-typhoidal salmonellosis in humans.

ACKNOWLEDGMENTS

This study was financially supported by The National Institutes of Health (NIH) Fogarty International Center (grant 043TW008650) to W.A.G./J.S.G. and WHO/AGISAR to T.E. We are grateful to Linda Cole, Shaun Kernaghan, Ketna Mistry, Ann Perets and Betty Wilkie of the Public Health Agency of Canada, National Microbiology Laboratory at Guelph for serotyping and phagetyping of the *Salmonella* isolates. Technical assistance of Mr. Nega Nigusie during sample collection is highly appreciated.

REFERENCES

1. Guibourdenche M, Roggentin P, Mikoleit M, Fields PI, Bockemühl J, Grimont PA, Weill FX. Supplement 2003–2007 (No. 47) to the White-Kauffmann-Le Minor scheme. Res Microbiol. 2010;161(1):26–9.
2. Hoelzer K, Moreno Switt AI, Wiedmann M. Animal contact as a source of human non-typhoidal salmonellosis. Vet Res. 2011;42:34.
3. Uzzau S, Brown DJ, Wallis T, Rubino S, Leori G, Bernard S, Casadesús J, Platt DJ, Olsen JE. Host adapted serotypes of Salmonella enterica. Epidemiol Infect. 2000;125(2):229–55.

4. Majowicz SE, Musto J, Scallan E, Angulo FJ, Kirk M, O'Brien SJ, Jones TF, Fazil A, Hoekstra RM, Studies ICoEDBoI. The global burden of nontyphoidal Salmonella gastroenteritis. Clin Infect Dis. 2010;50(6):882–9.
5. Bouchrif B, Le Hello S, Pardos M, Karraouan B, Perrier-Gros-Claude JD, Ennaji MM, Timinouni M, Weill FX. Ceftazidime-resistant Salmonella enterica, Morocco. Emerg Infect Dis. 2009;15(10):1693–5.
6. Fey PD, Safranek TJ, Rupp ME, Dunne EF, Ribot E, Iwen PC, Bradford PA, Angulo FJ, Hinrichs SH. Ceftriaxone-resistant salmonella infection acquired by a child from cattle. N Engl J Med. 2000;342(17):1242–9.
7. Bartholomew ML, Heffernan RT, Wright JG, Klos RF, Monson T, Khan S, Trees E, Sabol A, Willems RA, Flynn R, et al. Multistate outbreak of Salmonella enterica serotype enteritidis infection associated with pet guinea pigs. Vector Borne Zoonotic Dis. 2014;14(6):414–21.
8. Roy R, Higgins R, Fortin M, Tardif S. Salmonella give infection in 2 dairy herds. Can Vet J. 2001;42(6):468–70.
9. Wray C, Wary A. Salmonella in Domestic Animals. Wallingford: CABI Pub.; 2000.
10. Wells SJ, Fedorka-Cray PJ, Dargatz DA, Ferris K, Green A. Fecal shedding of Salmonella spp. by dairy cows on farm and at cull cow markets. J Food Prot. 2001;64(1):3–11.
11. Yilma Z, Faye B, Loiseau G. Occurrence and distribution of species of Enterobacteriaceae in selected Ethiopian traditional dairy products: a contribution to epidemiology. Food Control. 2007;18:1397–404.
12. Tigabu E, Asrat D, Kassa T, Sinmegn T, Molla B, Gebreyes W. Assessment of risk factors in milk contamination with Staphylococcus aureus in Urban and Peri-Urban small-holder dairy farming in Central Ethiopia. Zoonoses Public Health. 2015;62:637.
13. Adugna M, Asresie A. A review on microbiological quality of Ethiopian raw bovine milk. Food Sci Qual Manage. 2015;35:17–34.
14. Gupta A, Fontana J, Crowe C, Bolstorff B, Stout A, Van Duyne S, Hoekstra MP, Whichard JM, Barrett TJ, Angulo FJ, et al. Emergence of multidrug-resistant Salmonella enterica serotype Newport infections resistant to expanded-spectrum cephalosporins in the United States. J Infect Dis. 2003;188(11):1707–16.
15. Addis Z, Kebede N, Worku Z, Gezahegn H, Yirsaw A, Kassa T. Prevalence and antimicrobial resistance of Salmonella isolated from lactating cows and in contact humans in dairy farms of Addis Ababa: a cross sectional study. BMC Infect Dis. 2011;11:222.
16. Molla B, Sterman A, Mathews J, Artuso-Ponte V, Abley M, Farmer W, Rajala-Schultz P, Morrow WE, Gebreyes WA. Salmonella enterica in commercial swine feed and subsequent isolation of phenotypically and genotypically related strains from fecal samples. Appl Environ Microbiol. 2010;76(21):7188–93.
17. WHO. WHO Global Foodborne Infections Network Laboratory Protocol. In: Isolation of Salmonella spp. From Food and Animal Feace. 5th ed. 2010.
18. Cohen ND, Neibergs HL, McGruder ED, Whitford HW, Behle RW, Ray PM, Hargis BM. Genus-specific detection of salmonellae using the polymerase chain reaction (PCR). J Vet Diagn Invest. 1993;5(3):368–71.
19. Grimont PAD, Weill FX. Antigenic formulae of the Salmonella Serovars. 9th ed. Paris: Institut Pasteur; 2007.
20. Issenhuth-Jeanjean S, Roggentin P, Mikoleit M, Guibourdenche M, de Pinna E, Nair S, Fields PI, Weill FX. Supplement 2008–2010 (no. 48) to the White-Kauffmann-Le Minor scheme. Res Microbiol. 2014;165(7):526–30.
21. Ewing. Serological Identificaiton of Salmonella. 1986.
22. Shipp CR, Rowe B. A mechanised microtechnique for salmonella serotyping. J Clin Pathol. 1980;33(6):595–7.
23. Callow BR. A new phage-typing scheme for Salmonella typhi-murium. J Hyg (Lond). 1959;57:346–59.
24. Anderson ES, Ward LR, Saxe MJ, de Sa JD. Bacteriophage-typing designations of Salmonella typhimurium. J Hyg (Lond). 1977;78(2):297–300.
25. Rabsch W. Salmonella typhimurium phage typing for pathogens. Methods Mol Biol. 2007;394:177–211.

26. CLSI. Performance Standards for Antimicrobial Susceptibility Testing; Twenty-Third Informational SupplementM100-S23, vol. 33. 2013.
27. Branham LA, Carr MA, Scott CB, Callaway TR. E. coli O157 and Salmonella spp. in white-tailed deer and livestock. Curr Issues Intest Microbiol. 2005;6(2):25–9.
28. Callaway TR, Edrington TS, Anderson RC, Byrd JA, Nisbet DJ. Gastrointestinal microbial ecology and the safety of our food supply as related to Salmonella. J Anim Sci. 2008;86(14 Suppl):E163–72.
29. Fossler CP, Wells SJ, Kaneene JB, Ruegg PL, Warnick LD, Bender JB, Eberly LE, Godden SM, Halbert LW. Herd-level factors associated with isolation of Salmonella in a multi-state study of conventional and organic dairy farms I. Salmonella shedding in cows. Prev Vet Med. 2005;70(3–4):257–77.
30. Alemayehu D, Molla B, Muckle A. Prevalence and antimicrobial resistance pattern of Salmonella isolates from apparently healthy slaughtered cattle in Ethiopia. Trop Anim Health Prod. 2003;35(4):309–19.
31. Tarazi YH, Abo-Shehada MN. Herd- and individual-level prevalences of and risk factors for Salmonella spp. fecal shedding in dairy farms in Al-Dhulail Valley, Jordan. Trop Anim Health Prod. 2015;47(7):1241–8.
32. Bonde M, Sørensen JT. Salmonella infection level in Danish indoor and outdoor pig production systems measured by antbodies in meat juice and faecal shedding on-farm and at slaughter. Anim Health Anim Welf Biosecur Proc. 2007;2:729–34.
33. Hutchison ML, Walters LD, Avery SM, Munro F, Moore A. Analyses of livestock production, waste storage, and pathogen levels and prevalences in farm manures. Appl Environ Microbiol. 2005;71(3):1231–6.
34. Warnick LD, Kanistanon K, McDonough PL, Power L. Effect of previous antimicrobial treatment on fecal shedding of Salmonella enterica subsp. enterica serogroup B in New York dairy herds with recent clinical salmonellosis. Prev Vet Med. 2003;56(4):285–97.
35. Davison HC, Sayers AR, Smith RP, Pascoe SJ, Davies RH, Weaver JP, Evans SJ. Risk factors associated with the salmonella status of dairy farms in England and Wales. Vet Rec. 2006;159(26):871–80.
36. Blau DM, McCluskey BJ, Ladely SR, Dargatz DA, Fedorka-Cray PJ, Ferris KE, Headrick ML. Salmonella in dairy operations in the United States: prevalence and antimicrobial drug susceptibility. J Food Prot. 2005;68(4):696–702.
37. Cummings KJ, Warnick LD, Alexander KA, Cripps CJ, Gröhn YT, McDonough PL, Nydam DV, Reed KE. The incidence of salmonellosis among dairy herds in the northeastern United States. J Dairy Sci. 2009;92(8):3766–74.
38. Evans S, Davies R. Case control study of multiple-resistant Salmonella typhimurium DT104 infection of cattle in Great Britain. Vet Rec. 1996;139(23):557–8.
39. Cho YI, Yoon KJ. An overview of calf diarrhea - infectious etiology, diagnosis, and intervention. J Vet Sci. 2014;15(1):1–17.
40. Berge AC, Moore DA, Sischo WM. Prevalence and antimicrobial resistance patterns of Salmonella enterica in preweaned calves from dairies and calf ranches. Am J Vet Res. 2006;67(9):1580–8.
41. Okoro CK, Kingsley RA, Connor TR, Harris SR, Parry CM, Al-Mashhadani MN, Kariuki S, Msefula CL, Gordon MA, de Pinna E, et al. Intracontinental spread of human invasive Salmonella Typhimurium pathovariants in sub-Saharan Africa. Nat Genet. 2012;44(11):1215–21.
42. Bonalli M, Stephan R, Käppeli U, Cernela N, Adank L, Hächler H. Salmonella enterica serotype Virchow associated with human infections in Switzerland: 2004–2009. BMC Infect Dis. 2011;11:49.
43. Barton Behravesh C, Mody RK, Jungk J, Gaul L, Redd JT, Chen S, Cosgrove S, Hedican E, Sweat D, Chávez-Hauser L, et al. 2008 outbreak of Salmonella Saintpaul infections associated with raw produce. N Engl J Med. 2011;364(10):918–27.
44. Alemu S, Zewde BM. Prevalence and antimicrobial resistance profiles of Salmonella enterica serovars isolated from slaughtered cattle in Bahir Dar. Ethiopia Trop Anim Health Prod. 2012;44(3):595–600.

45. Eguale T, Marshall J, Molla B, Bhatiya A, Gebreyes WA, Engidawork E, Asrat D, Gunn JS. Association of multicellular behaviour and drug resistance in Salmonella enterica serovars isolated from animals and humans in Ethiopia. J Appl Microbiol. 2014;117(4):961–71.
46. Endrias Z, Poppe C. Antimicrobial resistance pattern of Salmonella serotypesisolated from food items and personnel in AddisAbaba, Ethiopia. Trop Anim Prod. 2009;41:241–9.
47. Eguale T, Gebreyes WA, Asrat D, Alemayehu H, Gunn JS, Engidawork E. Non-typhoidal Salmonella serotypes, antimicrobial resistance and co-infection with parasites among patients with diarrhea and other gastrointestinal complaints in Addis Ababa, Ethiopia. BMC Infect Dis. 2015;15:497.
48. Le Hello S, Hendriksen RS, Doublet B, Fisher I, Nielsen EM, Whichard JM, Bouchrif B, Fashae K, Granier SA, Jourdan-Da Silva N, et al. International spread of an epidemic population of Salmonella enterica serotype Kentucky ST198 resistant to ciprofloxacin. J Infect Dis. 2011;204(5):675–84.
49. Wright JG, Tengelsen LA, Smith KE, Bender JB, Frank RK, Grendon JH, Rice DH, Thiessen AM, Gilbertson CJ, Sivapalasingam S, et al. Multidrug-resistant Salmonella Typhimurium in four animal facilities. Emerg Infect Dis. 2005;11(8):1235–41.
50. Martin LJ, Fyfe M, Doré K, Buxton JA, Pollari F, Henry B, Middleton D, Ahmed R, Jamieson F, Ciebin B, et al. Increased burden of illness associated with antimicrobial-resistant Salmonella enterica serotype typhimurium infections. J Infect Dis. 2004;189(3):377–84.
51. Helms M, Simonsen J, Molbak K. Quinolone resistance is associated with increased risk of invasive illness or death during infection with Salmonella serotype Typhimurium. J Infect Dis. 2004;190(9):1652–4.
52. Varma JK, Molbak K, Barrett TJ, Beebe JL, Jones TF, Rabatsky-Ehr T, Smith KE, Vugia DJ, Chang HG, Angulo FJ. Antimicrobial-resistant nontyphoidal Salmonella is associated with excess bloodstream infections and hospitalizations. J Infect Dis. 2005;191(4):554–61.

CHAPTER 14

Contrasting clinical outcomes in two cohorts of cats naturally infected with feline immunodeficiency virus (FIV)

Paweł M. Bęczkowski[a,b,*], Annette Litster[c], Tsang Long Lin[d], Dominic J. Mellor[e], Brian J. Willett[a], Margaret J. Hosie[a]

[a] *MRC Centre for Virus Research, University of Glasgow, Glasgow, UK*

[b] *Small Animal Hospital, University of Glasgow, Glasgow, UK*

[c] *Department of Veterinary Clinical Sciences, Purdue University, West Lafayette, IN 47907, USA*

[d] *Indiana Animal Disease Diagnostic Laboratory and Department of Comparative Pathobiology, Purdue University, West Lafayette, IN 47907, USA*

[e] *School of Veterinary Medicine, University of Glasgow, Glasgow, UK*

ABSTRACT

Despite over 25 years of feline immunodeficiency virus (FIV) research, relatively little is known about the longitudinal course of FIV infection following natural infection. In contrast to published reports of experimental infections using lethal strains of the virus, clinical signs of naturally acquired FIV infection can be mild or inapparent, rather than life-threatening. In this prospective, longitudinal controlled study, based in Chicago, IL (n = 17) and Memphis, TN (n = 27), we investigated two cohorts of privately owned, naturally infected cats kept under different housing conditions. Cats in the Chicago cohort (Group 1) were kept in households of ≤2 cats, while the Memphis cohort (Group 2) comprised part of a large multi-cat household of over 60 cats kept indoors only, with unrestricted access to one another.

The majority of cats from Group 1 did not display clinical signs consistent with immunodeficiency during the 22-month observation period. In contrast, the outcome of infection in Group 2 was dramatically different; 17/27 (63%) of cats lost a median of 51.3% of their bodyweight ($P < 0.0005$) and died during the study period, with lymphoma being the most common cause of mortality.

Although the decrease in CD4+ T cell count between enrolment and terminal disease was significant ($P = 0.0017$), the CD4:CD8 ratio at the time of enrolment did not reliably distinguish FIV-positive cats classified as 'healthy' and 'not healthy' at either cohort. FIV load at enrolment was significantly lower in Group 1 than in Group 2 ($P < 0.0001$), but there were no significant differences at enrolment between healthy and not healthy cats at either group.

In conclusion, the results of this study suggest that management and housing conditions impact on disease progression and survival times of FIV-positive cats.

Keywords: FIV, Natural infection, Clinical outcome, FIV load, CD4:CD8 Lymphoma

1. INTRODUCTION

Feline immunodeficiency virus (FIV) is an important pathogen of domestic cats that is distributed worldwide. Prevalence estimates vary from 1% to 44%, depending on the geographical location as well as the entry criteria of each study (Hartmann, 1998; Ishida et al., 1989). Outdoor, adult, male, sick cats with bite wounds are at increased risk of infection, since transmission is mainly via bites from infected cats (Hartmann, 1998). Clinical signs are similar to those described for human immunodeficiency virus-acquired immune deficiency syndrome (HIV-AIDS) and include signs of immunosuppression, hematopoietic changes and neoplasia (Hartmann, 2011). The progression of experimental FIV infection can be monitored using the staging system used in HIV infection: acute, persistent generalized lymphadenopathy, asymptomatic carrier phase (AC), AIDS-related complex (ARC) and AIDS (English et al., 1994). While this classification is useful in well-controlled, experimental settings, it has limitations under field conditions, mainly because of strain-related variability in pathogenicity, variation in the ages of cats at the time of infection as well as possible exposure to secondary pathogens (Hartmann, 2012; Pedersen et al., 2001; Podell et al., 1997). Hence, there have been variable morbidity and mortality rates reported in naturally infected cats (Addie et al., 2000; Hofmann-Lehmann et al., 1997; Kohmoto et al., 1998; Liem et al., 2013).

Providing an accurate prognosis for FIV-positive cats is important so that optimal care can be provided by veterinarians and cat owners. Surrogate markers for HIV infection are well established and valuable in monitoring both responses to therapy and the likelihood of disease progression (Churchill, 1997; Katzenstein, 2003; Okonji et al., 2012; Shaunak and Teo, 2003; Smith and Stein, 2002). In contrast, to our knowledge, prospective controlled studies investigating the course of FIV infection and markers of disease progression have not been described. Consequently, the longitudinal course of naturally acquired FIV infection is poorly described and the variability in rates of progression to immunodeficiency are poorly understood.

In this controlled study, we examined clinical and laboratory parameters from two cohorts of cats living in different locations and kept under different housing conditions. The aim of the study was to report clinical findings, post-mortem findings, bodyweight, CD4+ T cell counts, FIV load and phylogenetic classification of viral envelope genes in two groups of cats naturally infected with FIV.

2. MATERIALS AND METHODS

2.1. Cats

Forty-four neutered, FIV-positive (SNAP FIV/FeLV Combo Test (IDEXX Laboratories, Westbrook, MN) cats were enrolled, with an equal number of age- and sex-matched FIV-negative cats from the same geographical locations. In order to obtain an adequate sample size FIV-positive cats and matched control cats during the 2-year enrolment period, it was necessary to source cats from two separate locations. FIV-positive results were further confirmed by virus isolation (Hosie et al., 2009). All cats were feline leukaemia virus (FeLV) antigen negative at enrolment. The number of cats enrolled into the study was capped at 44 due to the capacity of the laboratory at which molecular analyses were performed.

Seventeen of the 44 FIV-positive cats enrolled (Group 1) had been previously adopted from a large metropolitan adoption-guarantee shelter (PAWS Chicago) and lived in single-cat households in Chicago, IL, USA except for seven cats: two cats (C7 and C4) cohabited in a two-cat household; one cat (C13) lived in a two-cat household with a FIV-negative cat enrolled in the study; one cat (C9) was housed at PAWS Chicago for the first 11 weeks of the study and then was adopted into a house with an FIV-positive cat not enrolled in the study; and three cats (C2, C15 and C21) were housed at PAWS Chicago in a room containing up to three FIV-positive cats before they were each adopted into single cat households at 2, 14 and 58 weeks after enrolment, respectively.

The remaining twenty-seven FIV-positive cats enrolled (Group 2) were housed together in a large multi-cat household operating as an FIV-positive cat rescue in Memphis, TN, where a total of 53 FIV-positive and 10 FIV-negative cats were housed indoors with unrestricted access to one another. None of 10 FIV-negative cats from the Memphis cat rescue died during the study period. Since these cats were not age- or sex-matched for any of the enrolled FIV-infected cats from the same household (Group 2), their health status was not monitored using the study protocol. All enrolled FIV-negative cats at both locations lived in households of three cats or fewer.

The study and its aims were reviewed and approved by University of Glasgow Ethics Committee and the Purdue Animal Care and Use Committee. Cat owners provided written informed consent for their participation in the study.

2.2. Study timeline and collection dates

At enrolment, all cats underwent a general physical examination by a registered specialist in feline medicine (AL) and blood was collected for determination of FIV status and laboratory analysis. Oral examinations were conducted as part of the physical examination and a gingival score (0–3/3; Lobprise, 2007) was assigned. At the time of enrolment, FIV-positive cats were classified as 'healthy' if there were no abnormalities on physical examination and their gingival score was 0/3 or 1/3. Cats were classified as 'not healthy' if at least one clinical abnormality was detected, or if their gingival score was 2/3 or 3/3. The clinical classification at enrolment was based solely on physical examination findings and remained with each cat for the 22-month study period, regardless of any further changes. An entry criterion for age- and sex-matched FIV-negative cats was healthy status at the enrolment examination.

Serial physical examinations and specimen collections were made at 6-monthly intervals in FIV-positive cats over the study period. Physical examinations and specimen collections were performed at enrolment and

12 months later in FIV-negative cats. The date of the first known positive FIV ELISA test was obtained for all FIV-positive cats and the time between first diagnosis and study enrolment (months) was calculated. Abnormalities detected during the clinical examinations are listed in Table 1.

Table 1. Clinical abnormalities detected on physical examination in FIV-positive cats ($n = 44$) over the 22-month study period.

Description	**Number of cats with clinical abnormalities (%)**		
	Group 1 (*n* = 17)	**Group 2 (*n* = 27)**	**All FIV-positive cats (*n* = 44)**
Dental/oral cavity (faucitis)	12 (71)	19 (70)	31 (70)
Skin (alopecia, dermatitis)	10 (59)	12 (44)	22 (50)
Ears	3 (18)	4 (15)	7 (16)
Ocular disease	2 (12)	5 (19)	7 (16)
Cardiovascular	3 (18)	2 (7)	5 (11)
Digestive tract	0 (0)	4 (15)	4 (9)
Respiratory	0 (0)	2 (7)	2 (5)
Lymph nodes	0 (0)	0 (0)	0 (0)
Urogenital	0 (0)	0 (0)	0 (0)
Neurologic	0 (0)	0 (0)	0 (0)
Musculoskeletal	0 (0)	0 (0)	0 (0)
Weight loss/anorexia	0 (0)	22 (81)	22 (50)

2.3. Laboratory and post-mortem examinations

Flow cytometric analyses were performed as previously described (Lin and Litster, 2013). FIV loads were measured using a commercially available PCR test (IDEXX FIV RealPCR Test, IDEXX Laboratories, West Sacramento, CA). The assay detects the presence of viral nucleic acid, including both genomic DNA and viral RNA, in peripheral blood leukocytes with 80.5% sensitivity and 99.9% specificity (IDEXX Laboratories, West Sacramento, CA).

Post-mortem examinations were performed by a specialist veterinary pathologist (TLL).

2.4. Phylogenetic analysis

Maximum likelihood (ML) phylogenetic trees were constructed in MEGA5 (Tamura et al., 2011) under the HKY nucleotide substitution model, selected through jMODELTEST (Posada, 2008). Statistical support for the ML trees was estimated using 1000 bootstrap replicates (Efron et al., 1996). Multiple *env* sequences (n = 355), from cats enrolled in the study were subjected to rigorous recombination testing as described previously (Bęczkowski et al., 2014).

2.5. Statistical analysis

Graphing and statistical data analyses were performed using commercially available software (GraphPad Prism version 5.00, GraphPad Software). Descriptive data were shown as medians and interquartile range (IQR; median, 5th and 95th quartiles). Given the relatively small sample size, and after inspection of the data distributions, Mann–Whitney and Wilcoxon matched pairs tests were used to test hypotheses regarding differences in laboratory parameters between and within cat groups. Binary data were analyzed using Fisher's exact test. FIV load data were tested for correlation using Spearman correlation tests. Kaplan–Meier curves were compared using the Mantel–Cox 'log-rank' test and tested with the log-rank test for trends. Significance was set at $P < 0.05$.

3. RESULTS

3.1. Cats

Table 2 provides breed, gender, age and time since first FIV-positive test data for all FIV-positive cats and for cats from Group 1 and Group 2. Each matching pair of FIV-positive and FIV-negative cats was aged within 2 years of each other. All enrolled cats remained FeLV-negative over the study period and none of the FIV-negative cats seroconverted.

Table 2. Breed, gender, age and age at first known positive feline immunodeficiency virus (FIV) ELISA test at enrolment.

	Group 1 (*n* = 17)	Group 2 (*n* = 27)	All FIV-positive cats (*n* = 44)
Breed	DSH (94%), DLH (6%)	DSH (74%), DLH (15%), Siamese X (11%)	DSH (82%), DLH (11%), Siamese X (7%)
Gender	FS (18%), MN (82%)	FS (33.3%), MN (66.6%)	FS (27%), MN (73%)
Age (years; median/range)	4/1–9	5/2–10	4.5/1–10
Age at first known positive FIV ELISA test (years; median/range)	2.3/0.2–7.8	1.6/0–4.6	1.8/0–7.8
Time from first known positive FIV ELISA test to enrolment (years; median/range)	1.3/0–3.3	3/0.1–8	2/0–8

DSH, domestic shorthair; DLH, domestic longhair; X, cross; FS, female spayed; MN, male neutered.

3.2. Health status

In the FIV-positive group at the time of enrolment, there were equal numbers of healthy ($n = 22$) and not healthy animals ($n = 22$). In Group 1, 10 cats were classified as healthy (59%) and seven cats were classified as not healthy (41%); all but one cat remained in this classification during the observation period. In Group 2, 12 cats were initially classified as healthy (44%) and 15 cats that were classified as not healthy (56%) remained in these classifications throughout the study. However, 63% of cats (17/27) from Group 2 experienced severe weight loss and died during the study period. Half (6/12) of the cats classified as healthy and 73% (11/15) of cats classified as not healthy died during the study ($P = 0.26$). Almost half of cats diagnosed at post-mortem with lymphoma (44%; 4/9) had been classified as healthy at the time of enrolment (Table 3).

Table 3. Health status on the day of enrolment and post-mortem diagnoses for FIV-positive cats deceased during the study (Group 1, $n = 1$; Group 2, $n = 17$).

Cat #	Health status at enrolment	Post-mortem diagnosis
M3	NH	Lymphoma – retrobulbar
M5	NH	Lymphoma – bone marrow

M10	H	Lymphoma – bone marrow and lymph nodes
M11	H	Lymphoma – multicentric
M14	NH	Lymphadenomegaly; bone marrow hyperplasia
M16	NH	Pulmonary congestion and oedema with focal bacterial bronchopneumonia
M25	H	Upper and lower bacterial respiratory infection, probably secondary to viral infection
M26	NH	Chronic focal granulomatous nematode larval pneumonia; bone marrow hyperplasia
M30	NH	Lymphoid depletion; Aelurostrongylus abstrusus pneumonia; left front leg suppurative dermocellulitis; bone marrow hyperplasia, with erythroid and myeloid hyperplasia
M31	NH	Lymphoma – bone marrow
M33	NH	No post-mortem performed
M41	NH	Lymphoma – bone marrow and lymph nodes
M44	NH	Emaciation/nephritis/pancreatitis
M46	H	Lymphoma – bone marrow
M48	NH	Lymphoma – bone marrow and lymph nodes
M49	H	Lymphoma – bone marrow
M50	H	Undetermined/emaciation
C2	H	Cardiac failure, hypertrophic cardiomyopathy

C, Chicago, Group 1; M, Memphis, Group 2; H, healthy on the day of enrolment (no abnormalities detected on a physical examination); NH, not healthy on the day of enrolment (at least one abnormality detected on a physical examination).

3.3. Post-mortem findings

During the 22-month study period, 1/17 (5.9%) FIV-positive cats from Group 1 and 17/27 (63%) FIV-positive cats from Group 2 died. Fig. 1 illustrates a Kaplan–Meier survival plot of both cohorts. The cats that died were examined post-mortem and the findings are shown in Table 3. In Group 2, post-mortem data were not available for one cat (M33).

The most common pathological finding at post-mortem in Group 2 was lymphoma; various anatomical types were documented in 9/16 (56%) cats from Group 2 (Table 4). Four cats had lymphoma limited to a single site (retro-bulbar site in one case and bone marrow in three cases), while in five other cats, varying degrees of dissemination were observed. Bone marrow involvement was noted in all but one case. The second most common finding was nodal lymphoma, diagnosed in 5/9 cats (56%) and accompanied by splenic lymphoma in 3/9 cases (33%). One cat (M11) had a disseminated form involving the kidneys, liver, jejunum, heart, trachea and tongue. The median time since first FIV-positive ELISA test to death in cats with lymphoma was 5.3 years (range, 2.9–9 years).

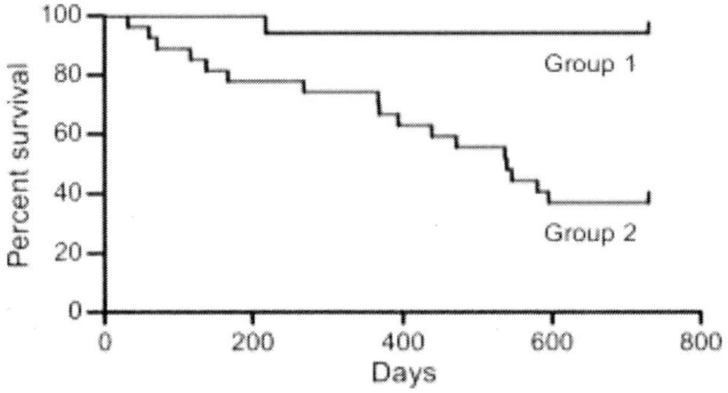

Figure. 1. Kaplan–Meier survival plots for FIV-positive cats from Group 1 (black; *n* = 17) and Group 2 (red; *n* = 27). Survival rates in Group 1 and Group 2 over the 22-month study period were 94% and 37%, respectively (*P* = 0.0006, Gehan–Breslow–Wilcoxon test). (For interpretation of the references to colour in this figure legend, the reader is referred to the web version of this article.)

Table 4. Organs with neoplastic lymphocyte infiltrates among FIV-positive cats diagnosed with lymphoma at post-mortem.

Cat #	Bone marrow	Lymph node	Spleen	Kidney	Liver	Jejunum	Tongue	Trachea	Heart	Eye
M3										+
M5	+	+	+							
M10	+	+								
M11	+	+	+	+	+	+	+	+	+	
M31	+									
M41	+	+	+							
M46	+									
M48	+	+								
M49	+									

Lymphoid hyperplasia and lymphoid depletion were noted concurrently in 13/16 (81%) cats. These findings were documented in various tissues, mostly in the spleen and lymph nodes, but in four cats, Peyer's patch atrophy was also diagnosed. Erythroid bone marrow hyperplasia was documented in 9/16 cats (56%), as was myeloid bone marrow hyperplasia. Multifocal lymphocytic interstitial nephritis was diagnosed in six cats (37.5%). In four cats, there were pathological signs of respiratory tract infection and pneumonia. Hypertrophic cardiomyopathy and signs of congestive heart failure were identified in the single FIV-positive cat from Group 1 that died.

None of the FIV-negative cats died during the study period.

3.4. Bodyweight

The bodyweight at the time of death for the single FIV-positive cat from Group 1 cohort that died was 15% heavier than the bodyweight recorded at enrolment (Fig. 2). The 17 FIV-positive cats from Group 2 that died lost a median of 51.3% of their bodyweight between enrolment (median 3.9 kg, range 2.27–5.54 kg) and death (median 1.9 kg, range 1.45–3.72 kg; $P < 0.0005$), over a median time span of 15 months (range 1.6–20 months). Those cats lost a median of 10.9% of their bodyweight between enrolment and 3 months before death (range −26.3% to +9.3%); median bodyweight loss in the last 3 months of life was 29.5% (range −55.4% to −8.9%).

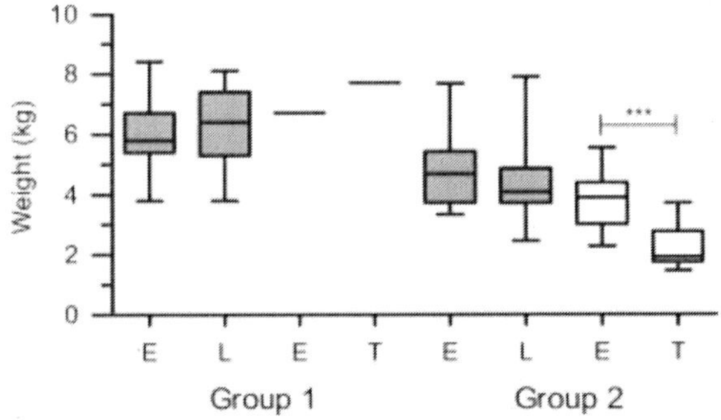

Figure. 2. Bodyweights of FIV-positive cats from Group 1 (*n* = 17) and Group 2 (*n* = 27). E, enrolment weight; L, weight at the end of the study period; T, terminal weight. Grey boxes denote cats alive at the end of the 22-month study period; White boxes denote cats dead at the end of the 22-month study period. (The central bar in the box denotes the median, while the top and bottom of the box represent the 75th and 25th centiles, respectively. The upper and lower bars represent the 95th and 5th centiles, respectively. ***$P < 0.0005$.)

The 16 FIV-positive cats from Group 1 that survived for the entire study period gained a median of 7.6% bodyweight over the 12 months following enrolment (enrolment, median 5.9 kg, range 3.8–8.4 kg; 12 months later 6.35 kg, range 3.9–8.1 kg; $P = 0.2089$). In contrast, the 10 FIV positive cats from Group 2 that remained alive lost a median of 12.8% of their bodyweight over the 12 months following enrolment (enrolment, median 4.68 kg, range 3.34–7.67 kg; 12 months later, median 4.08 kg, range 2.45–7.89 kg; $P = 0.2324$).

FIV-negative cats from Group 1 for which bodyweight data were available ($n = 16$) gained a median 16.8% bodyweight between enrolment (median 5.35 kg, range 4–8.5 kg) and 12 months later (median 6.25 kg, range 4.1–8 kg; $P = 0.0145$). Similarly, FIV negative cats from Memphis for which bodyweight data were available ($n = 24$) gained a median of 24.5% bodyweight between enrolment (median, 4.9 kg, range 2.7–9 kg) and 12 months later (median 6.1 kg, range 2.6–9.9 kg; $P = 0.0022$).

Available bodyweight data from Group 2 demonstrated that enrolment weights for FIV-positive cats from Group 2 that died were significantly lower than those for matched FIV-negative cats ($n = 17$; FIV-positive, median 3.9 kg, range 2.27–5.54; FIV-negative, median 5.6 kg, range 2.7–6.8 kg; $P = 0.0046$). Bodyweights were not statistically different when FIV-positive survivors from Group 2 and matching FIV-negative cats were compared at enrolment ($n = 10$; FIV-positive, median 5.12 kg, range 3.34–7.68 kg; FIV-negative median 5.8 kg, range 4.6–9.0 kg; $P = 0.1641$) and 12 months later (FIV-positive, median 4.63 kg, range 3.63–8.27 kg; FIV-negative, median 6.2 kg, range 4.9–9.9 kg; $P = 0.0977$).

3.5. CD4+ T cell counts and CD4:CD8 ratio

Fig. 3 illustrates the CD4:CD8 ratios obtained over 12 months for the FIV-positive cats from Group 2 with reference values acquired from FIV-negative cats classified as healthy from the same geographic location. The difference between FIV-positive (median 0.89, range 0.19–2.24) and FIV-negative cats (median 1.71, range 1.03–2.62) was statistically significant at the time of enrolment ($P < 0.0001$) and 12 months later (medians 0.69, range 0.11–1.54 and 1.48, range 0.88–2.77 for FIV-positive and FIV-negative cats, respectively; $P < 0.0001$). The CD4:CD8 ratio in surviving FIV-positive cats was maintained over 18 months at a consistently low level (median values: 0.89, 0.73, 0.69, and 0.81 for each time point).

CD4:CD8 ratio at the time of enrolment ($P < 0.0005$) was lower in the FIV-positive cats from Group 1 (median 0.77, range 0.08–1.27) than in matched FIV-negative cats (median 2.18, range 1.23–6.42; Fig. 3). CD4:CD8 ratios in the FIV-positive cats from Group 1 were maintained at a consistently low level over the 12 months following enrolment (median 0.65, range 0.2–

2.55) compared to age- and sex-matched FIV-negative cats (median 1.8, range 1.4–3.93; $P = 0.0054$). The CD4:CD8 ratios at the time of enrolment in the FIV-positive cats from Group 1 were lower (median 0.77, range 0.08–1.27) than those of the FIV-positive cats from Group 2 (median 0.89, range 0.19–2.23; Fig. 3).

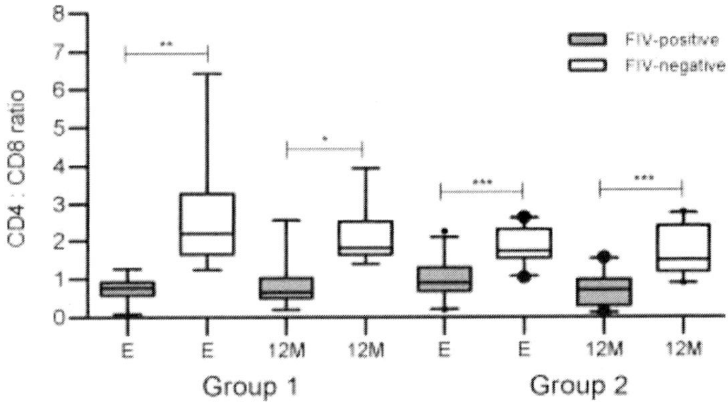

Figure. 3. CD4:CD8 ratios for FIV-positive cats from Group 1 and Group 2 (in grey) and age- and sex-matched FIV-negative cats from the same geographic location (in white). For Group 1, data represent 17 FIV-positive cats and 17 age- and sex-matched FIV-negative cats at enrolment (E) and 16 surviving FIV-positive and 16 FIV-negative cats at 12 months after enrolment (12M). In Group 2 at enrolment, data were collected from 27 FIV-positive cats and 27 FIV-negative cats; 12 months after enrolment, data were collected from 21 surviving FIV-positive cats and 21 FIV-negative cats. (The central bar in the box denotes the median, while the top and bottom of the box represent the 75th and 25th centiles, respectively. The upper and lower bars represent the 95th and 5th centiles, respectively. $^{*}P < 0.01$; $^{**}P < 0.001$; $^{***}P < 0.0001$.)

There were no statistically significant differences when the CD4:CD8 ratio was compared between cats classified as healthy and not healthy at enrolment at either group, or for all FIV-positive cats as a single group (Group 1, $P = 0.97$; Group 2, $P = 0.84$; all FIV-positive cats, $P = 0.99$; Fig. 4).

To determine the kinetics of the CD4:CD8 ratio inversion over the course of infection, we compared absolute CD4+ and CD8+ T lymphocyte numbers from 17 cats from Group 2 that died during the study (Fig. 5). The decrease in CD4+ T cell count was statistically significant when enrolment results (median 0.41 K/μL, range 0.13–1.38 K/μL) were compared with terminal results (median 0.18 K/μL, range 0.04–1.32 K/μL; $P = 0.0017$), but this was not the case for the decrease in CD8+ T cell count from enrolment (median 0.44 K/μL, range 0.18–1.49 K/μL) to terminal disease (median 0.38 K/μL, range 0.06–1.68 K/μL; $P = 0.35$). Therefore, the decrease in the CD4:CD8 ratio was attributed to a reduction in the CD4+ T cell count. Further analysis of total lymphocyte counts in deceased cats showed that the decrease in absolute lymphocyte numbers between enrolment (median 1.73 K/μL, range 1.15–4.03 K/μL) and terminal disease (median 1.25 K/μL, range 0.25–3.55 K/μL) reflected not only the loss of CD4+ T cells, but also depletion of CD21+ B cells (enrolment, median CD21+ B cell count 0.26 K/μL, range 0.12–0.89 K/μL; terminal specimens, median CD21+ B cell count 0.17 K/μL, range 0.04–0.37 K/μL; $P = 0.0061$).

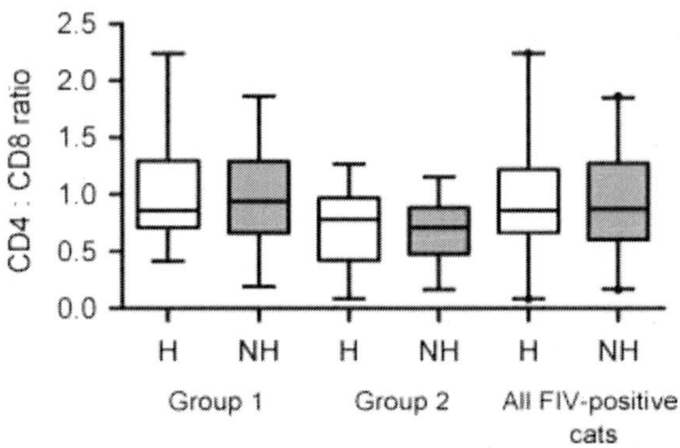

Figure. 4. CD4:CD8 ratio at enrolment for FIV-positive cats from Group 1 ($n = 17$), Group 2 ($n = 27$) and all FIV-positive cats ($n = 44$). Cats were classified as healthy (H) or not healthy (NH) based on physical examination findings at the time of enrolment. (The central bar in the box denotes the median, while the top and bottom of the box represent the 75th and 25th centiles, respectively. The upper and lower bars represent the 95th and 5th centiles, respectively.)

3.6. FIV load (FIV genomes/mL blood)

The FIV load at enrolment was significantly lower in Group 1 (median, 741 genomes/mL blood; range 43–18,796 genomes/mL blood) than Group 2 (median, 50,003 genomes/mL blood; range, 2540–3,792,000 genomes/mL blood; $P < 0.0001$; Fig. 6). There were no statistically significant differences in FIV load at enrolment when healthy and not healthy cats were compared in Group 1 (healthy, $n = 8$, median 717 genomes/mL blood; range, 43–18,796 genomes/mL blood; not healthy, $n = 7$, median 767 genomes/mL blood; range, 395–18,796 genomes/mL blood; $P > 0.05$) or Group 2 (healthy, $n = 11$, median 64,843 genomes/mL blood; range, 2540–1,667,000 genomes/mL blood; not healthy, $n = 15$, median 41,387 genomes/mL blood; range, 13,423–3,792,000 genomes/mL blood; $P > 0.05$). In Group 2, there were no significant differences in FIV load at enrolment when the following comparisons were made: cats still alive vs. cats that had died by the end of the study period (alive, $n = 9$, median 31,395 genomes/mL blood; range, 2540–41,8671 genomes/mL blood; dead, $n = 17$, median 176,551 genomes/mL blood; range, 7724–3,792,000 genomes/mL blood; $P > 0.05$); cats that had died by the end of the study period and were classified at enrolment as either healthy or not healthy (healthy, $n = 6$, median 241,757 genomes/mL blood, range, 7724–1,667,000 genomes/mL blood; not healthy, $n = 11$, median 176,551 genomes/mL blood; range, 13,423–3,792,000 genomes/mL blood; $P > 0.05$); and cats that were still alive by the end of the study period and were classified at enrolment as either healthy or not healthy (healthy, $n = 5$, median 56,867 genomes/mL blood, range, 2540–418,671 genomes/mL blood; not healthy, $n = 4$, median 27,771 genomes/mL blood, range 16,176–43,138 genomes/mL blood; $P > 0.05$).

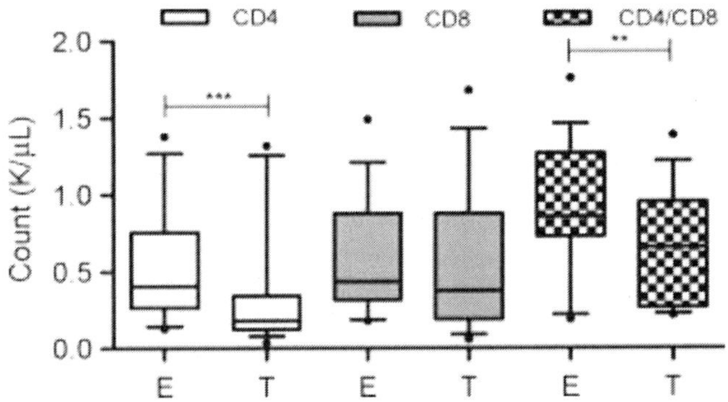

Figure. 5. CD4+ and CD8+ T cell count and CD4:CD8 ratio at enrolment (E) and terminally (T) from the FIV-positive cats from Group 2 that died during the study ($n = 17$). (The central bar in the box denotes the median, while the top and bottom of the box represent the 75th and 25th centiles, respectively. The upper and lower bars represent the 95th and 5th centiles, respectively. $^{**}P = 0.0034$; $^{***}P = 0.0017$.)

Statistical correlations between FIV load and CD4+ T cells, CD8+ T cells and the CD4:CD8 ratio were investigated for each group. There were statistically significant, or close to significant, negative correlations between the CD4:CD8 ratio and FIV load in Group 1 ($r = -0.5022$; $P = 0.0564$) and in Group 2 ($r = -0.5479$; $P = 0.0038$; Fig. 7).

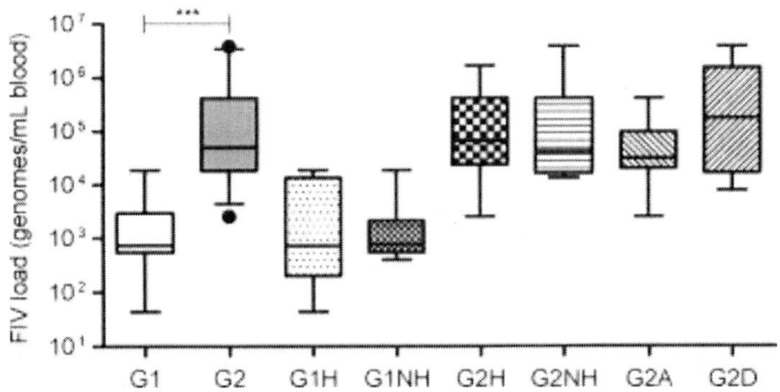

Figure. 6. FIV load at the time of enrolment (genomes/mL blood). Statistical comparisons were made between pairs of classifications across the graph. G1, Group 1 (n = 15); G2, Group 2 (n = 26); H, healthy on the day of enrolment (no abnormalities detected on a physical examination; Group 1, n = 8, Group 2, n = 11); NH, not healthy on the day of enrolment (at least one abnormality detected on a physical examination; Group 1, n = 7, Group 2, n = 15); A, alive at the end of the 22-month study period (Group 2, n = 9); D, cats dead at the end of the 22-month study period (Group 2, n = 17). ***P < 0.0001.

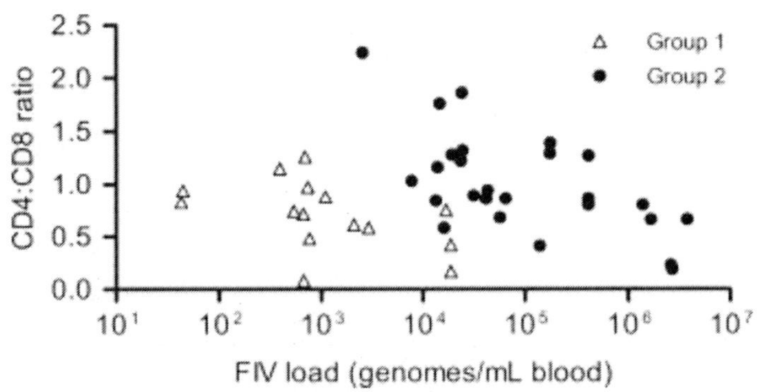

Figure. 7. Correlation between FIV load (FIV genomes/mL blood) and CD4:CD8 ratio for FIV-positive cats from Group 1 (n = 15; r = −0.5022, P = 0.0564) and Group 2 (n = 26; r = −0.5479, P = 0.0038).

3.7. Phylogenetic classification of FIV envelope genes

Full-length viral envelope (*env*) gene sequences from FIV-positive cats were examined (n = 355). In Group 1, clade B *env* viruses and clade A/B recombinant *env* viruses each accounted for 50% of infections. In Group 2,

69% of cats were infected with viruses bearing clade B *env*, while 31% of cats were infected with recombinant *env* viruses (Supplementary Table 3).

4. DISCUSSION

Here we describe contrasting clinical outcomes of naturally acquired FIV infection, demonstrating significant differences between cats enrolled from two separate cohorts. The relationship between FIV infection and clinical disease is unclear, with some studies reporting that infected cats are at increased risk of morbidity and mortality (Muirden, 2002), while others report no such association (Akhtardanesh et al., 2010; Liem et al., 2013). Considering all 44 FIV-positive cats, there was no relationship between mortality and morbidity and FIV infection; however, this was not the case when the two cohorts were examined separately. Our observations in Group 2 indicated that living conditions might have played a role in the more rapid disease onset. In contrast, cats from Group 1 were generally housed in single cat households and remained alive and in relatively good health over the study period. Therefore, it appears that FIV infection is more likely to progress in cats kept in crowded shelter conditions compared to those living in spacious environments. As a corollary, our results suggest that the FIV-positive cats in our study that lived in single or dual cat households remained free of clinical signs of illness over the study period; this can most likely be attributed to lower risks of exposure to opportunistic pathogens and reduced levels of environmental stress.

FIV infected cats are at greater risk of opportunistic infections and neoplasia (Hartmann, 2012). Regular assessment of their health status is essential and, according to European Advisory Board on Cat Diseases guidelines (Hosie et al., 2009), health checks are recommended every 6 months. The short time to disease progression leading to the deaths of 63% of cats in Group 2 within the 22-month follow-up period was unexpectedly high, especially in light of previously published studies that were unable to demonstrate a difference in lifespan between FIV-positive and FIV-negative cats (Addie et al., 2000; Ravi et al., 2010). The health status of cats from Group 1 was similar to that of cats in a household in which FIV infection did not reduce life expectancy over a period of 2 years (Addie et al., 2000). There are several clinical abnormalities associated with FIV infection, with gingivostomatitis being the most common. Although one study (Quimby et al., 2008) did not observe an association between FIV infection and gingivostomatitis, the high prevalence observed here was in agreement with previous reports (Bandecchi et al., 1992; Hofmann-Lehmann et al., 1995; Hosie et al., 2009; Knotek et al., 1999). As gingivostomatitis is rare in SPF cats experimentally infected with FIV, co-infections with other infectious agents, such as feline calicivirus, might have contributed to this syndrome in naturally infected cats (Reubel et al., 1994; Tenorio et al., 1991). It is perhaps relevant that severe acute on chronic upper respiratory tract disease was diagnosed in four cats from Group 2 at post-mortem. Other abnormalities previously attributed to FIV-induced dysregulation and impairment of immune surveillance include dermatitis, ocular disease, renal insufficiency, lower urinary tract infections and other opportunistic infections (Barlough et al., 1991; Reche et al., 2010). Although neurological abnormalities have been reported in natural and experimental, acute and chronic FIV infections (Abramo et al., 1995; Dow et al., 1990; Podell et al., 1997, 1999), neither behavioural nor motor abnormalities were observed in our study population.

The mortality rate of 63% in the FIV-positive cats from Group 2 was markedly higher than the FIV-positive cats from Group 1, in which only one cat died during the study period (6%). Post-mortem examinations identified several pathological changes in the FIV-infected cats. Although FIV infection is generally associated with immunosuppression, immune hyperstimulation is evident during the early stages of infection, including B cell activation, manifested as polyclonal hypergammaglobulinemia (Hopper et al., 1989; Takano et al., 2012), expansion of CD8+ T cells with an activated phenotype and subsequent CD8+ T lymphocytosis (Beatty et al., 1996; Willett et al., 1993). Lymph node follicular hyperplasia and dysplasia was documented post-mortem in this study, confirming the findings of other studies of acute and terminal stage disease (Matsumura et al., 1993; Shelton et al., 1989; Yamamoto et al., 1988). In addition, reactive changes in the bone marrow, observed in 56% of cats in this study, have been documented previously in HIV and FIV infections (Bain et al., 2009; Breuer et al., 1998; Geller et al., 1985).

Lymphoma was the most common condition diagnosed at post-mortem in Group 2 (9/16 cats; 56%), which raises questions regarding potential interactions between virus, host, environment, and other factors that could have resulted in a predisposition to neoplasia. FIV strain variability was examined, since the Memphis FIV strains (Group 2) might have had greater oncogenic potential than the strains infecting the Chicago cohort (Group 1). However, a comparison of full-length *env* gene and long terminal repeat sequences revealed no significant differences (unpublished observation). It is possible that the local cluster of lymphoma in Group 2 was associated with another infectious agent, such as FeLV, but this seemed unlikely since FIV/FeLV Snap test (IDEXX) results were negative for FeLV antigen. However, we cannot exclude the possibility of regressive FeLV infection (Hofmann-Lehmann et al., 2001), since PCR testing to detect integrated proviral DNA in bone marrow was not performed. Breed was unlikely to play a role, as the cats from Group 2 diagnosed with lymphoma and those from Group 1 were all classified as domestic shorthairs and came from a variety of sources within their geographic area. The shelter environment in which the cats from Group 2 lived was likely to play a role in the development of the various clinical manifestations of FIV, although exposure to potential carcinogens cannot be ruled out.

Marked weight loss between enrolment and terminal measurement was evident among the cats that died over the study period, whereas healthy FIV-infected cats maintained relatively stable bodyweights (Fig. 2). Although monitoring bodyweight regularly can be a valuable tool in the care of FIV-positive cats and appeared to have some prognostic value in our study of FIV-positive cats, declining bodyweight is a relatively non-specific finding, as it is not unusual to observe weight loss during the progression of various other diseases.

The hallmark of FIV infection is an inverted CD4:CD8 ratio, which, as demonstrated here and elsewhere (Hoffmann-Fezer et al., 1992; Novotney et al., 1990; Tompkins et al., 1991), is largely the consequence of a loss of CD4+ T cells. Here we provide further evidence that FIV-positive cats had significantly lower CD4:CD8 ratios than those for FIV-negative cats over the 22-month study period in each of the cohorts (Fig. 3). We also observed a statistically significant decline in the CD4:CD8 ratio between the time of enrolment and the terminal sampling in those cats that died over the study period (Fig. 5). However, there was not a significant difference in median CD4:CD8 ratio at enrolment when cats classified as either healthy or not healthy were compared (Fig. 4). Both groups of cats had relatively low CD4:CD8 ratios with median values less than 1, regardless of their health status. Indeed, the cat that had the lowest CD4+ T lymphocyte count (0.09 K/μL) and the lowest CD4:CD8 ratio (0.08) at enrolment was classified as healthy and is still alive. Similar observations have been reported previously in FIV (Novotney et al., 1990) and HIV infections (Levy, 1988), where seropositive individuals remained asymptomatic despite significant loss of CD4+ T lymphocytes. The variability in CD4:CD8 ratios among healthy and not healthy cats in this study population also limited their utility in the prediction of disease progression.

Extrapolating from HIV infection, the plasma viral load has been postulated as a potential prognostic indicator for FIV infection. A study focusing on disease progression in SPF cats experimentally infected with two strains of FIV of different virulence reported a correlation between disease progression and increased plasma viral load (Diehl and Hoover, 1995). Similarly, the relationship between plasma viral load, disease progression and survival time was found in a study of naturally infected cats in Japan (Goto et al., 2002). In the present study, the median FIV load at the time of enrolment from Group 2 was 67-fold higher than that from Group 1 ($P < 0.0001$) and there was a marked difference in health status and mortality rate between groups over the following 22 months. It is not possible to discern whether an increased viral load was a cause or a consequence of poor health status, since that would have required knowledge of viral load kinetics and changes in health status from the time of initial infection in the FIV-infected 'healthy' and 'not healthy' cats; in this prospective study the data were collected from the time of enrolment. However, statistically significant differences in FIV load between cats from Group 2 classified as 'healthy' and 'not healthy' and cats that remained alive or died during the study were not apparent. Observation of these parameters over longer periods of time, preferably with larger group sizes, would be required to fully validate their utility in predicting disease progression.

It has been hypothesized that FIV clade B viruses are evolutionarily older and more host-adapted, hence less pathogenic, than those of clade A (Sodora et al., 1994). One possible explanation for the high morbidity and mortality in Group 2 is that those cats were infected with highly virulent strains of FIV. However, this hypothesis was not supported by the phylogenetic analyses performed, as cats from both groups were infected with comparable numbers of clade B and recombinant A/B viruses. Furthermore, sequences isolated from serial 6-monthly blood collections demonstrated that highly monophyletic groups were formed in 96% of the cats from Group 2, suggesting that viral isolates were not transmitted between cats despite unrestricted cohabitation (Bęczkowski et al., 2014). Although we cannot exclude the possibility that cats from Memphis were infected with more virulent FIV isolates, based on thorough phylogenetic analyses of full-length *env* sequences, this seems unlikely.

A limitation of this study is the restriction of enrolments to two distinct geographic areas; the inclusion of cats from a wider range of locations could have increased the variability in virulence of the wild type FIVs examined and their possible contribution to the onset of disease. Additionally, for the purposes of this study of naturally infected cats either adopted from shelters or rescued, it was necessary to estimate the ages of the cats and to use the date that the cat first tested FIV-positive since the precise dates of infection were not known. It has been reported previously that age at time of first exposure to FIV can influence the course of infection (George et al., 1993) and can determine the outcome of disease (Akhtardanesh et al., 2010). From a biological and behavioural perspective, the age of approximately 1–2 years old at the time of first diagnosis reported in this study is consistent with young cats engaging in territorial fights, especially if they are entire. The cats in our study were originally adopted from shelters and many were entire and had outdoor access at the time of shelter intake, both known lifestyle risk factors for FIV (Levy et al., 2006). In our study population, infection was more common in male cats, confirming the results of previous studies (Grindem et al., 1989; Hosie et al., 1989; Ishida et al., 1989; Levy et al., 2006; Yamamoto et al., 1989). Additionally, FIV status and multi-cat housing status could have affected the health of the cats in our study; it was not possible to discern whether they acted independently or as cofactors. FIV-negative cats matched to enrolled FIV-positive cats on housing status, as well as age and sex, were not available during the study enrolment period. While an extra layer of matching based on housing status would have been ideal, it was not achievable in this real-world study.

Based on the relatively large number of cats enrolled and the striking clinical differences observed between the two cohorts, we conclude that keeping FIV-infected cats in overcrowded conditions can have a significant impact on the risk of disease progression, particularly in cats which already have their immune systems compromised by FIV infection. In contrast, FIV-positive cats remained in relatively good health when living in stable, single cat households. Clinical signs of disease were not observed at 6-monthly physical examinations over a 24-month period of the cats from Group 1 that were housed in single cat households. The average duration of the 'asymptomatic' phase of infection remains unknown and could only be estimated following observations over a longer period. In this study, the selection of study participants from different locations and housing conditions minimized potential selection bias. It is apparent that the conclusions of this study would have been very different indeed had efforts been focused on only one of the cohorts presented herein.

ACKNOWLEDGEMENTS

This study was supported by The Wellcome Trust and the Maddie's Fund. The Purdue Maddie's Shelter Medicine Programme is underwritten by a grant from Maddie's Fund, The Pet Rescue Foundation (www.maddiesfund.org), helping to fund the creation of a no-kill nation. We thank Kristen Hall CVT, Dr. Jui Ming Lin, Dr. Christian Leutenegger, PAWS Chicago, Drennan Animal Hospital, the Fitzhugh B. Crews FIV Cat Sanctuary and participating cat owners for their assistance with the study. We thank the IDEXX Corporation (Westbrook, MN and West Sacramento, CA) for providing the FIV load and flow cytometry results.

REFERENCES

1. Abramo et al., 1995 F. Abramo, S. Bo, M.G. Canese, A. PoliRegional distribution of lesions in the central nervous system of cats infected with feline immunodeficiency virusAIDS Res. Hum. Retrovir., 11 (1995), pp. 1247-1253
2. Addie et al., 2000 D.D. Addie, J.M. Dennis, S. Toth, J.J. Callanan, S. Reid, O. JarrettLong-term impact on a closed household of pet cats of natural infection with feline coronavirus, feline leukaemia virus and feline immunodeficiency virusVet. Rec., 146 (2000), pp. 419-424

3. Akhtardanesh et al., 2010 B. Akhtardanesh, N. Ziaali, H. Sharifi, S. RezaeiFeline immunodeficiency virus, feline leukemia virus and *Toxoplasma gondii* in stray and household cats in Kerman – Iran: seroprevalence and correlation with clinical and laboratory findingsRes. Vet. Sci., 89 (2010), pp. 306-310
4. Bain et al., 2009 B.J. Bain, D.M. Clark, B.S. WilkinsInfection and reactive changesBone Marrow Pathology, Wiley-Blackwell (2009), pp. 100-165
5. Bandecchi et al., 1992 P. Bandecchi, D. Matteucci, F. Baldinotti, G. Guidi, F. Abramo, F. Tozzini, M. BendinelliPrevalence of feline immunodeficiency virus and other retroviral infections in sick cats in ItalyVet. Immunol. Immunopathol., 31 (1992), pp. 337-345
6. Barlough et al., 1991 J.E. Barlough, C.D. Ackley, J.W. George, N. Levy, R. Acevedo, P.F. Moore, B.A. Rideout, M.D. Cooper, N.C. PedersenAcquired immune dysfunction in cats with experimentally induced feline immunodeficiency virus infection: comparison of short-term and long-term infectionsJ. Acquir. Immune Defic. Syndr., 4 (1991), pp. 219-227
7. Beatty et al., 1996 J.A. Beatty, B.J. Willett, E.A. Gault, O. JarrettA longitudinal study of feline immunodeficiency virus-specific cytotoxic T lymphocytes in experimentally infected cats, using antigen-specific inductionJ. Virol., 70 (1996), pp. 6199-6206
8. Bęczkowski et al., 2014 P.M. Bęczkowski, J. Hughes, R. Biek, A. Litster, B.J. Willett, M.J. HosieFeline immunodeficiency virus (FIV) env recombinants are common in natural infectionsRetrovirology, 11 (2014), p. 80
9. Breuer et al., 1998 W. Breuer, K. Stahr, M. Majzoub, W. HermannsBone-marrow changes in infectious diseases and lymphohaemopoietic neoplasias in dogs and cats—a retrospective studyJ. Comp. Pathol., 119 (1998), pp. 57-66
10. Churchill, 1997 D.R. ChurchillPrognostic markers and surrogate markers of clinical progression in HIV infectionInt. J. STD AIDS, 8 (1997), pp. 552-556quiz 557
11. Diehl and Hoover, 1995 L.J. Diehl, E.A. HooverDisease progression correlates with plasma viral-RNA load in an accelerated FIV modelAIDS Res. Hum. Retrovir., 11 (1995), p. S93
12. Dow et al., 1990 S.W. Dow, M.L. Poss, E.A. HooverFeline immunodeficiency virus: a neurotropic lentivirusJ. Acquir. Immune Defic. Syndr., 3 (1990), pp. 658-668
13. Efron et al., 1996 B. Efron, E. Halloran, S. HolmesBootstrap confidence levels for phylogenetic treesProc. Natl. Acad. Sci. U. S. A., 93 (1996), pp. 13429-13434
14. English et al., 1994 R.V. English, P. Nelson, C.M. Johnson, M. Nasisse, W.A. Tompkins, M.B. TompkinsDevelopment of clinical disease in cats experimentally infected with feline immunodeficiency virusJ. Infect. Dis., 170 (1994), pp. 543-552
15. Geller et al., 1985 S.A. Geller, R. Muller, M.L. Greenberg, F.P. SiegalAcquired immunodeficiency syndrome distinctive features of bone marrow biopsiesArch. Pathol. Lab. Med., 109 (1985), pp. 138-141
16. George et al., 1993 J.W. George, N.C. Pedersen, J. HigginsThe effect of age on the course of experimental feline immunodeficiency virus infection in catsAIDS Res. Hum. Retrovir., 9 (1993), pp. 897-905
17. Goto et al., 2002 Y. Goto, Y. Nishimura, K. Baba, T. Mizuno, Y. Endo, K. Masuda, K. Ohno, H. TsujimotoAssociation of plasma viral RNA load with prognosis in cats naturally infected with feline immunodeficiency virusJ. Virol., 76 (2002), pp. 10079-10083
18. Grindem et al., 1989 C.B. Grindem, W.T. Corbett, B.E. Ammerman, M.T. TomkinsSeroepidemiologic survey of feline immunodeficiency virus infection in cats of Wake County North CarolinaJ. Am. Vet. Med. Assoc., 194 (1989), pp. 226-228
19. Hartmann, 1998 K. HartmannFeline immunodeficiency virus infection: an overviewVet. J., 155 (1998), pp. 123-137
20. Hartmann, 2011 K. HartmannClinical aspects of feline immunodeficiency and feline leukemia virus infectionVet. Immunol. Immunopathol., 143 (2011), pp. 190-201
21. Hartmann, 2012 K. HartmannClinical aspects of feline retroviruses: a reviewViruses, 4 (2012), pp. 2684-2710
22. Hoffmann-Fezer et al., 1992 G. Hoffmann-Fezer, J. Thum, C. Ackley, M. Herbold, J. Mysliwietz, S. Thefeld, K. Hartmann, W. KraftDecline in CD4+ cell numbers in cats with naturally acquired feline immunodeficiency virus infectionJ. Virol., 66 (1992), pp. 1484-1488

23. Hofmann-Lehmann et al., 1995 R. Hofmann-Lehmann, E. Holznagel, A. Aubert, K. Bauer-Pham, H. LutzFIV vaccine studies II. Clinical findings, hematological changes and kinetics of blood lymphocyte subsetsVet. Immunol. Immunopathol., 46 (1995), pp. 115-125
24. Hofmann-Lehmann et al., 1997 R. Hofmann-Lehmann, E. Holznagel, P. Ossent, H. LutzParameters of disease progression in long-term experimental feline retrovirus (feline immunodeficiency virus and feline leukemia virus) infections: hematology, clinical chemistry, and lymphocyte subsetsClin. Diagn. Lab. Immunol., 4 (1997), pp. 33-42
25. Hofmann-Lehmann et al., 2001 R. Hofmann-Lehmann, J.B. Huder, S. Gruber, F. Boretti, B. Sigrist, H. LutzFeline leukemia provirus load during the course of experimental infection and in naturally infected catsJ. Gen. Virol., 82 (2001), pp. 1589-1596
26. Hopper et al., 1989 C.D. Hopper, A.H. Sparkes, T.J. Gruffydd-Jones, S.M. Crispin, P. Muir, D.A. Harbour, C.R. StokesClinical and laboratory findings in cats infected with feline immunodeficiency virusVet. Rec., 125 (1989), pp. 341-346
27. Hosie et al., 1989 M.J. Hosie, C. Robertson, O. JarrettPrevalence of feline leukaemia virus and antibodies to feline immunodeficiency virus in cats in the United KingdomVet. Rec., 125 (1989), pp. 293-297
28. Hosie et al., 2009 M.J. Hosie, D. Addie, S. Belák, C. Boucraut-Baralon, H. Egberink, T. Frymus, T. Gruffydd-Jones, K. Hartmann, A. Lloret, H. Lutz, F. Marsilio, M.G. Pennisi, A.D. Radford, E. Thiry, U. Truyen, M.C. HorzinekFeline immunodeficiency ABCD guidelines on prevention and managementJ. Feline Med. Surg., 11 (2009), pp. 575-584
29. Ishida et al., 1989 T. Ishida, T. Washizu, K. Toriyabe, S. Motoyoshi, I. Tomoda, N.C. PedersenFeline immunodeficiency virus infection in cats of JapanJ. Am. Vet. Med. Assoc., 194 (1989), pp. 221-225
30. Katzenstein, 2003 T.L. KatzensteinMolecular biological assessment methods and understanding the course of the HIV infectionAPMIS Suppl. (2003), pp. 1-37
31. Knotek et al., 1999 Z. Knotek, P. Hajkova, M. Svoboda, M. Toman, V. RaskaEpidemiology of feline leukaemia and feline immunodeficiency virus infections in the Czech RepublicZent. Veterinarmed. B, 46 (1999), pp. 665-671
32. Kohmoto et al., 1998 M. Kohmoto, K. Uetsuka, Y. Ikeda, Y. Inoshima, M. Shimojima, E. Sato, G. Inada, T. Toyosaki, T. Miyazawa, K. Doi, T. MikamiEight-year observation and comparative study of specific pathogen-free cats experimentally infected with feline immunodeficiency virus (FIV) subtypes A and B: terminal acquired immunodeficiency syndrome in a cat infected with FIV petaluma strainJ. Vet. Med. Sci., 60 (1998), pp. 315-321
33. Levy, 1988 J.A. LevyMysteries of HIV: challenges for therapy and preventionNature, 333 (1988), pp. 519-522
34. Levy et al., 2006 J.K. Levy, H.M. Scott, J.L. Lachtara, P.C. CrawfordSeroprevalence of feline leukemia virus and feline immunodeficiency virus infection among cats in North America and risk factors for seropositivityJ. Am. Vet. Med. Assoc., 228 (2006), pp. 371-376
35. Liem et al., 2013 B.P. Liem, N.K. Dhand, A.E. Pepper, V.R. Barrs, J.A. BeattyClinical findings and survival in cats naturally infected with feline immunodeficiency virusJ. Vet. Intern. Med., 27 (2013), pp. 798-805
36. Lin and Litster, 2013 J. Lin, A. LitsterFluorescence flow cytometry methodology to exclude platelet aggregate interference when measuring feline CD4 and CD8 lymphocyte countsVet. J., 198 (2013), pp. 275-278
37. Lobprise, 2007 H.B. LobpriseOral exam and chartingBlackwell's Five Minute Veterinary Consult Clinical Companion: Small Animal Dentistry (first ed.), Blackwell Publishing Ltd., Ames, IA, USA (2007)
38. Matsumura et al., 1993 S. Matsumura, T. Ishida, T. Washizu, I. Tomoda, S. Nagata, J. Chiba, T. KurataPathologic features of acquired immunodeficiency-like syndrome in cats experimentally infected with feline immunodeficiency virusJ. Vet. Med. Sci., 55 (1993), pp. 387-394
39. Muirden, 2002 A. MuirdenPrevalence of feline leukaemia virus and antibodies to feline immunodeficiency virus and feline coronavirus in stray cats sent to an RSPCA hospitalVet. Rec., 150 (2002), pp. 621-625

40. Novotney et al., 1990 C. Novotney, R.V. English, J. Housman, M.G. Davidson, M.P. Nasisse, C.R. Jeng, W.C. Davis, M.B. TompkinsLymphocyte population changes in cats naturally infected with feline immunodeficiency virusAIDS, 4 (1990), pp. 1213-1218
41. Okonji et al., 2012 J.A. Okonji, C. Zeh, P.J. Weidle, J. Williamson, B. Akoth, R. Masaba, M.G. Fowler, T.K. ThomasCD4, viral load response and adherence among antiretroviral-naive breastfeeding women receiving triple antiretroviral prophylaxis for prevention of mother-to-child transmission of HIV in Kisumu, KenyaJ. Acquir. Immune Defic. Syndr., 61 (2) (2012), pp. 249-257
42. Pedersen et al., 2001 N.C. Pedersen, C.M. Leutenegger, J. Woo, J. HigginsVirulence differences between two field isolates of feline immunodeficiency virus (FIV-apetaluma and FIV-cpgammar) in young adult specific pathogen free catsVet. Immunol. Immunopathol., 79 (2001), pp. 53-67
43. Podell et al., 1997 M. Podell, K. Hayes, M. Oglesbee, L. MathesProgressive encephalopathy associated with CD4/CD8 inversion in adult FIV-infected catsJ. Acquir. Immune Defic. Syndr. Hum. Retrovirol, 15 (1997), pp. 332-340
44. Podell et al., 1999 M. Podell, K. Maruyama, M. Smith, K.A. Hayes, W.R. Buck, D.S. Ruehlmann, L.E. MathesFrontal lobe neuronal injury correlates to altered function in FIV-infected catsJ. Acquir. Immune Defic. Syndr. Hum. Retrovirol, 22 (1999), pp. 10-18
45. Posada, 2008 D. PosadajMODELTEST: phylogenetic model averagingMol. Biol. Evol., 25 (2008), pp. 1253-1256
46. Quimby et al., 2008 J.M. Quimby, T. Elston, J. Hawley, M. Brewer, A. Miller, M.R. LappinEvaluation of the association of *Bartonella* species, feline herpesvirus 1, feline calicivirus, feline leukemia virus and feline immunodeficiency virus with chronic feline gingivostomatitisJ. Feline Med. Surg., 10 (2008), pp. 66-72
47. Ravi et al., 2010 M. Ravi, G.A. Wobeser, S.M. Taylor, M.L. JacksonNaturally acquired feline immunodeficiency virus (FIV) infection in cats from Western Canada: prevalence, disease associations, and survival analysisCan. Vet. J., 51 (2010), pp. 271-276
48. Reche et al., 2010 A. Reche Jr., A.G.T. Daniel, T.C.P. Lazaro Strauss, C.P. Taborda, S.A. Vieira Marques, K. Haipek, L.J. Oliveira, J.M. Monteiro, J.R. Kfoury Jr.Cutaneous mycoflora and CD4:CD8 ratio of cats infected with feline immunodeficiency virusJ. Feline Med. Surg., 12 (2010), pp. 355-358
49. Reubel et al., 1994 G.H. Reubel, J.W. George, J. Higgins, N.C. PedersenEffect of chronic feline immunodeficiency virus infection on experimental feline calicivirus-induced diseaseVet. Microbiol., 39 (1994), pp. 335-351
50. Shaunak and Teo, 2003 S. Shaunak, I. TeoMonitoring HIV disease with new and clinically useful surrogate markersCurr. Opin. Infect. Dis., 16 (2003), pp. 581-586
51. Shelton et al., 1989 G.H. Shelton, J.L. Abkowitz, M.L. Linenberger, R.G. Russell, C.K. GrantChronic leukopenia associated with feline immunodeficiency virus infection in a catJ. Am. Vet. Med. Assoc., 194 (1989), pp. 253-255
52. Smith and Stein, 2002 C.L. Smith, G.E. SteinViral load as a surrogate end point in HIV diseaseAnn. Pharmacother., 36 (2002), pp. 280-287
53. Sodora et al., 1994 D.L. Sodora, E.G. Shpaer, B.E. Kitchell, S.W. Dow, E.A. Hoover, J.I. MullinsIdentification of three feline immunodeficiency virus (FIV) env gene subtypes and comparison of the FIV and human immunodeficiency virus type 1 evolutionary patternsJ. Virol., 68 (1994), pp. 2230-2238
54. Takano et al., 2012 T. Takano, S. Hosoya, A. Shibao, B. Nagasaki, H. Yoshioka, R. Satoh, T. HohdatsuComparative study of the plasma globulin level CD21(−) B-cell counts and FOXP3 mRNA expression level in CD4(+) T-cells for different clinical stages of feline immunodeficiency virus infected catsRes. Vet. Sci., 92 (2012), pp. 157-161
55. Tamura et al., 2011 K. Tamura, D. Peterson, N. Peterson, G. Stecher, M. Nei, S. KumarMega5: molecular evolutionary genetics analysis using maximum likelihood, evolutionary distance, and maximum parsimony methodsMol. Biol. Evol., 28 (2011), pp. 2731-2739
56. Tenorio et al., 1991 A.P. Tenorio, C.E. Franti, B.R. Madewell, N.C. PedersenChronic oral infections of cats and their relationship to persistent oral carriage of feline calici-, immunodeficiency, or leukemia virusesVet. Immunol. Immunopathol., 29 (1991), pp. 1-14

57. Tompkins et al., 1991 M.B. Tompkins, P.D. Nelson, R.V. English, C. NovotneyEarly events in the immunopathogenesis of feline retrovirus infectionsJ. Am. Vet. Med. Assoc., 199 (1991), pp. 1311-1315
58. Willett et al., 1993 B.J. Willett, M.J. Hosie, J.J. Callanan, J.C. Neil, O. JarrettInfection with feline immunodeficiency virus is followed by the rapid expansion of a CD8+ lymphocyte subsetImmunology, 78 (1993), pp. 1-6
59. Yamamoto et al., 1988 J.K. Yamamoto, E. Sparger, E.W. Ho, P.R. Andersen, T.P. O'Connor, C.P. Mandell, L. Lowenstine, R. Munn, N.C. PedersenPathogenesis of experimentally induced feline immunodeficiency virus infection in catsAm. J. Vet. Res., 49 (1988), pp. 1246-1258
60. Yamamoto et al., 1989 J.K. Yamamoto, H. Hansen, E.W. Ho, T.Y. Morishita, T. Okuda, T.R. Sawa, R.M. Nakamura, N.C. PedersenEpidemiologic and clinical aspects of feline immunodeficiency virus infection in cats from the continental United States and Canada and possible mode of transmissionJ. Am. Vet. Med. Assoc., 194 (1989), pp. 213-220

CHAPTER 15

Multiresistant extended-spectrum β-lactamase-producing *Enterobacteriaceae* from humans, companion animals and horses in central Hesse, Germany

Judith Schmiedel[1], Linda Falgenhauer[1], Eugen Domann[1], Rolf Bauerfeind[2], Ellen Prenger-Berninghoff[2], Can Imirzalioglu[1] and Trinad Chakraborty[1]

[1]*Institute of Medical Microbiology, Justus Liebig University Giessen and German Center for Infection Research (DZIF), Partner site Giessen-Marburg-Langen, Schubertstrasse 81, 35392 Giessen, Germany.*

[2]*Institute of Hygiene and Infectious Diseases of Animals, Justus Liebig University Giessen, Frankfurter Strasse 85-89, 35392 Giessen, Germany.*

ABSTRACT

Background

Multiresistant Gram-negative bacteria producing extended-spectrum β-lactamases (ESBLs) are an emerging problem in human and veterinary medicine. This study focused on comparative molecular characterization of β-lactamase and ESBL-producing *Enterobacteriaceae* isolates from central Hesse in Germany. Isolates originated from humans, companion animals (dogs and cats) and horses.

Results

In this study 153 (83.6%) of the human isolates (n = 183) and 163 (91.6%) of the animal isolates (n = 178) were confirmed as ESBL producers by PCR and subsequent sequencing of the PCR amplicons. Predominant ESBL subtypes in human and animal samples were CTX-M-15 (49.3%) and CTX-M-1 (25.8%) respectively. Subtype *bla* $_{\text{CTX-M-2}}$ was found almost exclusively in equine and was absent from human isolates. The carbapenemase OXA-48 was detected in 19 ertapenem-resistant companion animal isolates in this study. The Plasmid-encoded quinolone resistance (PMQR) gene *aac('6)-Ib-cr* was the most frequently detected antibiotic- resistance gene present in 27.9% of the human and 36.9% of the animal ciprofloxacin-resistant isolates. Combinations of two or up to six different resistance genes (penicillinases, ESBLs and PMQR) were detected in 70% of all isolates investigated. The most frequent species in this study was *Escherichia coli* (74%), followed by *Klebsiella pneumoniae* (17.5%), and *Enterobacter cloacae* (4.2%). Investigation of *Escherichia coli* phylogenetic groups revealed underrepresentation of group B2 within the animal isolates.

Conclusions

Isolates from human, companion animals and horses shared several characteristics regarding presence of ESBL, PMQR and combination of different resistance genes. The results indicate active transmission and dissemination of multi-resistant *Enterobacteriaceae* among human and animal populations.

Keywords: ESBL, *Enterobacteriaceae,* Comparison, Human isolate, Animal isolates, One Health, concept

1. BACKGROUND

Extended-spectrum beta-lactamase (ESBL)-producing *Enterobacteriaceae* are isolated with increasing frequency from human and animal samples. In particular, the species *Escherichia (E.) coli* plays an important role as a source of the corresponding resistance genes [1]. ESBLs genes are commonly plasmid-encoded and can easily be transmitted by conjugation to other bacteria, even across species barriers. In addition, various resistance genes are located in or close to mobile genetic elements such as insertion sequences and transposons [2]. Lateral gene transfer and continuous DNA recombination make it extremely difficult to track transmission pathways of ESBL genes in bacterial populations. The transmission of ESBL-producing pathogens or ESBL genes between companion animals/livestock and owner/caretaker/consumer is currently a subject of intense and controversial discussion. Evidence has been presented for zoonotic spread [3, 4]. Several studies have addressed the risk of infection that could arise from keeping animals [5, 6, 7]. A recent study highlighted that the risk factors for ESBL carriage were travel to Greece or Africa within the last 12 months and keeping of pet animals, but not antibiotic consumption or recent hospitalization [8]. Until now, it is still difficult to estimate the amount of exchange and even more difficult to define the risk for human and animal health and also food safety.

The focus of this study was a comparative investigation of 361 ESBL-producing *Enterobacteriaceae* isolates obtained from animals and human patients presenting at a veterinary clinic and a hospital, respectively, both serving a similar catchment area. The prevalence of β-lactamase-, particularly ESBL-producing bacteria in companion animal, horse and human isolates of clinical origin was examined. Resistance genes (β-lactamase and plasmid-mediated quinolone resistance (PMQR)) and *E. coli* phylogenetic groups were investigated using molecular methods.

This work sheds light on shared populations of ESBL-producing *Enterobacteriaceae* in symptomatic companion animals, horses and humans in the geographical region of middle Hesse, Germany.

2. METHODS

2.1 Bacterial isolates

A collection of 513 *Enterobacteriaceae* isolates resistant to one or more third-generation cephalosporins (200 human isolates and 313 animal isolates) was examined. Identification was done using in-house biochemical tests. In case of ambiguous results confirmation was done using API 20E. Prior to further analysis, all isolates were grown on MacConkey agar supplemented with cefotaxime (2 mg/l) to promote selection of β-lactamase producers and ensure selection of *Enterobacteriaceae*. Afterwards isolates were tested for possible ESBL production by double disc synergy test (DDST) according to EUCAST guidelines for resistance mechanisms [9]. After performance of the DDST 183 human and 178 animal isolates were categorized as possible ESBL-producers and forwarded for a more detailed investigation. Isolates with a negative DDST result were not included in the study.

DDST-positive isolates from human clinical samples (n = 183) were taken from the strain collection of the Institute of Medical Microbiology, Giessen. Some of the isolates originate from routine screening for colonization with ESBL-producing *Enterobacteriaceae* at the University hospital Giessen. Additional isolates were collected from samples of clinically ill patients e.g. blood culture or urine samples in the same facility. All isolates were collected between 2009 and 2010. Approximately two-thirds of the human isolates originated from inpatients (n = 128) and one third were isolates from outpatients wards (n = 55).

DDST-positive isolates from animals (n = 178) were obtained during a survey for aerobic Gram-negative bacteria growing on MacConkey agar supplemented with cefotaxime (2 mg/l) among animal patients at the veterinary clinics of the Justus Liebig University (JLU) Giessen. All samples were clinically relevant infection related isolates. The isolates originated from horses (n = 100), dogs (n = 67) and cats (n = 11) and were collected between 2009 and 2011.

The study was approved by the ethics committee of medical faculty of the Justus Liebig University of Giessen and deemed exempt from informed consent.

2.2 Antibiotic susceptibility

Antibiotic susceptibility testing was done using the VITEK®2 compact system with AST N117 cards (Biomérieux) and Etest® stripes (Liofilchem®) containing ertapenem, cefepime, chloramphenicol and nalidixic acid. Results were evaluated according to CLSI guidelines for human pathogens (CLSI, 2012).

2.3 β-lactamase identification

For antibiotic resistance genes screening by PCR and identification by sequencing were performed as described previously [10]. Specific oligonucleotide primers for *bla* $_{TEM}$, *bla* $_{SHV}$[11], *bla* $_{CTX-M}$ genes [12] and *bla*$_{OXA-1}$[13] were used. The MAST carbapenemase detection set (MAST group, UK) was applied on carbapenem-resistant isolates. Furthermore, primers for *bla* $_{OXA-48}$ were used for carbapenem-resistant isolates [11]. With the exception of *bla* $_{OXA-1}$-positive strains, amplicons of PCR-positive strains were sequenced to identify the encoded ESBL allele in detail. Sequencing was performed using the automated sequencer ABI Prism® 3100 (Life technologies, USA). The blastn algorithm of NCBI (http://www.ncbi.nlm.nih.gov/blast/) was used for database searches to identify the resistance gene allele. The primers and the sequences they were derived from are presented in Table 1. Positive control strains from the Institute of Microbiology, Giessen, Germany, were *Klebsiella* (*K.*) *pneumoniae* H59 for *bla* $_{TEM}$, *bla* $_{SHV}$, *bla* $_{CTX-M}$, *bla* $_{OXA-1}$ and *aac(6')-Ib*, *Klebsiella pneumoniae* 714 for *bla* $_{OXA-48}$. For the *qnr*-genes, strains from the Institute of Hygiene and Infectious Diseases of Animals, Giessen, Germany (V148 for *qnrA*, V167 for *qnrB*, V60 for *qnrD* and V61 for *qnrS)*, were used as positive controls after sequencing and comparison with known sequences using DNASTAR software (DNASTAR Inc, Madison, USA) and the NCBI blastn algorithm.

Table 1 Pcr primers used to detect β-lactamase genes, pmqr genes and ***e. Coli*** phylogenetic groups

Primer	Sequence (5' - 3')	Target	Size of product (bp)	Reference
TEM_f	ATGAGTATTCAACATTTCCG	*bla* $_{TEM}$	851	[11]
TEM_r	TTAATCAGTGAGGCACCTAT			
SHV_f	GCAAAACGCCGGGTTATTC	*bla* $_{SHV}$	940	[11]
SHV_r	GGTTAGCGTTGCCAGTGCT			
CTX-M_f	TCTTCCAGAATAAGGAATCCC	*bla* $_{CTX-M}$	909	[12]
CTX-M_r	CCGTTTCCGCTATTACAAAC			
OXA-1_f	AGCAGCGCCAGTGCATCA	*bla* $_{OXA-1}$	700	[13]
OXA-1_r	ATTCGACCCCAAGTTTCC			
OXA-48_f	AAATCACAGGGCGTAGTTGTG	*bla* $_{OXA-48}$	555	[11]
OXA-48_r	GACCCACCAGCCAATCTTAG			
qnrA_f	AGAGGATTTCTCACGCCAGG	*qnrA*	619	[14]
qnrA_r	GCAGCACTATKACTCCCAAGG			
qnrB_f	GGMATHGAAATTCGCCACTG	*qnrB*	264	[16]
qnrB_r	TTTGCYGYYCGCCAGTCGAA			
qnrC_f	GGGTTGTACATTTATTGAATC	*qnrC*	447	[17]
qnrC_r	TCCACTTTACGAGGTTCT			
qnrD_f	CGAGATCAATTTACGGGGAATA	*qnrD*	582	[18]
qnrD_r	AACAAGCTGAAGCGCCTG			
qnrS_f	GCAAGTTCATTGAACAGGCT	*qnrS*	428	[16]
qnrS_r	TCTAAACCGTCGAGTTCGGCG			

qepA_f	CTGCAGGTACTGCGTCATG	*qepA*	403	[15]
qepA_r	CGTGTTGCTGGAGTTCTTC			
aac_f	TTGCGATGCTCTATGAGTGGCTA	*aac(6')-Ib*	482	[19]
aac_r	CTCGAATGCCTGGCGTGTTT			
chuA.1	GACGAACCAACGGTCAGGAT	*chuA*	279	[21]
chuA.2	TGCCGCCAGTACCAAAGACA			
YjaA.1	TGAAGTGTCAGGAGACGCTG	*yajA*	211	[21]
YjaA.2	ATGGAGAATGCGTTCCTCAAC			
TspE4C2.1	GAGTAATGTCGGGGCATTCA	TspE4C2	152	[21]
TspE4C2.2	CGCGCCAACAAAGTATTACG			

2.4 Detection of Plasmid-encoded quinolone resistance (PMQR)

Detection for PMQR by PCR was performed on all isolates with a minimum inhibitory concentration (MIC) ≥ 1 µg/ml for ciprofloxacin (Human isolates: n = 140, animal isolates: n = 122). Oligonucleotide primers for detection of *qnrA, qnrB, qnrC, qnrD, qnrS, qepA and aac(6')-Ib* were used [14, 15, 16, 17, 18, 19]. Identification of the *aac(6')-Ib-cr* variant was performed according to Jones et al. 2008 [20].

2.5 Phylogenetic groups of ESBL-producers

Distribution of *E. coli* phylogenetic groups was analysed by PCR targeting genes *chuA*, *yjaA* and the DNA fragment TSPE4.C2 to assign the strains to the groups A, B1, B2 and D according to Clermont et al. 2000 [21].

2.6 Analysis of resistance genes patterns

Information regarding the source of isolates, species and the resistance genes detected were assembled into an Excel file. The data was subsequently analysed using GENE-E [22] which enables the clustering of data with matrix visualization and analysis to support visual data exploration.

3. RESULTS

3.1 Overview of bacterial species

The majority of the isolates studied were *E. coli* (74%), followed by *K. pneumoniae* (17.5%) and *Enterobacter cloacae* (4.2%). Other species detected were *Klebsiella oxytoca, Enterobacter intermedius, Citrobacter freundii, Providencia stuartii, Morganella morganii* and *Proteus mirabilis*. The percentages of *E. coli* and *K. pneumoniae* isolates among the human and animal isolates were very similar.

3.2 Resistance rates

All isolates (n = 361) were categorised as multidrug resistant (MDR) according to the international expert proposal for interim standard definitions for acquired resistance promoted by the ECDC [23]. Both human (n = 183) and animal isolates (n = 178) revealed resistance against ampicillin and trimethoprim/sulfmethoxazole. Slight differences could be observed concerning the tested third generation cephalosporins ceftazidime, cefotaxime and cefepime. Of the human isolates 52.5% (n = 96) showed resistance against ceftazidime, 78.7% (n = 144) resistance against cefotaxime and 98.4% (n = 180) resistance against cefepime. Of the animal isolates 75.8% (n = 135) displayed resistance against ceftazidime and 99.4% (n = 177) to cefotaxime and cefepime. Of the human isolates 2.2% (n = 4), and 1.1% (n = 2) of the animal isolates revealed resistance against imipenem. Resistance against the other tested carbapenem (ertapenem) was detected in 24.5% (n = 45) of human isolates and in 19.7% (n = 35) of animal isolates. The aminoglycosides gentamicin and amikacin differed in their results. Only 5.1% (n = 9) of animal isolates exhibited resistance against amikacin, as compared to 14.8% (n = 27) of the human isolates. The resistance rates against gentamicin were higher in both cases with 35% (n = 64) in human isolates and 62.4% (n = 111) in animal isolates. High resistance rates in both groups were also revealed for the fluoroquinolone ciprofloxacin (76.5% (n = 140) of the human isolates and 68.5% (n = 122) of the animal isolates).

3.3 Distribution of β-lactamases

Almost all isolates carried at least one beta-lactamase (*bla*)-gene (Table 2). In all, 91.6% of the animal isolates and 83.6% of the human isolates were confirmed as ESBL producers by PCR and sequencing. The remaining isolates (7.5%) were negative for the presence of *bla* $_{TEM}$, *bla* $_{SHV}$, *bla* $_{OXA-1}$, *bla* $_{OXA-48}$ and *bla* $_{CTX-M}$.

Table 2 Distribution of β-lactamase genes in ***enterobacteriaceae*** isolates (n = 361) of human and animal origin

Source	**Human**		**Animal**			
	Outpatients	**Inpatients**	**Dogs**	**Cats**	**Horses**	**Total no. of isolates**
NO. OF ISOLATES [A]	55	128	67	11	100	361
PENICILLINASES [B]						
bla $_{TEM-1}$	58.2 (32)	41.4 (53)	62.7 (42)	45.5 (5)	67 (67)	55.1 (199)
bla $_{TEM-190}$			1.5 (1)			0.3 (1)
bla $_{SHV-1}$		2.3 (3)	16.4 (11)		2 (2)	4.4 (16)
bla $_{SHV-11}$		1.6 (2)				0.6 (2)
bla $_{OXA-1}$	23.6 (13)	32.8 (42)	59.7 (40)	45.5 (5)	29 (29)	35.7 (129)
ESBL [B]						
bla $_{TEM-52}$		2.3 (3)				0.8 (3)
bla $_{SHV-2}$			1.5 (1)	9.1 (1)		0.6 (2)
bla $_{SHV-5}$		0.8 (1)				0.3 (1)
bla $_{SHV-28}$			3(2)		1(1)	0.8 (3)
bla $_{CTX-M-1}$	32.7 (18)	45.3 (24)	16.4 (11)	27.3 (3)	37 (37)	25.8 (93)
bla $_{CTX-M-2}$				9.1 (1)	15 (15)	4.4 (16)
bla $_{CTX-M-9}$			3 (2)	9.1 (1)	1 (1)	1.1 (4)
bla $_{CTX-M-14}$			1.5 (1)			0.3 (1)
bla $_{CTX-M-15}$	54.5 (30)	51.6 (66)	59.7 (40)	36.4 (4)	38 (38)	49.3 (178)
bla $_{CTX-M-32}$	1.8 (1)	0.8 (1)				0.6 (2)
bla $_{CTX-M-38}$		0.8 (1)				0.3 (1)

bla$_{\text{CTX-M-79}}$	1.8 (1)	2.3 (3)			1(1)	1.4 (5)
bla$_{\text{CTX-M-97}}$	1.8 (1)				3(3)	1.1 (4)
bla$_{\text{CTX-M-117}}$		0.8 (1)				0.3 (1)
CARBAPENEMASES [B]						
bla$_{\text{OXA-48}}$			23.9 (16)	18.2 (2)	1(1)	5.3 (19)

[a]numerous isolates encoded for more than one β-lactamase gene.

[b]% of β-lactamase genes (no. of isolates).

3.4 CTX-M

Subtypes of the ESBL CTX-M were by far the most frequently encoded ESBLs in both groups (Table 2 and Figure 1). Among the human isolates 23% encoded CTX-M-1 and 52.5% CTX-M-15. Some human isolates carried *bla*$_{\text{CTX-M-79}}$, *bla*$_{\text{CTX-M-32}}$, *bla*$_{\text{CTX-M-97}}$, *bla*$_{\text{CTX-M-38}}$ or *bla*$_{\text{CTX-M-117}}$. The *bla*$_{\text{CTX-M-1}}$ gene was found in 32.7% of human outpatient isolates and in 45.3% of inpatient isolates, while prevalence rates of *bla*$_{\text{CTX-M-15}}$were high and similar in both groups (outpatients 54.5%, inpatients 51.6%). Animal isolates revealed similar results as 28.7% of these isolates carried *bla*$_{\text{CTX-M-1}}$ and 46.1% *bla*$_{\text{CTX-M-15}}$. Other detected *bla*$_{\text{CTX-M}}$ genes were *bla*$_{\text{CTX-M-2}}$, *bla*$_{\text{CTX-M-9}}$, *bla*$_{\text{CTX-M-97}}$, *bla*$_{\text{CTX-M-79}}$, and *bla*$_{\text{CTX-M-14}}$. The presence of *bla*$_{\text{CTX-M-2}}$, *bla*$_{\text{CTX-M-9}}$, and *bla*$_{\text{CTX-M-14}}$ was restricted to animal isolates, while *bla*$_{\text{CTX-M-32}}$, *bla*$_{\text{CTX-M-38}}$, *bla*$_{\text{CTX-M-117}}$ were found only among the human isolates. *Bla*$_{\text{CTX-M-1}}$ was the dominant resistance subtype in horses (37%) but it was less frequently isolated in dogs (16.4%). In contrast, *bla*$_{\text{CTX-M-15}}$ occurred more often in dog isolates (59.7%) than in isolates from horses (38%) or cats (36.4%). In addition, *bla*$_{\text{CTX-M-15}}$ was also the dominant CTX-M subtype among *K. pneumoniae* isolates whereas among the *E. coli* isolates both subtypes were present to the same degree. Altogether, *bla*$_{\text{CTX-M-1}}$ and *bla*$_{\text{CTX-M-15}}$ were the prominent β-lactamase genes present in 25.8% and 49.3% of all investigated isolates.

3.5 TEM

Non-ESBL gene *bla*$_{\text{TEM-1}}$ was the most frequently identified β-lactamase gene among all investigated isolates. It was detected in 46.4% of the human isolates and in 64% of all animal isolates. It was present in 199 of 361 isolates (55%). The highest detection rates were observed in outpatients (58.2%), inpatients (41.4%), dog isolates (62.7%) and horse isolates (67%). TEM-52 was the only TEM-type ESBL that occurred in this study.

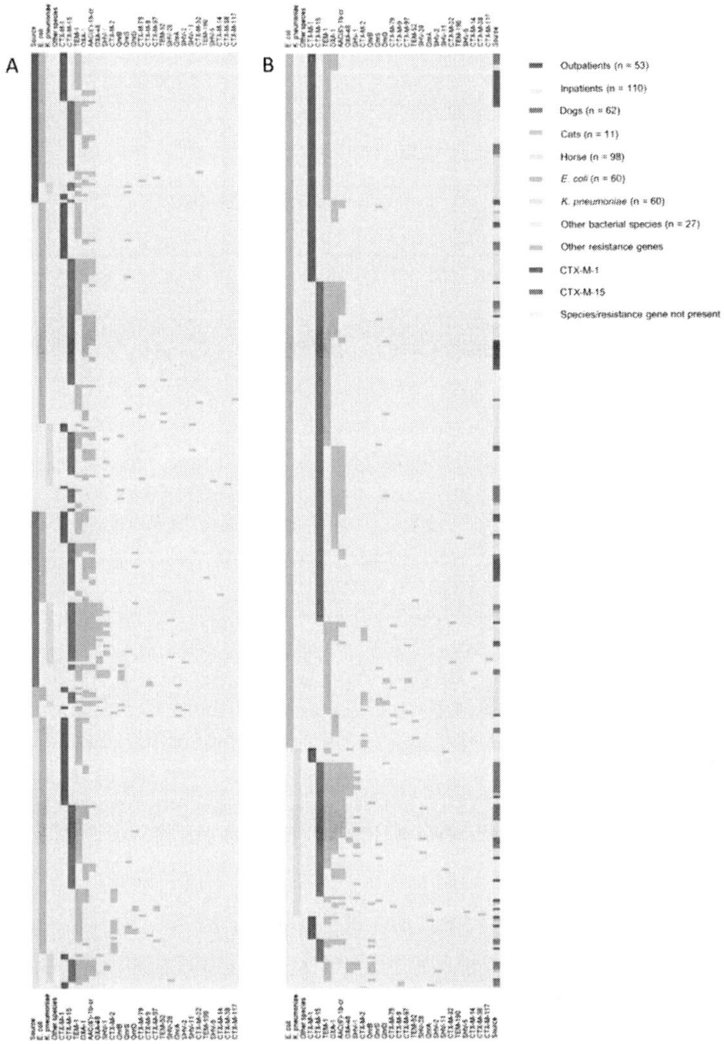

Figure 1 Heat maps generated from identified resistance genes and bacterial species among human and animal isolates. Identified resistance genes and bacterial species are listed according to frequency (except for CTX-M-1 and CTX-M-15) on the right and left side of the figures. Source = origin of isolates (outpatients, inpatients, dogs, cats, horses). Not included are isolates without detectable resistance gene (n = 28). The term "Other species" includes *Enterobacter cloacae* (n = 12), *Klebsiella oxytoca* (n = 9), *Enterobacter intermedius* (n = 2), *Enterobacter gergoviae* (n = 1), *Citrobacter freundii* (n = 1) and *Proteus mirabilis* (n = 1). **A** is focussing on the origin of isolates whereas the **B** emphasizes the involved bacterial species.

It was identified in three isolates from human inpatients. Results are shown in Table 2.

3.6 SHV

Prevalence of *bla* $_{SHV}$ was 17.2% in human and 52.9% in animal *K. pneumoniae. Bla* $_{SHV-1}$ was identified in 10.3% of the human and in 29.4% of the animal *K. pneumoniae* isolates. Several ESBL variants could be identified in addition (Table 2). Most of the latter isolates originated from dogs. Detected SHV-type ESBL genes were *bla* $_{SHV-2}$, *bla* $_{SHV-5}$, *bla* $_{SHV-11}$ and *bla* $_{SHV-28}$. *Bla* $_{SHV-5}$ and *bla* $_{SHV-11}$ were restricted to isolates from human inpatients. ESBL genes *bla* $_{SHV-2}$ and *bla* $_{SHV-28}$ were detected among the animal isolates only.

3.7 OXA-1

The presence of the penicillinase gene *bla* $_{OXA-1}$ could be demonstrated in 30.1% of the human isolates. Among the animal isolates *bla* $_{OXA-1}$ was identified in 41.6% of all strains. Detection rates were highest in isolates from dogs (59.7%), followed by cat (45.5%), outpatient (32.8%) and horse (29%) isolates. However, *bla*$_{OXA-1}$ was also found in inpatient isolates (23.6%) (Table 2).

3.8 OXA-48

All carbapenem resistant isolates (human isolates: 45, animal isolates: 40) were tested for presence of the carbapenemase OXA-48. None of the tested human isolates harboured *bla* $_{OXA-48}$. In contrast, 19 (5.3%) of the animal isolates harboured this gene. The isolates harbouring *bla* $_{OXA-48}$ originated from dogs (16), cats (2) and a horse (Table 2).

3.9 PMQR

Distribution and presence of PMQR genes are illustrated in Table 3. The most frequently detected PMQR gene in ciprofloxacin-resistant isolates (n = 262) was *aac('6)-Ib-cr*. Prevalence was 27.9% (n = 39) in the human isolates, 36.9% (n = 45) in the animal isolates and 32.1% (n = 84) in all investigated isolates. None of the isolates carried *qnrC* or *qepA*. Among the human isolates *qnrB* was found in three and *qnrS* in five isolates. Two more *qnr* variants were detected among the animal isolates. Gene *qnrA* was carried by three, *qnrB* by seven, *qnrD* by eight and *qnrS* by five animal isolates.

Most *qnr* type genes (*qnrA, qnrB, qnrD, qnrS*) were detected in dog isolates. Inpatient isolates carried PMQR genes at higher rates (37.8%) than isolates from outpatients (23.8%). A similar observation could be made among the animal isolates. Highest identification rates (49.5%) were observed in isolates originating from animals tested during stay in the small animal clinic. In summary, most PMQR genes were present in dog isolates (65.1%). Overall, PMQR genes could be demonstrated in 43.9% (n = 115) of all investigated isolates.

Table 3 Distribution of pmqr genes in ciprofloxacin-resistant isolates (n = 262) of human and animal origin

		% of PMQR genes (no. of isolates)						
Source	**No. of isolates[b]**	***qnrA***	***qnrB***	***qnrC***	***qnrD***	***qnrS***	***qepA***	***aac(6')-Ib-cr***
HUMAN								
Outpatients	42					4.8 (2)		19 (8)
Inpatients	98		3.1 (3)			3.1 (3)		31.6 (31)
Total	140		2.1 (3)			3.6 (5)		27.9 (39)
ANIMAL								
Dogs	63	1.6 (1)	7.9 (5)		4.8 (3)	1.6 (1)		49.2 (31)
Cats	10	20 (2)	20 (2)					40 (4)
Horses	49				10.2 (5)	8.2 (4)		20.4 (10)
Total	122	2.5 (3)	5.7 (7)		6.6 (8)	4.1 (5)		36.9 (45)
TOTAL NO. OF ISOLATES	262	1.1 (3)	3.8 (10)		3.1 (8)	3.8 (10)		32.1 (84)

[b]some isolates encoded for more than one PMQR gene.

3.10 Analysis of resistance gene patterns

Preliminary detection and visualization of clusters and/or resistance patterns in context to isolate source and bacterial species was performed using Figure 1 which was generated by means of the GENE-E program. Data was then confirmed by analysis of absolute and relative numbers/ percentages as displayed in Table 4.

Combinations of various plasmid encoded resistance genes (ESBL and non-ESBL β-lactamases and PMQR) were observed in the majority of isolates (70.1%) included in this study (Table 4 and Figure 1). Most frequently detected (26.3%) was the combination of a TEM type penicillinase with a CTX-M type ESBL. Also regularly observed was a penicillinase (TEM or OXA-1) in combination with CTX-M and plasmid-mediated quinolone resistance (PMQR) genes.

These combinations were predominantly found among *E. coli* isolates. Other isolates encoded for several combinations of a TEM penicillinase, CTX-M enzymes, the penicillinase *bla*$_{OXA-1}$, the penicillinase *bla* $_{SHV}$ and PMQR genes.

The carbapenemase OXA-48 was mainly associated with numerous resistance genes such as a TEM penicillinase, CTX-M and PMQR. In addition, it was exclusively identified in *K. pneumoniae* and *Enterobacter cloacae* isolates from animals. Another frequently observed combination was that of *bla* $_{CTX-M-15}$ and *aac('6)-Ib-cr*. Other combinations were detected only once or to a much lesser extent than those mentioned above. Combination of more than two resistance genes could be observed more often among human *E. coli* isolates and especially among *K. pneumoniae* animal isolates (Figure 1).

3.11 Phylogenetic grouping of β-lactamase-producing *E. coli*

Phylogenetic group typing was applied to 141 human and 126 animal ESBL-producing *E. coli* isolates (Table 5 and Figure 2). Both groups differed from each other in the frequencies of known phylogenetic groups. In general, the human isolates were almost evenly distributed among the phylogenetic groups A (30.5%), B1 (27.7%) and B2 (26.2%) and to a lesser extent with group D (15.6%). In contrast, most of the analysed animal isolates belonged to group A (43.7%) and a considerably smaller number of isolates to B1 (35.7%).

Approximately 17.5% of all animal isolates could be assigned to group D, whereas only 2.4% were classified as group B2.

Table 4 Observed combinations of β-lactamase and pmqr genes according to bacterial species

Combination of resistance genes[a]	Outpatients					Inpatients					Dogs					Cats					Horses				
	% of bacterial species (no. of isolates)																								
	EC	*KP*	*KO*	*EN*	*OS*[b]	*EC*	*KP*	*KO*	*EN*	*OS*[b]	*EC*	*KP*	*KO*	*EN*	*OS*[b]	*EC*	*KP*	*KO*	*EN*	*OS*[b]	*EC*	*KP*	*KO*	*EN*	*OS*[b]
TEM_{Pen}/CTX-M	50 (24)					26.9 (25)	12 (3)				17.1 (6)					20 (1)					40.7 (35)				25 (1)
TEM_{Pen}/CTX-M/ PMQR		25 (1)				1.2 (1)			25 (1)		2.9 (1)					20 (1)			33.3 (1)		8.1 (7)				
TEM_{Pen}/CTX-M/ OXA_{Pen}	2.1 (1)		33.3 (1)			2.2 (2)					14.3 (5)										12.8 (11)	10 (1)			
TEM_{Pen}/CTX-M/ OXA_{Pen}/PMQR	4.2 (2)	25 (1)				8.6 (8)	8 (2)				8.6 (3)	18.2 (4)	100 (1)	12.5 (1)		20 (1)					1.2 (1)	40 (4)			
TEM_{Pen}/SHV_{Pen}/ CTX-M/OXA_{Pen}/ PMQR							4 (1)					9.1 (2)										20 (2)			
TEM_{Pen}/SHV_{Pen}/ CTX-M/OXA_{Pen}/ OXA_{Carba}/PMQR												22.7 (5)													
TEM_{Pen}/CTX-M/ OXA_{Pen}/OXA_{Carba}/ PMQR												27.3 (6)					50 (1)								
CTX-M/OXA_{Pen}/ PMQR	8.3 (4)					14 (13)	12 (3)				17.1 (6)					20 (1)					3.5 (3)				
TEM_{Pen}/SHV_{Pen}/ OXA_{Carba}/PMQR														37.5 (3)											
CTX-M/PMQR						2.2 (2)		20 (1)			5.7 (2)			12.5 (1)					33.3 (1)						
CTX-M/OXA_{Pen}	8.3 (4)					6.5 (6)	8 (2)				2.9 (1)					20 (1)					8.1 (7)				
Other combinations	2.1 (1)	25 (1)				3.2 (3)	16 (4)					22.7 (5)		12.5 (1)			50 (1)		33.3 (1)			20 (2)			
No resistance gene detected	4.2 (2)					14 (13)	12 (3)		50 (2)		8.6 (3)			12.5 (1)	100 (1)						2.3 (2)				
Single resistance gene	20.8 (10)	25 (1)	66.7 (2)			21.5 (20)	28 (7)	80 (4)	25 (1)	100 (1)	22.9 (8)			12.5 (1)		20 (1)				1	23.3 (20)	10 (1)			75 (3)
Total no. of isolates	48	4	3	0	0	93	25	5	4	1	35	22	1	8	1	5	2	0	3	1	86	10	0	0	4

EC, Escherichia coli; KP, Klebsiella pneumoniae; KO, Klebsiella oxytoca; EN, Enterobacter cloacae; OS, Other species.

[a] TEM_{Pen}, TEM Penicillinase, OXA_{Pen}, OXA Pencillinase, OXA_{Carba}, OXA Carbapenemase, SHV_{Pen}, SHV Penicillinase.

[b] Other species include *Enterobacter intermedius* (n = 2). *Enterobacter gergoviae* (n = 1), *Citrobacter freundii* (n = 1), and *Proteus mirabilis* (n = 1), *Providencia stuartii* (n = 1), *Morganella morganii* (n = 1).

Table 5 Distribution of ***e. Coli*** phylogenetic groups among human and animal isolates (n = 267)

		% of phylogenetic groups (no. of isolates)			
Source	**No. of isolates**	**A**	**B1**	**B2**	**D**
HUMAN					
Outpatients	48	31.3 (15)	29.2 (14)	25 (12)	14.6 (7)
Inpatients	93	30.1 (28)	26.9 (25)	26.9 (25)	16.1 (15)
Total	141	30.5 (43)	27.7 (39)	26.2 (37)	15.6 (22)
ANIMAL					
Dogs	35	48.6 (17)	22.9 (8)	2.9 (1)	25.7 (9)
Cats	5	60 (3)	20 (1)		20 (1)
Horses	86	40.7 (35)	41.9 (36)	2.3 (2)	14 (12)
Total	126	43.7 (55)	35.7 (45)	2.4 (3)	17.5 (22)
TOTAL NO. OF ISOLATES	267	36.7 (98)	31.5 (84)	14.9 (40)	16.5 (44)

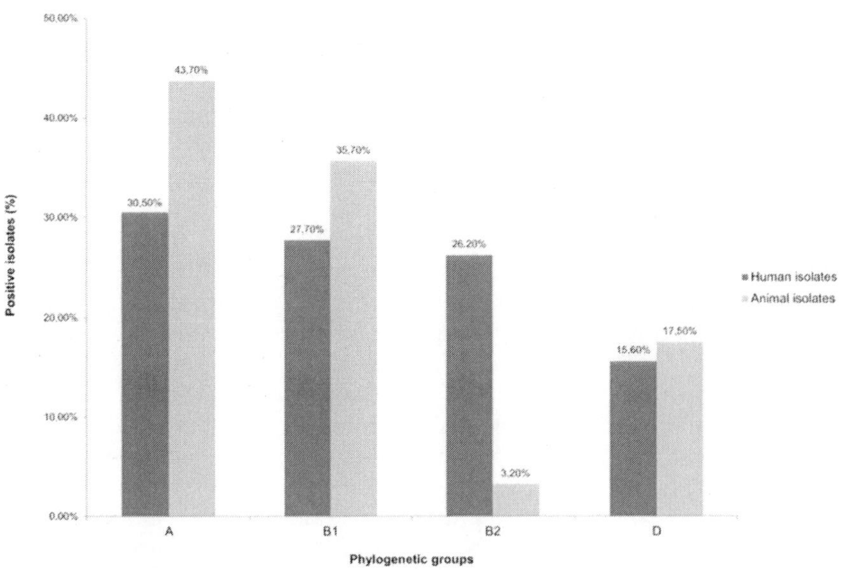

Figure 2 Distribution of ***e. coli*** phylogenetic groups among human and animal esbl-producing isolates. A total of 267 isolates from human (n = 141) and animal (n = 126) sources were assigned to *E. coli* phylogenetic groups A, B1, B2 and D.

4. DISCUSSION

The global spread of multi-resistant ESBL-producing *Enterobacteriaceae* can in part be explained by mobility of the resistance bearing genetic elements e.g. plasmids, transposons and insertion sequence elements. This fact is a major concern in terms of epidemiology and infection control. Transmission pathways between humans and animals are currently a subject of active discussion [8, 24, 25]. This study was designed to investigate the distribution of β-lactamases, particularly ESBL, and PMQR genes among isolates from human, companion animal and horse samples in a defined geographical area of Germany. The key question was whether similar resistance characteristics are currently present among both groups. Indeed, our studies here demonstrate that isolates from human and animal samples share numerous characteristics.

Antimicrobial susceptibility testing revealed common and disturbing patterns. High resistance rates for all β-lactams and other antibiotic classes were observed. In particular animal isolates harbouring resistance to the last-resort antibiotics ertapenem and impenem were detected. The results for ertapenem were particularly alarming, with 16.3% resistance among animal isolates and 24% among the human isolates. In Germany, the use of carbapenems in livestock animals is prohibited and its use for treating companion animals is strongly restricted. Nevertheless, almost all ertapenem resistant strains among the animal isolates (95.6%) originated from dogs and cats. This would suggest human to animal transmission of resistant strains by their owners. Alternatively, the use of carbapenems for treatment of companion animals cannot be excluded even though these pharmaceuticals are expensive and their use is uncommon even in pet animal practice. Carbapenemase activity in bacteria from companion animals and livestock has been already identified recently [26, 27, 28]. To investigate this phenomenon in more detail, all carbapenem resistant isolates from this study were tested for the presence of *bla* $_{OXA-48}$. Surprisingly, a substantial difference could be observed regarding the molecular mechanisms underlying this resistance phenotype. None of the 45 tested human isolates was positive for *bla* $_{OXA-48}$, though 26 (57.8%) were identified with a possible combination of an AmpC enzyme and porin deficiency. This mechanism has been observed previously in AmpC-producing *K. pneumoniae* by Shi et al. They stated that the combination of AmpC production and porin loss can result in reduced susceptibility to carbapenems [29]. In contrast to these results, 19 (47.5%) of the 40 examined animal isolates harboured an OXA-48 carbapenemase.

Among 361 isolates resistant to one or more third generation cephalosporins 84% harboured a CTX-M type enzyme. Predominant ESBL genes in human and animal samples were *bla* $_{CTX-M-15}$ and *bla* $_{CTX-M-1}$. In the past *bla* $_{CTX-M-15}$ was merely associated with isolates from humans [30] and CTX-M-1 was the major CTX-M subtype in cattle and pigs in Europe [31]. In addition *bla* $_{CTX-M-1}$ was also detected in companion animal and poultry isolates [32, 33]. In the meantime, attribution of the above-named CTX-M types as specific to either human or animal isolates is no longer tenable. An increasing number of studies have identified *bla* $_{CTX-M-1}$ and *bla* $_{CTX-M-15}$ in both populations [3, 34] and our study confirmed this development. Among our isolates *bla* $_{CTX-M-15}$ was the most prevalent ESBL gene which occurred in approximately 50% of the strains investigated (human: 52.5%, animal: 46.1%). The number for *bla* $_{CTX-M-1}$ was considerably lesser but this gene was still relatively frequently present (23%, 28.7%). Isolates collected from horses seemed to hold a special position since rates of *bla* $_{CTX-M-1}$- and *bla* $_{CTX-M-15}$-positive isolates were almost equal (*bla* $_{CTX-M-1}$ 37%; *bla* $_{CTX-M-15}$ 38%). Previous studies have mainly identified CTX-M-1 as the predominant ESBL subtype among horse isolates [35, 36], which is also the most common type among livestock. In Germany horses are mostly used as recreational animals, normally situated close to livestock or in livestock-like environments. However, they usually have closer contact to humans and in case of disease medical care for horses is much more intense than in agricultural animal species. This could be a possible explanation for the particularly high detection rates of *bla* $_{CTX-M-15}$ reported here. Interestingly, *bla* $_{CTX-M-2}$ was found almost exclusively in equine *E. coli* isolates and was not present among human isolates. A study from Belgium identified *bla* $_{CTX-M-2}$ as major CTX-M subtype among diseased horses [37]. These animals had received prolonged treatment with several antibiotics and the *bla* $_{CTX-M-2}$ gene was part of the novel complex class 1 integron IS*CR* $_1$. Ferreira et al. could demonstrate the same linkage between *bla* $_{CTX-M-2}$ and IS*CR* $_1$ however, within the chromosome of *E. coli* isolates from healthy broiler chickens in Brazil [38]. Further investigation of the region surrounding the *bla* $_{CTX-M-2}$ gene may be necessary to identify possible associations with IS*CR* $_1$ among the isolates examined in this work. However, the CTX-M-2 subtype is not restricted to animal or even horse isolates. It was previously found in *Salmonella enterica* isolates from the UK and *E. coli* isolates from Brazil originating from clinically ill human patients [39, 40].

Ciprofloxacin-resistant animal isolates (55.7%) showed higher prevalence of PMQR-encoding genes compared to human isolates (33.6%). In addition, four *qnr*-variants (*qnrA, qnrB, qnrD* and *qnrS*) were identified in the group of animal isolates, compared to only *qnrB* and *qnrS* among the human isolates. All identified *qnr* variants have already been identified in both groups, however no study is consistent with the results found here [41, 42].

Another interesting result is the low prevalence of *bla* $_{SHV-1}$ among the *K. pneumoniae* isolates (10.3% of the human and in 29.4% of the animal *K. pneumoniae* isolates). Even though it is widely considered a plasmid-encoded enzyme, it has been found at high frequencies in *K. pneumoniae*.

Further the *bla* $_{SHV-1}$ origin is suspected in the chromosome of *K. pneumoniae*[43]. As some authors indicate high frequencies of up to 90% in *K. pneumoniae* isolates, our data does not seem to fit into the picture [43]. Yet there are other studies showing also low prevalences [44]. Based on these data it could be a local phenomenon.

The β-lactamase genes *bla* $_{SHV-1}$, *bla* $_{SHV-5}$, *bla* $_{SHV-11}$ and *bla* $_{TEM-52}$ were restricted to isolates from inpatients. This fact could be explained by their ancestry, since these genes are frequently found in isolates from hospitals [45].

Combination of various resistance genes was the rule rather than the exception (Table 4). In 70.1% of all isolates two or more genes responsible for β-lactam or quinolone resistance were identified. Combinations of several β-lactamase genes are commonly observed regularly within genomes of Gram-negative bacterial pathogens [46, 47] and co-selection of β-lactamase genes and determinants for resistance against other antibiotics e.g. fluoroquinolones has been postulated as a possible mechanism responsible for the widespread distribution of those combined genes [48]. The frequently observed combination of *bla* $_{CTX-M-15}$and *aac('6)-Ib-cr* could provide genetic support for this theory, which has been noted before [49, 50]. For this study the presence of *bla* $_{OXA-48}$ in combination with at least two penicillinase genes e.g. TEM, SHV or OXA, a PMQR gene and even sometimes with a CTX-M gene is alarming. Even more worrying is the fact that almost all isolates identified with the OXA-48 carbapenemase originate from companion animals. However, other carbapenemases could be present among the human and animal isolates, which were not tested for. The carbapenem-resistant animal isolates were mainly *K. pneumoniae* and *Enterobacter cloacae* isolates. An unrelated study conducted in our geographic region also detected carbapenem-resistant *Klebsiella* spp. among isolates from dogs [26].

E. coli and *K. pneumoniae* are the main species associated with β-lactamase production among isolates resistant to third generation cephalosporins in Germany and other members of the European Union [51]. Data from this study underlines this predominance of both bacterial species as a carrier of ESBL-encoding resistance genes with 74% *E. coli* and 17.5% *K. pneumoniae* isolates in total. Interestingly, the phylogenetic group B2 was strongly underrepresented among the ESBL-*E. coli* isolates from animals. Compared to the human isolates (26.2%) it was only present in 2.4%. The origins of primary sample material could provide an explanation for this discrepancy. Several studies suggest dominance of certain phylogenetic groups in particular specimen and types of infection [52, 53]. Zhang et al. demonstrated clustering of phylogenetic group B2 among urinary *E. coli* isolates from human patients [52]. The fact that 53% of the human isolates in this study had been recovered from urine while only 19% were from rectal swabs suggests an association between higher numbers of B2-*E. coli* isolates and the urinary tract of humans. The main source for the animal isolates was faeces (51%) from animals suffering from enteritis, diarrhoea and colic. It is not known whether the ESBL isolates detected were indeed causative for the syndromes observed as they could also be commensal strains that were selectively isolated solely on the basis of their resistance phenotype. This could explain the strong presence of phylogenetic group A/B1 and almost complete lack of group B2. Nevertheless, the pathogenic potential of ESBL-encoding isolates remains uncertain and needs further research. Studies performed on *E. coli* harbouring β-lactamases revealed several virulence-associated traits such as toxins, fimbriae, siderophores, polysaccharide coatings and invasins [54, 55]. In addition, patients with bacteremic urinary tract infections (UTIs) caused by ESBL-producing *Enterobacteriaceae* also have prolonged hospitalization and higher antibiotic costs [56].

Our study has limitations with regard to the selection of isolates. Human isolates originate from screening of patients upon admission for carrier-status as well as from clinically ill patients, whereas animal isolates only derive from diseased animals. In addition, selection criteria for ESBL- or PMQR-carrying isolates might have introduced bias. A screening breakpoint of >1 mg/L is recommended for cefotaxime in accordance with the guidelines issued by EUCAST [9]. Usage of the higher concentration of 2 mg/L might have led to loss of some ESBL types. PMQR-genes typically induce low-level quinolone-resistance with ciprofloxacin MIC of ≥ 0.25 μg/mL [57]. For this reason some PMQR may not have been detected due to the MIC of 1 μg/mL used.

Notwithstanding, the study gives an overall view on current dissemination of ESBL-encoding *Enterobacteriaceae* isolates within the region investigated. Further studies should be directed to examining specific transmission pathways assessing the exchange between the human and animal populations in owner-pet studies. In order to address this aspect in more detail, simultaneous screening of pets and/or owners positive for ESBL-producing *Enterobacteriaceae* should be carried out in the future. Another important factor for possible transmission or spread is the environment. Multiresistant bacteria are isolated with increasing frequencies from soil or water and wildlife [58]. Some studies implicate ESBL-producing *E. coli* already as a form of environmental pollution [59]. Our findings are supported by other non-regional studies that confirm the data presented [4, 60, 61]. In line with this work several studies describe presence of ESBL producing *Enterobacteriaceae* among human and animal samples with dominance of the CTX-M subtype [1, 2, 3].

Other aspects such as the colonisation and persistence in humans, animals and the environment as well as host species adaption and interspecies transmissibility of the bacteria involved must be taken into account. As plasmids are epidemiologically relevant agents in ESBL transfer, their role and association with sequence types (STs) need to be examined. Transfer of similar plasmids from human to animal bacterial populations and vice versa has been shown in broilers and broiler farmers in the Netherlands [62] and suggest that a similar situation could also exist between pet and owner.

5. CONCLUSIONS

The results from this study substantiate the alarming occurrence and ongoing spread of various ESBL-producing, multiresistant *Enterobacteriaceae* strains in the human and animal populations within a defined geographic region in central Hesse, Germany. Our data confirm that companion animals represent novel reservoirs of carbapenemase-producing strains of *K. pneumoniae* and *Enterobacter cloacae*. The study strongly supports the One Health concept which expands interdisciplinary collaborations and communications in all aspects pertaining to the health care for humans and animals [63].

ETHICAL APPROVAL

The study was approved by the ethics committee of medical faculty of the Justus Liebig University of Giessen (AZ: 95/11).

ACKNOWLEDGEMENTS

We thank Christina Gerstmann, Anja Schwanitz and David Dippel for excellent technical assistance.

This study was supported by grants from the Federal Ministry of Education and Research (BMBF, Germany) within the framework of the RESET research network (contract no. 01KI1013G) and the German Centre for Infection research (DZIF / grant number 8000 701–3 [HZI]) to TC and CI.

REFERENCES

1. Liebana E, Carattoli A, Coque TM, Hasman H, Magiorakos AP, Mevius D, Peixe L, Poirel L, Schuepbach-Regula G, Torneke K, Torren-Edo J, Torres C, Threlfall J: Public health risks of

enterobacterial isolates producing extended-spectrum-lactamases or AmpC-lactamases in food and food-producing animals: An EU perspective of epidemiology, analytical methods, risk factors, and control options. Clin Infect Dis. 2013, 56: 1030-1037.

2. Dolejska M, Villa L, Hasman H, Hansen L, Carattoli A: Characterization of IncN plasmids carrying blaCTX-M-1 and qnr genes in *Escherichia coli* and *Salmonella* from animals, the environment and humans. J Antimicrob Chemother. 2013, 68: 333-339.
3. Ewers C, Bethe A, Semmler T, Guenther S, Wieler LH: Extended-spectrum β-lactamase-producing and AmpC-producing *Escherichia coli* from livestock and companion animals, and their putative impact on public health: a global perspective. Clin Microbiol Infect. 2012, 18: 646-655.
4. Wu G, Day MJ, Mafura MT, Nunez-Garcia J, Fenner JJ, Sharma M, van Essen-Zandbergen A, Rodríguez I, Dierikx C, Kadlec K: Comparative analysis of ESBL-positive *Escherichia coli* isolates from animals and humans from the UK, the Netherlands and Germany. PLoS One. 2013, 8: e75392
5. Ewers C, Grobbel M, Stamm I, Kopp PA, Diehl I, Semmler T, Fruth A, Beutlich J, Guerra B, Wieler LH, Guenther S: Emergence of human pandemic O25:H4-ST131 CTX-M-15 extended-spectrum-β-lactamase-producing *Escherichia coli* among companion animals. J Antimicrob Chemother. 2010, 65: 651-660.
6. Ewers C, Bethe A, Stamm I, Grobbel M, Kopp PA, Guerra B, Stubbe M, Doi Y, Zong Z, Kola A, Schaufler K, Semmler T, Fruth A, Wieler LH, Guenther S: CTX-M-15-D-ST648 *Escherichia coli* from companion animals and horses: another pandemic clone combining multiresistance and extraintestinal virulence?. J Antimicrob Chemother. 2014, 69: 1224-1230.
7. Friese A, Schulz J, Laube H, von Salivati C, Hartung J, Roesler U: Faecal occurrence and emissions of livestock-associated methicillin-resistant *Staphylococcus aureus* (laMRSA) and ESBL/AmpC-producing *E. coli* from animal farms in Germany. Berl Munch Tierarztl Wochenschr. 2013, 126: 175-180.
8. Meyer E, Gastmeier P, Kola A, Schwab F: Pet animals and foreign travel are risk factors for colonisation with extended-spectrum β-lactamase-producing *Escherichia coli*. Infection. 2012, 40: 685-687.
9. EUCAST: 2012 EUCAST guidelines for detection of resistance mechanisms and specific resistances of clinical and/or epidemiological importance. [http://www.eucast.org/fileadmin/src/media/PDFs/EUCAST_files/Consultation/EUCAST_guidelines_detection_of_resistance_mechanisms_121222.pdf]
10. Mshana SE, Imirzalioglu C, Hossain H, Hain T, Domann E, Chakraborty T: Conjugative IncFI plasmids carrying CTX-M-15 among *Escherichia coli* ESBL producing isolates at a university hospital in Germany. BMC Infect Dis. 2009, 9: 97-
11. Grobner S, Linke D, Schutz W, Fladerer C, Madlung J, Autenrieth IB, Witte W, Pfeifer Y: Emergence of carbapenem-non-susceptible extended-spectrum –beta-lactamase-producing *Klebsiella pneumoniae* isolates at the university hospital of Tubingen, Germany. J Med Microbiol. 2009, 58: 912-922.
12. Kiratisin P, Apisarnthanarak A, Laesripa C, Saifon P: Molecular characterization and epidemiology of extended-spectrum- β-lactamase-producing *Escherichia coli* and *Klebsiella pneumoniae* isolates causing health care-associated infection in Thailand, where the CTX-M family is endemic. Antimicrob Agents Chemother. 2008, 52: 2818-2824.
13. Castillo B, Vinué L, Román EJ, Guerra B, Carattoli A, Torres C, Martínez-Martínez L: Molecular characterization of multiresistant *Escherichia coli* producing or not extended-spectrum β-lactamases. BMC Microbiol. 2013, 13: 84-
14. Chen X, Zhang W, Pan W, Yin J, Pan Z, Gao S, Jiao X: Prevalence of qnr, aac(6')-Ib-cr, qepA, and oqxAB in *Escherichia coli* Isolates from humans, animals, and the environment. Antimicrob Agents Chemother. 2012, 56: 3423-3427.
15. Cattoir V, Poirel L, Nordmann P: Plasmid-mediated quinolone resistance pump QepA2 in an *Escherichia coli* isolate from France. Antimicrob Agents Chemother. 2008, 52: 3801-3804.

16. Cattoir V, Poirel L, Rotimi V, Soussy CJ, Nordmann P: Multiplex PCR for detection of plasmid-mediated quinolone resistance qnr genes in ESBL-producing enterobacterial isolates. J Antimicrob Chemother. 2007, 60: 394-397.
17. Wang M, Guo Q, Xu X, Wang X, Ye X, Wu S, Hooper DC: New plasmid-mediated quinolone resistance gene, qnrC, found in a clinical isolate of *Proteus mirabilis*. Antimicrob Agents Chemother. 2009, 53: 1892-1897.
18. Cavaco LM, Hasman H, Xia S, Aarestrup FM: qnrD, a novel gene conferring transferable quinolone Resistance in *Salmonella enterica* serovar Kentucky and Bovis morbificans strains of human origin. Antimicrob Agents Chemother. 2009, 53: 603-608.
19. Park CH, Robicsek A, Jacoby GA, Sahm D, Hooper DC: Prevalence in the United States of aac(6')-Ib-cr encoding a ciprofloxacin-modifying enzyme. Antimicrob Agents Chemother. 2006, 50: 3953-3955.
20. Jones GL, Warren RE, Skidmore SJ, Davies VA, Gibreel T, Upton M: Prevalence and distribution of plasmid-mediated quinolone resistance genes in clinical isolates of *Escherichia coli* lacking extended-spectrum beta-lactamases. J Antimicrob Chemother. 2008, 62: 1245-1251.
21. Clermont O, Bonacorsi S, Bingen E: Rapid and simple determination of the *Escherichia coli* phylogenetic group. Appl Environ Microbiol. 2000, 66: 4555-4558.
22. GENE-E. [http://www.broadinstitute.org/cancer/software/GENE-E/] []
23. Magiorakos AP, Srinivasan A, Carey RB, Carmeli Y, Falagas ME, Giske CG, Harbarth S, Hindler JF, Kahlmeter G, Olsson-Liljequist B, Paterson DL, Rice LB, Stelling J, Struelens MJ, Vatopoulos A, Weber JT, Monnet DL: Multidrug-resistant, extensively drug-resistant and pandrug-resistant bacteria: An international expert proposal for interim standard definitions for acquired resistance. Clin Microbiol Infect. 2012, 18: 268-281.
24. Wieler LH, Ewers C, Guenther S, Walther B, Lübke-Becker A: Methicillin-resistant staphylococci (MRS) and extended-spectrum beta-lactamases (ESBL)-producing *Enterobacteriaceae* in companion animals: Nosocomial infections as one reason for the rising prevalence of these potential zoonotic pathogens in clinical samples. Int J Med Microbiol. 2011, 301: 635-641.
25. Rodriguez I, Barownick W, Helmuth R, Mendoza MC, Rodicio MR, Schroeter A, Guerra B: Extended-spectrum β-lactamases and AmpC β-lactamases in ceftiofur-resistant *Salmonella enterica* isolates from food and livestock obtained in Germany during 2003–07. J Antimicrob Chemother. 2009, 64: 301-309.
26. Stolle I, Prenger-Berninghoff E, Stamm I, Scheufen S, Hassdenteufel E, Guenther S, Bethe A, Pfeifer Y, Ewers C: Emergence of OXA-48 carbapenemase-producing *Escherichia coli* and *Klebsiella pneumoniae* in dogs. J Antimicrob Chemother. 2013, 68: 2802-2808.
27. Woodford N, Wareham DW, Guerra B, Teale C: Carbapenemase-producing *Enterobacteriaceae* and non-*Enterobacteriaceae* from animals and the environment: an emerging public health risk of our own making?. J Antimicrob Chemother. 2014, 69: 287-291.
28. Fischer J, Rodriguez I, Schmoger S, Friese A, Roesler U, Helmuth R, Guerra B: *Salmonella enterica* subsp. *enterica* producing VIM-1 carbapenemase isolated from livestock farms. J Antimicrob Chemother. 2013, 68: 478-480.
29. Shi W, Li K, Ji Y, Jiang Q, Wang Y, Shi M, Mi Z: Carbapenem and cefoxitin resistance of Klebsiella pneumoniae strains associated with porin OmpK36 loss and DHA-1 β-lactamase production. Braz J Microbiol. 2013, 44: 435-442.
30. Livermore DM: CTX-M: changing the face of ESBLs in the UK. J Antimicrob Chemother. 2005, 56: 451-454.
31. Geser N, Stephan R, Hächler H: Occurrence and characteristics of extended-spectrum β-lactamase (ESBL) producing Enterobacteriaceae in food producing animals, minced meat and raw milk. BMC Vet Res. 2012, 8: 21-
32. Ilse O: Extended-spectrum β-lactamase genes of *Escherichia coli* in chicken meat and humans, the Netherlands. Emerg Infect Dis. 2011, 17: 1216-1222.
33. Hordijk J, Schoormans A, Kwakernaak M, Duim B, Broens E, Dierikx C, Mevius D, Wagenaar JA: High prevalence of fecal carriage of extended spectrum β-lactamase/AmpC-producing *Enterobacteriaceae* in cats and dogs. Front Microbiol. 2013, 4: 242-

34. Poirel L, Nordmann P, Ducroz S, Boulouis HJ, Arne P, Millemann Y: Extended-spectrum-β-lactamase CTX-M-15-producing *Klebsiella pneumoniae* of sequence type ST274 in companion animals. Antimicrob Agents Chemother. 2013, 57: 2372-2375.
35. Johns I, Verheyen K, Good L, Rycroft A: Antimicrobial resistance in faecal *Escherichia coli* isolates from horses treated with antimicrobials: A longitudinal study in hospitalised and non-hospitalised horses. Vet Microbiol. 2012, 159: 381-389.
36. Dierikx CM, van Duijkeren E, Schoormans AHW, van Essen-Zandbergen A, Veldman K, Kant A, Huijsdens XW, van der Zwaluw K, Wagenaar JA, Mevius DJ: Occurrence and characteristics of extended-spectrum-β-lactamase- and AmpC-producing clinical isolates derived from companion animals and horses. J Antimicrob Chemother. 2012, 67: 1368-1374.
37. Smet A, Boyen F, Flahou B, Doublet B, Praud K, Martens A, Butaye P, Cloeckaert A, Haesebrouck F: Emergence of CTX-M-2-producing *Escherichia coli* in diseased horses: evidence of genetic exchanges of blaCTX-M-2 linked to ISCR1. J Antimicrob Chemother. 2012, 67: 1289-1291.
38. Ferreira JC, Penha Filho RAC, Andrade LN, Berchieri A, Darini ALC: Cutler: Detection of chromosomal blaCTX-M-2 in diverse *Escherichia coli* isolates from healthy broiler chickens. Clin Microbiol Infect. 2014, doi:10.1111/1469-0691.12531. published ahead of print
39. Berman H, Barberino MG, Moreira ED, Riley L, Reis JN: Distribution of strain type and antimicrobial susceptibility of *Escherichia coli* causing meningitis in a large urban setting in Brazil. J Clin Microbiol. 2014, doi:10.1128/JCM.03104-13. JCM.03104-13; published ahead of print
40. Burke L, Hopkins KL, Meunier D, de Pinna E, Fitzgerald-Hughes D, Humphreys H, Woodford N: Resistance to third-generation cephalosporins in human non-typhoidal *Salmonella enterica* isolates from England and Wales, 2010–12. J Antimicrob Chemother. 2014, 69: 977-981.
41. Yang T, Zeng Z, Rao L, Chen X, He D, Lv L, Wang J, Zeng L, Feng M, Liu JH: The association between occurrence of plasmid-mediated quinolone resistance and ciprofloxacin resistance in *Escherichia coli* isolates of different origins. Vet Microbiol. 2014, 170: 89-96.
42. Poirel L, Cattoir V, Nordmann P: Plasmid-Mediated Quinolone Resistance; Interactions between Human, Animal, and Environmental Ecologies. Front Microbiol. 2012, 3: 24-
43. Chaves J, Ladona MG, Segura C, Coira A, Reig R, Ampurdanés C: SHV-1 beta-lactamase is mainly a chromosomally encoded species-specific enzyme in *Klebsiella pneumoniae*. Antimicrob Agents Chemother. 2001, 45: 2856-2861.
44. Mshana SE, Hain T, Domann E, Lyamuya EF, Chakraborty T, Imirzalioglu C: Predominance of Klebsiella pneumoniae ST14 carrying CTX-M-15 causing neonatal sepsis in Tanzania. BMC Infect Dis. 2013, 13: 466-
45. Leistner R, Sakellariou C, Gürntke S, Kola A, Steinmetz I, Kohler C, Pfeifer Y, Eller C, Gastmeier P, Schwab F: Mortality and molecular epidemiology associated with extended-spectrum β-lactamase production in Escherichia coli from bloodstream infection. Infect Drug Resist. 2014, 7: 57-62.
46. Shahid M, Singh A, Sobia F, Rashid M, Malik A, Shukla I, Khan HM: BlaCTX-M, blaTEM, and blaSHV in *Enterobacteriaceae* from North-Indian tertiary hospital: high occurrence of combination genes. Asian Pacific J Tropical Med. 2011, 4: 101-105.
47. Munday CJ: Predominance and genetic diversity of community- and hospital-acquired CTX-M extended-spectrum β-lactamases in York, UK. J Antimicrob Chemother. 2004, 54: 628-633.
48. Cantón R, Coque TM: The CTX-M β-lactamase pandemic. Curr Opin Microbiol. 2006, 9: 466-475.
49. Sabtcheva S, Kaku M, Saga T, Ishii Y, Kantardjiev T: High prevalence of the aac(6')-Ib-cr gene and its dissemination among *Enterobacteriaceae* Isolates by CTX-M-15 plasmids in Bulgaria. Antimicrob Agents Chemother. 2009, 53: 335-336.
50. Shibl AM, Al-Agamy MH, Khubnani H, Senok AC, Tawfik AF, Livermore DM: High prevalence of acquired quinolone-resistance genes among *Enterobacteriaceae* from Saudi Arabia with CTX-M-15 β-lactamase. Diagn Microbiol Infect Dis. 2012, 73: 350-353.

51. European center for disease prevention and control (2012): Antibiotic resistance surveillance in Europe 2012. [http://www.ecdc.europa.eu]
52. Zhang L, Foxman B, Marrs C: Both urinary and rectal *Escherichia coli* isolates are dominated by strains of phylogenetic group B2. J Clin Microbiol. 2002, 40: 3951-3955.
53. Picard B, Garcia JS, Gouriou S, Duriez P, Brahimi N, Bingen E, Elion J, Denamur E: The Link between phylogeny and virulence in *Escherichia coli* extraintestinal infection. Infect Immun. 1999, 67: 546-553.
54. Karisik E, Ellington MJ, Livermore DM, Woodford N: Virulence factors in *Escherichia coli* with CTX-M-15 and other extended-spectrum -lactamases in the UK. J Antimicrob Chemother. 2007, 61: 54-58.
55. Valat C, Auvray F, Forest K, Metayer V, Gay E, Garam C, Madec JY, Haenni M: Phylogenetic grouping and virulence potential of extended-spectrum-β-lactamase-producing *Escherichia coli* strains in cattle. Appl Environ Microbiol. 2012, 78: 4677-4682.
56. Yang YS, Ku CH, Lin JC, Shang ST, Chiu CH, Yeh KM, Lin CC, Chang FY: Impact of extended-spectrum β-lactamase-producing *Escherichia coli* and *Klebsiella pneumoniae* on the outcome of community-onset bacteremic urinary tract infections. J Microbiol Immunol Infect. 2010, 43: 194-199.
57. Martínez-Martínez L, Pascual A, Jacoby GA: Quinolone resistance from a transferable plasmid. Lancet. 1998, 351: 797-799.
58. Hartmann A, Locatelli A, Amoureux L, Depret G, Jolivet C, Gueneau E, Neuwirth C: Occurrence of CTX-M producing *Escherichia coli* in soils, cattle, and farm environment in France (Burgundy region). Front Microbiol. 2012, 3: 83-
59. Guenther S, Ewers C, Wieler LH: Extended-spectrum beta-lactamases producing *E. coli* in wildlife, yet another form of environmental pollution?. Front Microbio. 2011, 2: 246-
60. Huber H, Zweifel C, Wittenbrink MM, Stephan R: ESBL-producing uropathogenic *Escherichia coli* isolated from dogs and cats in Switzerland. Vet Microbiol. 2013, 162: 992-996.
61. Dahmen S, Haenni M, Chatre P, Madec JY: Characterization of blaCTX-M IncFII plasmids and clones of *Escherichia coli* from pets in France. J Antimicrob Chemother. 2013, 68: 2797-2801.
62. Dierikx C, van der Goot J, Fabri T, van Essen-Zandbergen A, Smith H, Mevius D: Extended-spectrum-β-lactamase- and AmpC-β-lactamase-producing *Escherichia coli* in Dutch broilers and broiler farmers. J Antimicrob Chemother. 2013, 68: 60-67.
63. Calistri P, Iannetti S, Danzetta M, Narcisi V, Cito F, Di Sabatino D, Bruno R, Sauro F, Atzeni M, Carvelli A, Giovannini A: The components of 'one world - one health' approach. Transbound Emerg Dis. 2013, 60: 4-13.

CHAPTER 16

Comparative Evaluation of the Effects of Velogenic Newcastle Disease Virus Infection on the Hematology of Ducks and Chickens

Chekwube Paul Eze[1], John Osita Arinze Okoye[1], Innocent Okonkwo Ogbonna[2], Wilfred Sunday Ezema[1], Didacus Chukwuemeka Eze[1], Emmanuel Chukwudi Okwor[1], Christian Okorie-Kanu[3], Kalu Idika Idika[4]

[1]*Department of Veterinary Pathology and Microbiology, University of Nigeria, Nsukka, Nigeria*

[2]*Microbiology Unit, Department of Biological Sciences, University of Agriculture, Makurdi, Nigeria*

[3]*Department of Veterinary Pathology and Microbiology, University of Agriculture, Umudike, Nigeria*

[4]*Department of Veterinary Parasitology and Entomology, University of Nigeria, Nsukka, Nigeria*

ABSTRACT

The hematological lesions consequent upon velogenic Newcastle disease virus (NDV) infection were investigated in 6-week-old ducks and chickens. Following intramuscular inoculation, the results indicated significantly lower ($p < 0.05$) packed cell volume (PCV) in infected chickens (IC) on days 3 - 9 post inoculation (PI) and in infected ducks (ID) on days 3 - 15 PI. The hemoglobin concentrations were significantly lower ($p < 0.05$) in IC on days

3, 6 and 15 PI while in the ID, they were significantly lower ($p < 0.05$) on days 3, 9 and 15 PI. The total erythrocyte counts were significantly lower ($p < 0.05$) in IC on days 3, 9 and 15 PI and in ID, they were significantly lower ($p < 0.05$) on days 3 and 9 PI. The mean corpuscular values indicated macrocytic hypochromic anemia in IC and macrocytic normochromic anemia in ID. The leucogram showed leucopenia in IC and initial leucopenia followed by leucocytosis in ID. The hematological pictures of the velogenic NDV in this experiment indicate less susceptibility of ducks when compared with the chickens. The severity of this virus infection in chickens and the mild clinical signs and lesions presented by ducks showed that ducks are far less susceptible than chickens.

Keywords: Velogenic Newcastle Disease, Experimental Infection, Hematology, Ducks, Chickens

1. INTRODUCTION

Newcastle disease (ND) is a highly contagious and infectious disease that affects almost all avian species including poultry, cage and wild life bird species [1] -[3] . It is widespread affecting many continents of the world namely; Asia, Africa, and America [4] and causes considerable economic losses, not only due to high flock mortality but also through the economic impact that may arise from trade restrictions and embargoes [5] [6] . It has been reported to affect other birds such as guinea fowls, quails, turkeys, pheasants and peacock [1][6] . Mortality due to ND ranges from negligible to as high as 100% depending on the form or pathogenicity of the virus [1].

Kudu 113, a strain of NDV isolated from apparently healthy ducks and characterized by Echeonwu et al. [7] in Nigeria, has been studied in chickens and guinea fowls [6] [8]. This strain is frequently seen in outbreaks in Nigeria [9] . Ducks and other anseriformes are suspected to have maintained the endemicity of the disease around the world [4] [10] [11] . Scanty information is available on the basic performance of ducks in Nigeria [12] [13]. There is also a paucity of information on the effects of vNDV infection on the hematology of ducks and chickens. In this article, the comparative evaluation of the effects of vNDV infection on the hematology of ducks and chickens was done.

2. MATERIALS AND METHODS

2.1. Flock History

One hundred and seventy day old birds hatched the same day were obtained comprising 70 ducklings from poultry section of National Veterinary Research Institute (NVRI) Vom and 100 cockerel chicks from Zartech Hatchery. Brooding was done separately for the chicks and ducklings on deep litters under the same environmental condition. The cockerel received IBD vaccine by intraocular route at days 10 and 24 post hatch (PH). Chicks' mash was given ad libitum to the birds from day old to 8 weeks post hatch (PH). Growers' mash was given ad libitum also from 9 weeks PH until the end of the experiment and water was allowed free choice.

2.2. The NDV Inoculum

The vNDV strain, Kudu-113 isolated from the cloacal swab of an apparently healthy ducks in Kuru, Plateau State, biologically characterized by Echeonwu et al. [7] was used in the challenge experiment.

2.3. Experimental NDV Challenge

At six weeks of age, the chicks and ducklings were each randomly assigned into two groups of infected chicks (IC), uninfected chicks (UC) and infected ducks (ID), uninfected ducks (UD). The inoculum was reconstituted to give embryo lethal dose ($ELD_{50)}$ titre of $10^{6.36}$ per ml. The chicks and ducks in infected groups were inoculated intramuscularly (IM) with 0.2 ml of the inoculum (infected groups). The chicks and ducks in uninfected groups received 0.2 ml of PBS IM (control groups). The infected and control groups were housed at different locations and maintained on deep litter system.

2.4. Hematology

At days 3, 6, 9, 12, 15 and 21 PI, six birds from each group were randomly selected and phlebotomy was done, through the wing vein, using sterile syringe. 3 ml of blood was collected in EDTA impregnated sample bottles for hematology.

2.4.1. Packed Cell Volume (PCV)

The PCV was determined by the haematocrit method [14], using micro capillary tubes, micro haematocrit centrifuge and reader (Hawskey®, England). The PCV values were determined using the reader and recorded.

2.4.2. Hemoglobin Concentration (HbC)

The hemoglobin concentration (HbC) was determined by the cyanomethemoglobin method [15] . Twenty microliter of the blood sample was added to 5 ml of Drabkins hemoglobin reagent in a clean test tube and allowed to react for 20 minutes. The absorbance of the mixture was read at 540ηm wavelength against the blank reagent using a spectrophotometer. Standards were prepared, using rat blood and the wavelength determined. The HbC was calculated by multiplying the spectrophotometer reading with a calibrating factor (14.5) obtained from the absorbance and concentration of the standards.

HbC = K × Sample reading/Rat standard reading;

where **K** is a constant (**14.5**).

2.4.3. Total Erythrocyte Count (TEC)

The red blood cell count (RBC) was obtained by the hemocytometer method [16] , using an improved Neubauer counting chamber (NCC) (Hawskey, England) and avian RBC diluting fluid [14].

2.4.4. Mean Corpuscular Volume (MCV)

The results of the TEC, and PCV were used to calculate the average erythrocyte and its hemoglobin concentration, using the formula;

MCV (fL) = PCV × 10/TEC.

It was used in the definition of morphologic characteristic of RBC [17].

2.4.5. Mean Corpuscular Hemoglobin (MCH)

The results of the TEC and HbC were used to calculate the average corpuscular hemoglobin, using the formula;

MCH (pg) = HbC × 10/TEC [17].

2.4.6. Mean Corpuscular Hemoglobin Concentration (MCHC)

The results of the PCV and HbC were used to calculate the average corpuscular hemoglobin concentration, using the formula;

MCHC (g/dL) = HbC × 100/PCV [17].

2.4.7. Total Leucocyte Count (TLC)

The TLC was done by the hemocytometer method [16] , using an improved Neubauer counting chamber (Hawskey, England) and avian white blood cell diluting fluid [14] composed of aqueous phloxine, propylene, glycol, and sodium carbamate.

2.4.8. Differential Leukocyte Counts (DLC)

The DLC was done using Leishman technique [14]. The total of 100 cells was counted, by means of DLC tally counter and the number recorded. The results of each cell type were expressed as a percentage of the TLC and converted to absolute values.

2.5. Data Analysis

The hematological values between and within groups were subjected to statistical analysis using independent sample t-test and the level of significance was determined and accepted at $p \leq 0.05$ for all the results using statistical products for service and solution (SPSS) version 16.0 computer software. The mean ± standard error of mean (SEM) of the results obtained in the experiment were calculated and presented in tables and graphs.

3. RESULTS

The results of the PCV are shown in Table 1. The haematocrit values in the infected chickens were significantly lower ($p < 0.05$) than those of the control on days 3 to 9 PI, whereas in the infected ducks, they were significantly lower ($p < 0.05$) than those of the control ducks on days 3 to 15 PI.

The results of the HbC are shown in Table 2. The infected chickens recorded significantly lower ($p < 0.05$) values than the control on days 3 - 15 PI while in the infected ducks, they were significantly lower ($p < 0.05$) than those of the control on days 3 and 9 to 15 PI. The results of the TEC are shown in Table 3. The infected chickens had significantly lower TEC ($p < 0.05$) than the control group on days 3, 9 - 15 PI. In the infected ducks the TEC were significantly lower ($p < 0.05$) than the control group on days 3 - 12 PI.

The results of the MCV are shown in Table 4. The MCV of the infected chickens were significantly higher ($p < 0.05$) on days 3 and 9 - 15 PI, whereas the infected ducks recorded significantly higher ($p < 0.05$) values than the control on days 3 and 9 PI.

The results of the MCH are shown in Table 5. The infected chickens had significantly lower MCH ($p < 0.05$) on day 6 PI whereas, on day 9 PI, the infected chickens presented significantly higher ($p < 0.05$) values than in **Figure 1**. The infected chickens had significantly lower TLC ($p > 0.05$) the control. On the other hand, the infected ducks recorded significantly higher ($p < 0.05$) values than the control on days 3, 9 and 21 PI and significantly lower ($p < 0.05$) values than the control from days 12 - 15 PI.

Table 1. Results of mean packed cell volume ± SEM (%).

Days	IC	UC	ID	UD
0	30.17 ± 0.79	30.17 ± 0.79	37.33 ± 1.26	37.33 ± 1.26
3	24.92 ± 2.74*	33.33 ± 2.99	33.67 ± 1.02*	41.67 ± 1.05
6	29.00 ± 0.63*	34.17 ± 1.35	34.17 ± 1.30*	37.83 ± 1.05
9	27.83 ± 0.70*	32.17 ± 1.14	32.17 ± 0.65*	35.50 ± 1.18
12	28.83 ± 0.60	30.50 ± 1.38	32.17 ± 0.87*	39.67 ± 0.76
15	29.00 ± 0.86	31.33 ± 1.15	35.00 ± 2.04*	42.00 ± 1.59
21	29.17 ± 1.33	31.50 ± 0.56	38.17 ± 2.04	37.67 ± 1.15

NB* Means values significantly different at $p < 0.05$. Normal PCV range for birds (30.0 - 55.0).

Table 2. Results of mean hemoglobin concentration ± SEM (g/dL).

Days	IC	UC	ID	UD
0	09.70 ± 0.27	09.70 ± 0.27	12.52 ± 0.28	12.52 ± 0.28
3	08.24 ± 0.84*	13.36 ± 1.15	10.67 ± 0.32*	12.81 ± 0.18
6	06.16 ± 0.47*	11.33 ± 0.24	10.44 ± 0.54	11.34 ± 0.53
9	09.38 ± 0.83*	12.33 ± 0.61	12.58 ± 0.69*	13.52 ± 0.46
12	07.00 ± 0.33*	10.17 ± 0.58	10.01 ± 0.59*	13.78 ± 0.87
15	06.86 ± 0.33*	10.70 ± 0.78	10.21 ± 0.65*	13.05 ± 0.93
21	10.76 ± 0.54	11.06 ± 0.29	11.62 ± 1.06	12.78 ± 0.73

NB* Means values significantly different at $p < 0.05$. Normal HbC range for birds (8.0 - 13.0).

The results of the MCHC are shown in Table 6. The infected chickens presented lower significant ($p < 0.05$) values on days 6, 12 and 15 PI. In the infected ducks, the MCHC were significantly lower ($p < 0.05$) than the control only on day 12 PI.

The results of TLC are presented in **Figure 1**. The infected chickens had significantly lower TLC ($p > 0.05$) than the control from days 6 - 21 PI, while the TLC of the infected ducks presented significantly lower ($p < 0.05$) values than the control only on day 3 PI, but significantly higher ($p < 0.05$) values than the control on days 9 - 21 PI.

Table 3. Results of mean total erythrocyte counts (TEC) ± SEM (×106/μL).

Days	IC	UC	ID	UD
0	3.21 ± 0.32	3.21 ± 0.32	2.94 ± 0.21	2.94 ± 0.21
3	1.63 ± 0.03*	2.99 ± 0.04	1.90 ± 0.03*	3.19 ± 0.30
6	3.17 ± 0.40	3.07 ± 0.22	2.42 ± 0.12*	2.88 ± 0.14
9	1.70 ± 0.09*	2.88 ± 0.94	3.06 ± 0.13*	4.45 ± 0.08
12	1.97 ± 0.10*	2.84 ± 0.13	3.35 ± 0.09*	3.71 ± 0.15
15	1.93 ± 0.08*	2.89 ± 0.05	3.63 ± 0.16	3.41 ± 0.10
21	2.71 ± 0.07	2.82 ± 0.09	3.32 ± 0.80	3.01 ± 0.38

NB* Means values significantly different at $p < 0.05$. Normal TEC range for birds (2.8 - 4.5).

Table 4. Results of mean corpuscular volume ± SEM (fL).

Days	IC	UC	ID	UD
0	093.99 ± 11.18	093.99 ± 11.18	126.97 ± 06.02	126.97 ± 06.02
3	183.82 ± 15.39*	111.44 ± 13.62	177.21 ± 07.73*	130.63 ± 03.19
6	098.83 ± 11.98	111.30 ± 08.03	141.20 ± 11.45	131.35 ± 07.03
9	156.67 ± 03.09*	111.70 ± 12.76	105.13 ± 05.52*	079.78 ± 03.02
12	148.33 ± 08.71*	107.39 ± 07.64	096.03 ± 04.01	106.93 ± 04.15
15	155.33 ± 07.39*	108.41 ± 04.40	096.42 ± 03.89	109.09 ± 08.01
21	117.83 ± 13.10	111.70 ± 03.25	118.91 ± 06.88	098.61 ± 04.94

NB* Means values significantly different at $p < 0.05$. Normal MCV range for birds (100 - 130).

Table 5. Results of mean corpuscular hemoglobin ± SEM (pg).

Days	IC	UC	ID	UD
0	30.22 ± 2.06	30.22 ± 2.06	42.59 ± 3.57	42.59 ± 3.57
3	50.55 ± 4.72	44.68 ± 4.15	56.16 ± 1.89*	40.16 ± 0.33
6	19.43 ± 1.08*	36.91 ± 1.55	43.14 ± 2.29	39.38 ± 2.16
9	55.18 ± 5.25*	42.81 ± 2.10	41.11 ± 3.38*	30.38 ± 1.25
12	35.53 ± 1.33	35.17 ± 4.33	29.88 ± 2.29*	37.14 ± 1.63
15	35.54 ± 0.99	37.02 ± 2.20	28.13 ± 1.96*	33.90 ± 3.07
21	39.70 ± 1.50	39.22 ± 1.23	36.20 ± 3.33*	33.46 ± 2.69

NB* Means values significantly different at $p < 0.05$. Normal MCH range for birds (28.57 - 46.43).

Table 6. Results of mean corpuscular hemoglobin concentration ± SEM (g/L).

Days	IC	UC	ID	UD
0	32.15 ± 1.77	32.15 ± 1.77	33.54 ± 0.56	33.54 ± 0.56
3	33.07 ± 8.75	40.08 ± 0.88	31.69 ± 1.54	30.74 ± 0.85
6	21.12 ± 1.28*	33.16 ± 1.16	30.55 ± 2.03	29.98 ± 1.56
9	33.70 ± 3.64	38.33 ± 2.03	39.10 ± 2.01	38.08 ± 1.69
12	24.28 ± 1.11*	33.33 ± 2.45	31.12 ± 1.25*	37.14 ± 1.79
15	23.64 ± 0.90*	34.15 ± 3.52	29.17 ± 2.61	34.16 ± 2.57
21	36.87 ± 2.79	35.11 ± 1.17	30.04 ± 3.14	33.46 ± 2.46

NB* Means values significantly different at $p < 0.05$. Normal MCHC (26.70 - 33.47).

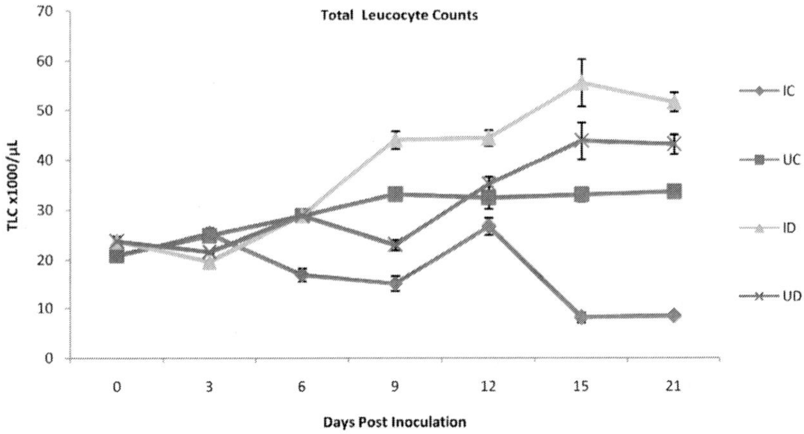

Figure 1. Results of total leukocyte counts ± SEM (103/µL).

The average differential leukocyte counts (DLC) results are shown in Table 7. There was significantly lower ($p < 0.05$) heterophil counts in infected chickens than the control from days 3 PI and consistently up till day 21 PI, whereas the infected ducks had significantly higher ($p < 0.05$) heterophil counts than the control on days 3 - 9 PI. By the day 12 the value was significantly lower ($p < 0.05$) than the control, but later increased significantly higher ($p < 0.05$) than the control group at day 21 PI (Table 7).

Table 7. Average absolute differential leukocyte counts in chickens and ducks (Mean ± SEM/µL).

Day	Cell	IC	UC	ID	UD
	H	5408 ± 300	5408 ± 300	6050 ± 245	6050 ± 245
	L	15.392 ± 854	15.392 ± 854	16.987 ± 690	16.987 ± 690
D0PI	M	0	0	233 ± 009	233 ± 009
	E	0	0	0	0
	B	0	0	0	0
	H	2271 ± 103*	2980 ± 117	2925 ± 053*	2152 ± 029
	L	21,193 ± 964	21,602 ± 847	14,235 ± 258*	18,722 ± 251
D3PI	M	1766 ± 080*	248 ± 010	1755 ± 032*	646 ± 009
	E	0	0	195 ± 007	0
	B	0	0	0	215 ± 003
	H	675 ± 103*	8634 ± 117	2306 ± 072*	1046 ± 024
	L	15,352 ± 1210*	17,844 ± 847	24,785 ± 776*	23,274 ± 535
D6PI	M	844 ± 067*	2302 ± 010	1729 ± 054*	1308 ± 030
	E	0	0	0	262 ± 006
	B	0	0	0	262 ± 006
	H	1356 ± 138*	10,582 ± 180	3697 ± 161*	1607 ± 074

	L	8114 ± 1453*	20.069 ± 680	38.654 ± 1408*	23.902 ± 986
D12PI	M	386 ± 069*	2913 ± 079	1777 ± 065*	2109 ± 087
	E	0	0	0	0
	B	0	0	0	352 ± 015
	H	970 ± 124*	6257 ± 624	4992 ± 426	4824 ± 400
	L	6626 ± 848*	26.015 ± 1381	48.259 ± 4116*	34.203 ± 2838
D15PI	M	485 ± 062*	659 ± 027	2219 ± 189*	3070 ± 255
	E	0	0	0	0
	B	0	0	0	0
	H	1445 ± 020*	6684 ± 114	5163 ± 191*	3870 ± 175
	L	6120 ± 086*	25.399 ± 434	40.268 ± 1490*	36.550 ± 1655
D21PI	M	765 ± 011*	1337 ± 023	3614 ± 134*	129 ± 006
	E	085 ± 001	0	103 ± 004	129 ± 006
	B	085 ± 001	0	154 ± 006	0

H = Heterophil, L = Lymphocyte, M = Monocyte E = Eosinophil and B = Basophil. *Means values significantly different at $p < 0.05$.

The lymphocyte counts were significantly lower ($p < 0.05$) in infected chickens than the control on days 6 and consistently remained lower up till 21 PI (Table 7).

In the infected ducks, the lymphocyte counts presented significantly lower ($p < 0.05$) values than the control only on day 3 PI and consistently recorded higher significant ($p < 0.05$) values than the control from days 6 to 21 PI (Table 7). The monocyte counts in infected chickens were significantly higher ($p < 0.05$) than the control only on day 3 PI, but consistently showed significantly lower ($p < 0.05$) values in infected than the control from days 6 to 15 PI (Table 7). In the infected ducks, there were bouts of monocytosis. The monocyte counts were significantly higher ($p < 0.05$) than the control from days 3 to 9 PI and 21 PI, but had significantly lower ($p < 0.05$) values than the control on days 12 and 15 PI (Table 7). The basophil and eosinophil counts were not consistent for both infected and control chickens and ducks (Table 7).

4. DISCUSSION

The erythrocyte response in the chickens and ducks had varying patterns in this study, even though there was significant depression of TEC, PCV and HbC in both infected chickens and duck when compared with the controls indicating anemia [18] . Blood samples were not collected from moribund birds to forestall polycythemia attributable to hemo-concentration occasioned by reduced water intake and the associated dehydration [19] . However, the values in infected ducks were within the reference normal range for birds in our teaching hospital. The severe anemia observed in the infected chickens may likely be due to possible intravascular hemolysis and blood loss usually seen in form of proventricular hemorrhages and GI ulcers in VND [20] . The mean corpuscular values in chickens indicated that circulating RBC at those points was mainly immature red cells (reticulocytosis) with less hemoglobin better described as macrocytic hypo-chromic anemia while in the ducks, the circulating RBC at those points was mainly mature RBC with adequately saturated hemoglobin, described as macrocytic normo-chromic anemia [21] -[23].

The variations in WBC—leukocytopenia in chickens and leucocytosis in ducks—may be due to heteropenia, lymphopenia and monocytopenia, and heterophilia, lymphocytosis and monocytosis respectively, demonstrating the marked depression and elevation reactions of the white corpuscles of chickens and ducks during ND respectively [14]. The lymphopenia recorded in chicken may be associated with depletion of lymphocytes in the bursa of Fabricus, spleen and thymus as reported by Ezema et al. [24]. Monocytopenia is a form of leukopenia associated with a deficiency of monocytes and which may be caused by acute infections, and other numerous disease syndromes [25]. The monocytes in the blood are in transit between the marrow and tissues, where they are transformed into tissue macrophages [26] . They participate virtually in all inflammatory and immune disorders such as severe viral and bacterial infections for mopping up of necrotic debris [27].

The leukocytic alterations recorded in this study indicate that vND exert marked depression on leucogram in chickens, as reported by Calderon et al. [20]. The bone marrow may have responded to tissue damage with initial leukocytosis and heterophilia but later depressed to heteropenia as observed in this study. Heterophilia and concurrent lymphopenia are observed in viral disease of birds [28] and lymphopenia of acute infection may have complex mechanisms which include endogenous corticosterone release with temporary lymphocyte redistribution, and trapping of circulating lymphocytes within lymphoid tissue, especially, during viral infection [29]. A sequential study in chickens after vND was made by Harrison et al. [25] who detected that the virus induced programmed cell death principally in mononuclear cells—macrophages and lymphocytes. Lymphopenia is common in acute inflammatory responses, because inflammatory mediators stimulate the diapedetic migration of heterophils and lymphocytes from the blood and lymphoid tissues to the points of inflammation where they are eventually destroyed [28]. This may account for the immunosuppression previously reported in vND in chickens [24]. The monocytosis and then monocytopenia in the infected chicken and monocytosis in duck may be related to the transit period of monocytes from bone marrow to the tissues as macrophages [27] and the variations between the 2 species may be related to the extent of tissue damage [27]. Eosinophil, and basophil absolute values in both infected birds, in this study, showed no consistent and relevant changes. This is in line with previous report by Calderon et al. [20] and the basic functions of eosinophils and basophils are in parasitic infections and hypersensitivity reactions respectively [30].

Chekwube Paul Eze, John Osita Arinze Okoye, Innocent Okonkwo Ogbonna, Wilfred Sunday Ezema, Didacus Chukwuemeka Eze, Emmanuel Chukwudi Okwor, Christian Okorie-Kanu, Kalu Idika Idika Generally, the hematology in this study showed the leucogram common in the VNDV challenged chickens and ducks as leukocytopenia and leucocytosis respectively. This may be the cause of variations in the severity of the disease in both species.

REFERENCES

1. Alexander, D.J. and Gough, R.E. (2003) Newcastle Disease and Other Avian Paramyxovirus Infections. In: Saif, Y.M., Barnes, H.J., Glisson, J.R., Fadly, A.M., McDougal, L.R. and Swayne, D.E., Eds., Disease of Poultry, 11th Edition, Iowa State University Press, Ames, 1A, 63-87.
2. Terregino, C., Cattoli, G., Grossele, B., Bertoli, E., Tisato, E. and Capua, I. (2003) Characterization of Newcastle Disease Virus Isolates Obtained from Eurasian Collared Doves (Streptopelia decaocto) in Italy. Avian Pathology, 32, 63-68. http://dx.doi.org/10.1080/0307 945021000070732
3. Vidanovic, D., Sekler, M., Asanin, R., Milic, N., Nisavic, J., Petrovic, T. and Savic, V. (2011) Characterization of Ve-logenic Newcastle Disease Viruses Isolated from Dead Wild Birds in Serbia during 2007. Journal of Wildlife Diseases, 47, 433-444. http://dx.doi.org/10.7589/0090-3558-47.2.433
4. Office International des Epizooties (OIE) (2009) Manual of Diagnostic Tests and Vaccines for Terrestial Animals. Newcastle Disease. OIE Standard Commission Publication, Office International Epizooties, 2009 Version, Part 2, Section 2.1, Chapter 2.3.14., 576-589.
5. Leslie, J. (2000) Newcastle Disease: Outbreak Losses and Control Policy Costs. Veterinary Record (Journal of British Veterinary Association), 146, 603-606. http://dx.doi.org/10.1136/vr.146.21.603
6. Igwe, A.O., Ezema, W.S., Ibu, J.I., Eze, J.I. and Okoye, J.O.A. (2013) Comparative Study of the Haematology and Persistence of Velogenic Newcastle Diseases in Chickens and Guinea Fowls. Research Opinion in Animal and Veteri-nary Sciences, 3, 136-142. http://www.roavs.com/pdf-files/Issue-5-2013/136-14...
7. Echeonwu, G.O.N., Ireogbu, C.U. and Emeruwa, A.C. (1993) Recovery of Velogenic Newcastle Disease Virus from Dead and Healthy Free Roaming Birds in Nigeria. Avian Pathology, 22, 383-387. http://dx.doi.org/10.1080/03079459308418928
8. Okoye, J.O.A., Agu, A.O., Chineme, C.N. and Echeonwu, G.O.N. (2000) Pathological Characterization in Chicken of a Velogenic Newcastle Disease Virus Isolate from Guinea Fowl. Revue d'élevage et de Médecine Vétérinaire des pays Tropicaux, 53, 325-330.
9. Ibu, O.J., Okoye, O.A., Adulugba, E.P., Chah, K.F., Shoyinka, S.V.O., Salihu, E., Chukwuedo, A.A. and Baba, S.S. (2009) Prevalence of Newcastle Disease Viruses in Wild and Captive Birds in Central Nigeria. International Journal of Poultry Science, 8, 574-578. http://dx.doi.org/10.3923/ijps.2009.574.578
10. Ikani, E.I. (2001) Duck Production in Nigeria (Extension Bulletin no 133) Poultry Series no 7. (National Agricultural Extension and Research Liaison Services). Ahmadu Bello University, Zaria, 1-32.
11. Sa'idu, L., Abdu, P.A., Tekdek, L.B., Umoh, J.U., Usman, M. and Oladele, S.B. (2006) Newcastle Disease in Nigeria. Nigerian Veterinary Journal, 2, 23-32.
12. Okeude, N.J., Okoli, I.C. and Igwe, G.O.F. (2003) Hematological Characteristics of Ducks (Cairina moschata) of South Eastern Nigeria. Tropicultura, 21, 61-65.

13. Alfred, S.D.Y. and Agbede, J.O. (2012) Influencing Factors of Duck Production in the Southwest of Nigeria. African Journal of Agricultural Research, 7, 3498-3505.
14. Coles, E.H. (1986) Veterinary Clinical Pathology. 4th Edition, W.B. Saunders Company, Philadelphia, 17-19.
15. Kachmar, J.F. (1970) Determination of Blood Hemoglobin by the Cyanomethaemoglobin Procedure. In: Tietz, N.W., Ed., Fundamentals of Clinical Chemistry, W S Saunders Company, Philadelphia, 268-269.
16. Schalm, O.M., Jain, N.C. and Carrol, E.J. (1975) Veterinary Hematology. 3rd Edition, Lea and Fabingers, Philadelphia.
17. Jain, N.C. (1993) Essentials of Veterinary Hematology. Lea and Febringers, Philadelphia, 365-372.
18. Velguth, K.E., Payton, M.E. and Hoover, J.P. (2010) Relationship of Hemoglobin Concentration to Packed Cell Volume in Avian Blood Samples. Journal of Avian Medicine and Surgery, 24, 115-121. http://dx.doi.org/10.1647/2008-042.1
19. Ihedioha, J.I. and Chineme, C.N. (2005) Fundamental of Systemic Veterinary Pathology, Volume 2, Great A P Express.
20. Calderon, N.L., Galindo-Muñiz, F., Ortiz, M., Lomniczi, B., Fehervari, T. and Paasch, L.H. (2005) Thrombocytopenia in Newcastle Disease: Haematological Evaluation and Histological Study of Bone Marrow. Acta Veterinaria Hungarica, 53, 507-513. http://dx.doi.org/10.1556/AVet.53.2005.4.11
21. Campbell, T.W. (2004) Clinical Chemistry of Birds. In: Thrall, M.A., Ed., Veterinary Hematology and Clinical Chemistry, Lippincott Williams and Wilkins, Philadelphia, 479-492.
22. Vajpayee, N., Graham, S.S. and Bem, S. (2007) Basic Examination of Blood and Bone Marrow. In: McPherson, R.A. and Pincus, M.R., Eds., Henry's Clinical Diagnosis and Management by Laboratory Methods, 22nd Edition, Elsevi-er/Saunders, Philadelphia, 30.
23. Elghetany, M.T. and Banki, K. (2011) Erythrocytic Disorders. In: McPherson, R.A. and Pincus, M.R., Eds., Henry's Clinical Diagnosis and Management by Laboratory Methods, 22nd Edition, Elsevier/Saunders, Philadelphia, 32. http://dx.doi.org/10.1016/B978-1-4377-0974-2.00032-4
24. Ezema, W.S., Okoye, J.O.A. and Nwanta, J.A. (2009) LaSota Vaccination May Not Protect against the Lesions of Velogenic Newcastle Disease in Chicken. Tropical Animal Health and Production, 41, 477-484. http://dx.doi.org/10.1007/s11250-008-9210-x
25. Harrison, L., Brown, C., Afonso, C., Zhang, J. and Susta, L. (2011) Early Occurrence of Apoptosis in Lymphoid Tissues from Chickens Infected with Strains of Newcastle Disease Virus of Varying Virulence. Journal of Comparative Pathology, 145, 327-335. http://dx.doi.org/10.1016/j.jcpa.2011.03.005
26. Grogan, K.B., Fernandez, R.J., Barranon, R.F.J. and Espinosa, H.G. (2008) Avian Immune System: A Brief Review. Merial Select, Gainesville, 1-12.
27. Thomson, R.G. (1984) General Veterinary Pathology.
28. Ng, W.L., Chu, C.M., Wu, A.K.L., Cheng, V.C.C. and Yuen, K.Y. (2005) Lymphopenia at Presentation Is Associated with Increased Risk of Infections in Patients with Systemic Lupus Erythematosus. Quarterly Journal of Medicine, 99, 37-47. http://dx.doi.org/10.1093/qjmed/hci155
29. Schmidt, E.M.S., Paulill, A.C., Caron, L.F., Filho, T.F., Augustini, M., Ventura, H.L.B. and Locatelli-Dittrich, R. (2008) Evaluation of Experimental Vaccination against Newcastle Disease and the Blood Proteinogram in Ring-Necked Pheasants (Phasiannus coichicus) during Breeding Season. International Journal of Poultry Science, 7, 661-664. http://dx.doi.org/10.3923/ijps.2008.661.664
30. McGavin, M.D. and Zachary, J.F. (2006) Pathologic Basis of Veterinary Disease. 4th Edition, Chapter 8, Elsevier, Mosby.

CHAPTER 17

Metagenomics of rumen bacteriophage from thirteen lactating dairy cattle

Elizabeth M Ross[1,2,3*], Steve Petrovski[1,2], Peter J Moate[1] and Ben J Hayes[1,2,3]

[1]*Department of Environment and Primary Industries, Bundoora, VIC 3086, Australia*

[2]*Dairy Futures Cooperative Research Centre, Bundoora, VIC 3086*

[3]*La Trobe University, Bundoora, VIC 3086, Australia.*

ABSTRACT

Background

The bovine rumen hosts a diverse and complex community of *Eukarya, Bacteria, Archea* and viruses (including bacteriophage). The rumen viral population (the rumen virome) has received little attention compared to the rumen microbial population (the rumen microbiome). We used massively parallel sequencing of virus like particles to investigate the diversity of the rumen virome in thirteen lactating Australian Holstein dairy cattle all housed in the same location, 12 of which were sampled on the same day.

Results

Fourteen putative viral sequence fragments over 30 Kbp in length were assembled and annotated. Many of the putative genes in the assembled contigs showed no homology to previously annotated genes, highlighting the large amount of work still required to fully annotate the functions encoded in viral genomes. The abundance of the contig sequences varied widely between animals, even though the cattle were of the same age, stage of lactation and fed the same diets. Additionally the twelve animals which were co-habited shared a number of their dominant viral contigs. We compared the functional characteristics of our bovine viromes with that of other viromes, as well as rumen microbiomes. At the functional level, we found strong similarities between all of the viral samples, which were highly distinct from the rumen microbiome samples.

Conclusions

Our findings suggest a large amount of between animal variation in the bovine rumen virome and that co-habiting animals may have more similar viromes than non co-habited animals. We report the deepest sequencing to date of the rumen virome. This work highlights the enormous amount of novelty and variation present in the rumen virome.

Keywords: Virome, Rumen, Bacteriophage, Metagenomics

1. BACKGROUND

The value of domestic ruminants (cattle, sheep and goats) comes from their ability to convert, by means of rumen fermentation, low-quality forages into high quality, high protein products (milk and meat) suitable for human consumption [1]. The rumen, the first section of the ruminant stomach, contains symbiotic microorganisms that breakdown ingested food. These microorganisms include *Eukarya, Bacteria, Archaea*[2], and large numbers ($>10^7$ pfu ml^{-1}) of virus including bacteriophage [3]. The entire bacteriophage or 'phage' population in a sample is termed a virome.

Massively parallel sequencing (MPS) is a technology which allows the generation of several million DNA sequences in parallel. MPS has been used to investigate the *Bacterial* and *Archaeal* populations of the bovine (*Bos taurus*) rumen [4, 5, 6, 7]. However, MPS studies on the rumen virome are scant. One study [8] sequenced 425 Mbp of viral DNA from the rumen of three animals, one lactating, one non lactating and one culled animal. The MPS study [8] and earlier studies such as [9], both suggest that there are large amounts of variation in the bovine rumen virome. However some of the variation may have been due to external sources (diet, lactation state, location, time of sampling).

Since the recent rumen virome study [8], there have been technological advances that have allowed the volume of sequence data to increase by an order of magnitude. These advances now enable the sequencing of viromes from a greater numbers of animals, resulting in the ability to gain an insight into natural variation between animals. MPS virome studies have made it possible to study viral genomes even when the host bacterium cannot be cultured. However, there are significant challenges involving analysis of MPS virome data [10]. Tools for the analysis of virome samples [11, 12, 13] are becoming more readily available allowing estimates of virome characteristics such as species richness. The combination of published analysis techniques, along with the increased ability to generate data has allowed the field to switch from descriptive analyses to quantitative investigations of how various factors (such as host effects or temporal variation) affect the virome [14].

To access the amount of variation in the rumen virome we extracted virus like DNA from the rumen fluid of 13 lactating dairy cattle. Twelve of the 13 cows were housed together, fed the same diet, had a shared location and were sampled on the same day, allowing us to investigate the magnitude of natural variation in the rumen virome.

2. RESULTS AND DISCUSSION

2.1 Assembly and annotation

We sequenced virus like particles from the rumen fluid of 13 lactating Australian Holstein dairy cattle on the HiSeq2000. Each library contained between 2 and 57 million read pairs (Table 1). After a combined assembly of all samples, 14 contigs were greater than 30 Kbp in length. These contigs were manually inspected for even coverage and annotated (Figure 1). Three of the contigs have all the genes necessary for a complete functional phage genome (contigs A, H and K). We mapped our sequence reads back to the complete set of contigs. While up to 57.5% of reads mapped to the whole assembly, only 0.84% of reads mapped to contigs greater than 10 Kbp (Figure 2a). This would suggest that although we have assembled up to 57% of the bovine rumen virome, most of what has been assembled is still highly fragmented, and contained in small contigs.

Five of the 14 largest contigs were observed in all samples (≥10 reads mapping back to contig), with all 14 observed in the sample from animal 2202 (which had the deepest sequence coverage; (Table 1).

Table 1 Virome sequence volume, alignments and richness

Sample	Data trimmed[T] (Read pairs)	Alignments to other viromes (%)						Richness by CatchAll			
		Cow 6993 [8]	Cow 7664 [8]	Cow 7887 [8]	Swine faeces [18][+]	Human saliva [19][+]	Pond [20][+]	Best model[B]	Estimated total species ± S.E.	Best discounted model[D]	Estimated total species ± S.E.
Cow 2202[C]	52,239,452	3.22	3.25	1.86	0.07	<0.01	<0.01	2MixedExp	5550786 ± 465864.1	SingleExp	18412.6 ± 1544.2
Cow 657	2,116,650	4.86	4.91	3.12	0.05	<0.01	<0.01	3MixedExp	4679342 ± 676180.6	2MixedExp	26795.8 ± 3871.9
Cow 1995	3,410,438	3.88	4.69	2.69	0.04	<0.01	<0.01	2MixedExp	5952188.7 ± 295239.5	SingleExp	5313.4 ± 263.5
Cow 2028	6,956,332	3.57	4.24	2.30	0.04	<0.01	<0.01	3MixedExp	6515260.9 ± 338382	2MixedExp	3369.8 ± 175
Cow 2042	4,048,214	3.19	3.71	2.22	0.03	<0.01	<0.01	2MixedExp	7904757.2 ± 622624.3	SingleExp	9023.1 ± 707.6
Cow 2165	4,362,890	4.57	5.14	3.20	0.04	<0.01	<0.01	4MixedExp	10353681.8 ± 119769.7	3MixedExp	1376492.9 ± 15923
Cow 3060	3,899,392	4.53	5.01	3.35	0.05	<0.01	<0.01	2MixedExp	5805999.9 ± 388938	SingleExp	11428.7 ± 765.5
Cow 4570	4,365,420	3.41	4.05	2.53	0.04	<0.01	<0.01	4MixedExp	22410379.6 ± 511083.2	3MixedExp	4126756.3 ± 94113.3
Cow 5679	3,681,922	3.70	4.84	2.67	0.04	<0.01	<0.01	3MixedExp	6098575.2 ± 404293.2	2MixedExp	6772.1 ± 448.5
Cow 6833	3,455,882	3.58	4.18	2.55	0.05	<0.01	<0.01	3MixedExp	7427673 ± 2447886.4	2MixedExp	30174.7 ± 9943.8
Cow 6857	8,998,900	4.18	4.71	2.77	0.04	<0.01	<0.01	3MixedExp	7916340 ± 2025360.6	2MixedExp	23466.2 ± 6003
Cow 6870	3,906,564	4.49	5.88	3.60	0.03	<0.01	<0.01	2MixedExp	4794540.3 ± 304303.4	SingleExp	13438.6 ± 852.7
Cow 7939	5,119,322	3.45	5.02	2.84	0.03	<0.01	<0.01	2MixedExp	6757358.4 ± 455261	SingleExp	7504.3 ± 495.6
Average*	8,197,029	3.89	4.59	2.75	0.04	<0.01	<0.01	-	7,858,991	-	435,304

*Average of all samples from this study.
[C]Cannulated animal.
[T]Amount of sequence remaining after removal of poor quality bases.
[+]Multiple samples from study combined.
[B]Best Model as assessed by the CatchALL program.
[D]Best Model that has been corrected using the assumption of overrepresented singletons.

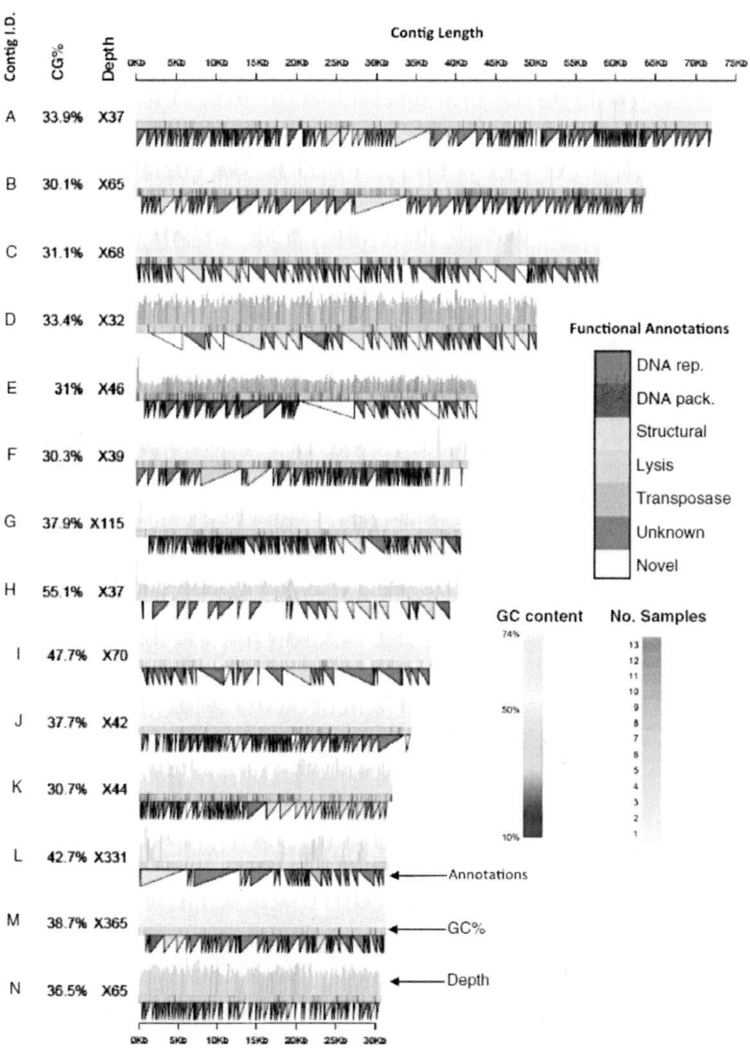

Figure 1 Virome contigs greater than 30 kbp. The median coverage of the contig and the GC% are shown to the left of each contig. The colour of the centre bar represents the GC% in a 100 bp sliding window. The grey lines on top of each contig represent the sequence depth, as determined by aligning reads to the contig using BWA. The shade of grey represents the number of different samples that have reads mapping to that position. The triangles below the contig show the function of the annotation at that position. DNA rep. = DNA replication and repair, DNA pack. = DNA packaging, Unknown = all significant BLAST hits were to unannotated subjects, Novel = no significant BLAST hit.

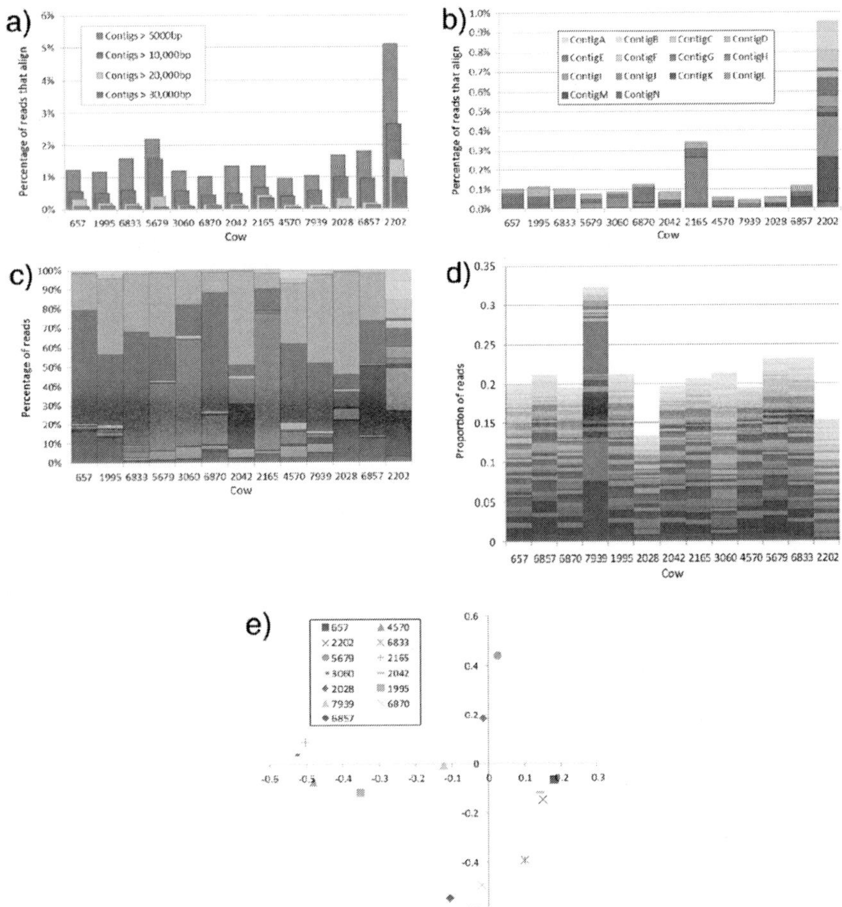

Figure 2 Between animal variation in alignments to contigs. **a)** Percentage of reads that map back to assembled contigs of different sizes. **b)** Distribution of reads that map to the 14 contigs annotated in Figure 1. **c)** Distribution of reads that map to the 14 contigs annotated in Figure 1 normalised to a total of 100%. **d)** Distribution of reads that map to the 100 most abundant (in our dataset) contigs assembled from [8]. **e)** Principle component analysis of reads that map to the 50 most abundant contigs assembled from [8]. PC1 (X axis) explains 15.42% of the variation, PC2 (Y axis) explains 12.73% of the variation.

Despite over a third of the most common contigs being represented in all samples, the most abundant contig only accounted for 0.24% of sequence reads (contig H in animal 2165; Figure 2b). There was a strong correlation between the sequencing depth of the sample and the percentage of reads that mapped back to the assembly (contigs > 5 Kbp; $r = 0.952$), partly reflecting the fact that the contigs were assembled from the same data. As sequence depth has a strong impact on contig size, it is likely that we have only assembled genomes from the most dominant phage species in the bovine rumen.

Of the 815 putative genes in the 14 largest contigs, 61.5% were assigned an unknown function because all of the significant BLAST hits were themselves unannotated. Genes involved with either DNA replication or repair consisted 13.3% of putative genes, 9.2% of putative genes were structural and 1.8% appear to be involved in lysis. Completely novel (no significant BLAST hits to the public nt database) putative genes made up an additional 13.0%. These novel and unannotated genes may encode unknown functions including those involved in cell lysis or lytogenic reproduction of the phage.

Phage may be integrated into the genomes of their host prokraryotes. To look for evidence of this we aligned sequence data from the microbiomes of the same samples from [15] to the 14 largest contigs. All of the largest contigs were observed in one or more microbiome samples. This could be explained by viral contamination of the microbiome. Such contamination would result in the same proportions of viral contigs in the microbiome as are observed in the virome. Therefore the proportions of reads that map to the viral contigs in the microbiome were compared to the virome. The abundance of two contigs, E and M, were significantly higher in the microbiome than expected under the viral contamination model ($p < 0.05$ after a Bonferroni correction). This is evidence that these two phage are integrating into the prokaryote genomes. However this hypothesis that cannot be validated until the host genomes are sequenced or the host and phage and isolated and cultured. Additionally, contigs A, B, C, E, F, G and N show homology to genes involved in lysis, suggesting that they may be lytic, however this is again speculation until the phage are isolated.

2.2 Variation in contig abundance between animals

The twelve co-habited animals' viromes appear to be dominated by contigs D, E and J, while animal 2202's virome shows a much more even coverage between contigs (Figure 2b). However these contigs only account of <0.34% of the viromes of the co-habited animals. There also appears to be a heterogeneity in the distributions of contigs between animals (Figure 2b-c). Contig H has a high abundance in cow 2165 (10,480 reads mapped back to the contig), but is almost below detectable limits in the deepest sequenced animal (2202, only 10 reads mapped back to the contig) and is undetectable in cow 657 (0 reads mapped back to the contig). The evenness observed in the virome of animal 2202 possibly hints that the rumen viral community is not dominated by a single species, but has a large number of taxa with a reasonably even distribution. This evenness of distribution was likely not observed in our other samples due to their limited sequence depth resulting in many unassembled species. To assess the distribution on an independent set of contigs we aligned our reads to contigs assembled from the independent set of animals from [8]. Between 0.22% and 0.70% of our reads aligned to the contigs. The distribution of contigs was much more even when these independent contigs were used (Figure 2d). A principle component analysis did not separate the independently housed animal from the co-house animal, although as less than 1% of the reads from any virome aligned, this should be interpreted with caution. The evenness of the phage population concurs with the findings of [8] that even the most abundant phage species represent only a small proportion of the rumen virome. Our finding that the most common of phage contigs only represent 0.24% of the virome is consistent with this. The higher degree of similarity between the viromes of the 12 co-habited animals and animal 2202 may suggest that there is transfer of viral species between the hosts within a herd, although 2202 was also cannulated, this confounding effect is not possible to resolve without further virome data. Still the variation between co-habited animals (for example contig H in animals 2165 and 657) illustrates that any transfer between animals does not maintain a completely homogeneous herd. Whether the within herd variation is related to traits that are related to rumen microbiome functions, such as methane emission levels or bloat susceptibility is yet to be seen, however this dataset shows that there is natural within herd virome variation which should be investigated. Future studies may explore the relationship between this variation and host traits such as disease susceptibility and methane emissions.

2.3 Species richness

Based on our sequence data, it is clear that estimations of rumen phage diversity using electron microscopy of less than 40 species per host summerised in [16] are dramatic underestimations, and the true number of species likely lies closer to the estimate in [8] of between 4 and 27 thousand. To estimate the number of species present in each of our samples, we used CatchAll [13]. The estimated total number of species were between 4.6 and 22.4 million per sample (Table 1). Over estimations of richness can be caused by spurious singletons, and the biological and technical causes for these have been outlined in [13]. To account for spurious singletons we also used a model which corrects for such artefacts.

This model reported between 3,370 and 4,126,756 (Table 1) species per sample. The divergence between these results and those reported in [8] is not surprising given that the authors of CatchAll consistently report richness higher than PHACCS [12], and claim that PHACCS underestimates richness. Our assembly contains 272 Mbp of assembled sequence, if we assume an average phage genome size of 30 Kbp this results in approximately 9073 genomes (assuming little overlap between contigs) worth of sequence in our assembly. Given that we have only assembled half of the sequence in our samples, and the assembled half would be the less diverse than the unassembled half, approximately 18 thousand genomes could be expected as a lower bound for number of species estimations. This rough calculation concurs with the CatchAll discounted model estimate for cow 2202, and together they tend to suggest that previous assessments of rumen phage richness are indeed underestimations. In any case, it is clear that there is a large amount of variation in the species that make up the *B. taurus* rumen virome. Also the large range of estimates, even under the reduced model, suggests that the viromes of some animals are much more species rich than others. This could be an indication of a more diverse prokaryote population in some animals than others. This large range should be considered in future experimental designs, as deeper sequencing of particularly diverse samples will be needed to capture a snapshot of the viral community.

2.4 Comparison to other viromes

There are a number of host associated and non host associated viromes sequenced to date. To compare these virome sequence profiles to ours we obtained reads from publically available datasets derived from published studies. Here we have only isolated viral DNA (not RNA), therefore we were unable to compare our results to studies such as that by [17], which isolated RNA viruses. We selected four different types of viromes for comparison with our data. The viromes included samples from cattle rumen, human saliva, swine faeces and non host-associated environments [8, 18, 19, 20]. The reads from the published viromes were aligned to our contigs. The other bovine rumen viromes had only a small percentage (between 0.70% and 1.56%; Table 2) of reads that mapped to our contigs. When we compared our reads to reads in the virome obtained from the rumen fluid of cattle as previously reported [8] only 2-5% of our reads mapped (Table 1). This would suggest that there are substantial differences between the viromes observed by [8] and ourselves. These differences could be a result of the phage isolation technique [21], or possibly the sequencing technology used. It is also possible that the differences were due to real variations, as there was large amounts of variation observed within our samples and it could be expected that cows from other locations and on different diets might be even more distinct.

Table 2 Virome alignments to assembled contigs

Sample	All contigs	Contigs >5000 bp	Contigs >10,000 bp	Contigs >20,000 bp	Contigs >30,000 bp
Average Rumen Virome*	37.1%	1.66%	0.84%	0.33%	0.17%
Cow 6993 [8]	0.70%	0.03%	0.01%	<0.01%	0.00%
Cow 766 [8]	0.56%	0.03%	0.01%	<0.01%	<0.01%
Cow 7887 [8]	1.56%	0.03%	<0.01%	0.00%	0.00%
Swine Faeces [18]⁺	0.01%	<0.01%	<0.01%	0.00%	0.00%
Human Saliva [19]⁺	0.00%	0.00%	0.00%	0.00%	0.00%
Pond [20]⁺	<0.01%	0.00%	0.00%	0.00%	0.00%

⁺Multiple samples from study combined.
*Average of all samples from this study.

We then compared our reads to the non-bovine viromes. The swine faeces virome [18], the human saliva virome [19] and aquaculture pond derived virome [20] did not have any reads mapping to our contigs (Table 2). We then attempted to map our reads to the 454 reads of the above mentioned databases. Less than 0.1% of our reads mapped to the swine faeces database, while less than 0.01% of our reads mapped to the human saliva or pond viromes (Table 1). There is an observable cline in the amount of our reads that map to these databases as they become less similar to our sample type, from highest from our own assembly, followed by the same sample type (bovine rumen), then an alternative species digestive tract (swine faeces), then non gut derived viromes.

This cline in reads that align to contigs from different habitats may be a result of the similar prokaryote populations within gut samples. Both swine faeces and bovine rumen are dominated by Firmicutes and Bacteroidetes phyla [6, 22], while the salivary microbiome is dominated by Proteobacteria [23]. Hence the viral species in gut samples likely prey mostly on Firmicutes and Bacteroidetes, which would result in more similarities between the gut derived viromes then between the rumen and non-gut derived viromes.

2.5 Functional characteristics of the rumen virome

We also investigated the hypothesis that the functional characteristics of the bovine rumen virome were conserved between animals despite the large amount of sequence variation observed. We assigned out reads to KEGG second level pathways. Of the assigned second level KEGG pathways, Nucleotide Metabolism and Replication and Repair combined made up 57% of virome assignments. This is in contrast to the prokaryote metagenomes of rumen fluid from animals from the same location [6] (Figure 3a).

Evenness of second level KEGG pathways was significantly lower in the phage sequences compared to the microbe data (t-test, $p < 0.001$; Table 3). This unevenness in the phage dataset is likely due to the lack of genes coding for functions such as carbohydrate metabolism, which the phage relies on the host cell to provide. We then compared the functional characteristics of our sequences to other viromes. A principle component analysis revealed that all viromes clustered, and all microbial metagenomes clustered based on assignments to KEGG second level pathways (Figure 3b). These analyses reveal how the virome of a sample is functionally distinct from the microbial metagenome, and that the functional characteristics of viromes are more similar to each other than to microbial metagenomes, even when the viromes are taken from different sample types (e.g. aquaculture ponds versus rumen fluid). This is likely a reflection of the limited number of annotatable functions encoded for in the phage metagenome. Interestingly the functions coded for in our viromes are not as variable as that observed in human faeces [14]. Some of the human faeces variation may be caused by the small number of reads which could be used in the KEGG analysis due to the limited percentage of alignments and the starting data volume.

The relatively narrow functional characteristics observed in the rumen fluid virome may be in part due to the lack of homology to annotated genes in the databases used for annotation (Figure 1). The potential of novel bacteriophage derived genes to control the bacterial populations of the host will likely be of interest in both a biomedical sense (for diseases which are associated with bacterial populations such as inflammatory bowel disease) and, in the case of livestock, reducing enteric methane emissions from the rumen.

2.6 Rumen virome importance

Domestic ruminants are important production animals to people worldwide. Their ability to use microbial fermentation to provide energy for food (meat and milk), fibre (wool) and power (bullocks used as draught animals) has seen their numbers soar to 3.6 billion [24]. Because of this, researchers have long been interested in the rumen microbiome, however the virome has been neglected due to the lack of appropriate marker genes and analysis benchmarks. It is imperative that the virome is studied along with the microbiome and the host animal. The virome, being pathogens of the microbiome, can substantially alter rumen function, which in turn may hinder the ability of the host to utilise feed efficiently. Additionally, as the microbiome in ruminants is much more closely linked to major nutrient acquisition than in humans and mice, a manipulation of the virome (and therefore the microbiome) may enable a more substantial manipulation of nutrient acquisition than is possible in monogastric animals.

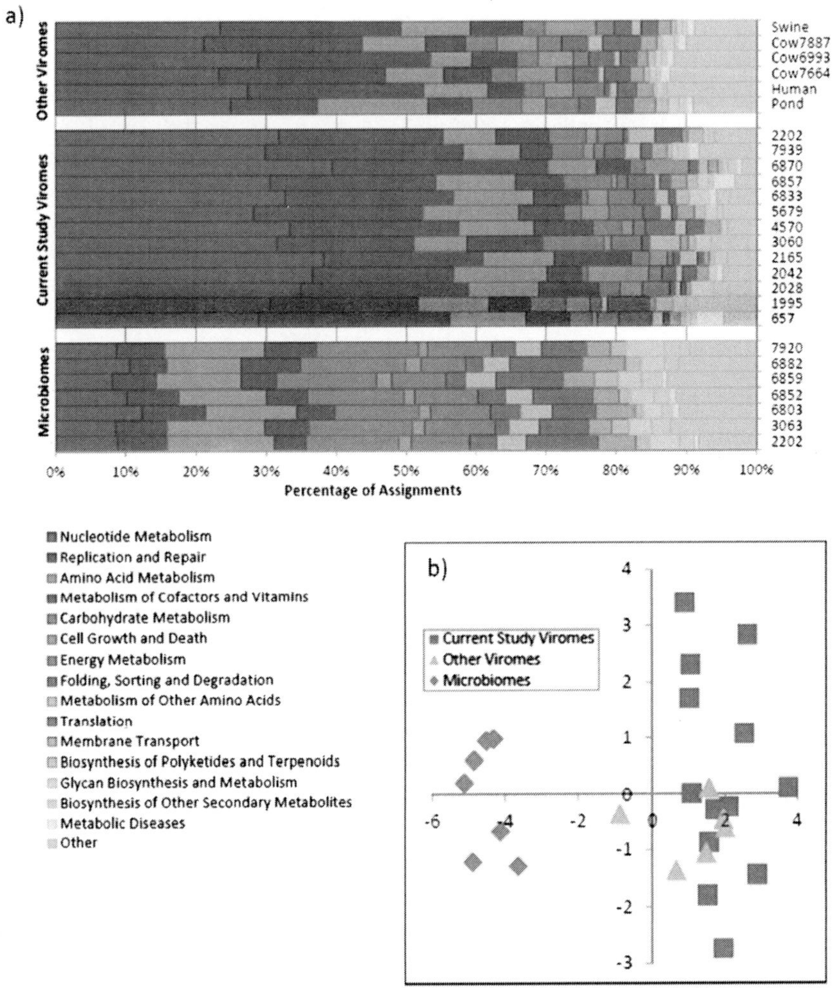

Figure 3 Between animal variation in KEGG pathways. a) Second level KEGG assignments of sequence reads. The "microbes" samples are microbial metagenomes [6]; other viromes include representatives from cattle, human saliva, swine faeces and non host-associated environments [8, 18, 19, 20]. b) Principle component analysis of the second level KEGG pathways. PC1 (X axis) explains 53.1% of the variation, PC2 (Y axis) explains 13.1% of the variation.

Table 3 Diversity and evenness of KEGG pathways

Metagenome source	Sample	Metagenome type	Categories observed[N]	Shannon diversity Index (H')	Evenness (E)
This study	Cow 2202[C]	Virome	20	2.192	0.732
	Cow 657	Virome	21	2.162	0.710
	Cow 1995	Virome	22	2.299	0.744
	Cow 2028	Virome	20	2.052	0.685
	Cow 2042	Virome	17	2.011	0.710
	Cow 2165	Virome	16	1.911	0.689
	Cow 3060	Virome	16	2.150	0.775
	Cow 4570	Virome	15	2.023	0.747
	Cow 5679	Virome	18	2.184	0.756
	Cow 6833	Virome	19	2.125	0.722
	Cow 6857	Virome	18	2.142	0.741
	Cow 6870	Virome	16	1.756	0.633
	Cow 7939	Virome	21	2.200	0.722
	Average*	Virome	18.4	2.093	0.720

Other Studies	Cow 6993 [8]	Virome	36	2.425	0.677
	Cow 766 [8]	Virome	34	2.553	0.724
	Cow 7887 [8]	Virome	35	2.580	0.726
	Swine Faeces [18]+	Virome	33	2.398	0.686
	Human Saliva [19]+	Virome	36	2.221	0.620
	Pond [20]+	Virome	31	2.537	0.739
	Cow 2202C [6]	Microbial	27	2.731	0.829
	Cow 3063C [6]	Microbial	30	2.794	0.822
	Cow 6803C [6]	Microbial	30	2.778	0.817
	Cow 6852C [6]	Microbial	29	2.809	0.834
	Cow 6859C [6]	Microbial	29	2.860	0.849
	Cow 6882C [6]	Microbial	24	2.752	0.866
	Cow 7920C [6]	Microbial	28	2.762	0.829

*Average of all samples from this study.
CCannulated animal.
NNumber of second level KEGG categories observed in sample's data.
H' – Shannon diversity Index.
E – Evenness (H' / H_{max}).
+Multiple samples from study combined.

Phages have been suggested as a possible methane mitigation strategy through "bacteriophage therapy" [25]. This strategy requires identification of phage species which target the dominant methane producing Archaea in the rumen. Such approaches will benefit from metagenomic virome sequencing by the availability of additional information regarding the dominant phage species in the rumen virome, and will help establish an understanding of the stability or variability of the rumen virome over time, which may affect such therapies. Furthermore, comparison of the virome and microbiome of each rumen sample may shed light on the ecological relationship between phage and bacterium. For now, this work has illustrated the vast amount of species richness and between animal variation in the rumen virome. Identifying which of these sequences may belong to Archaea pathogens is a significant challenge for future work.

Studies in other species have found that phages can have a protective role for their host [26], and studies investigating control of bacterial gut populations with phages have been successful [27]. The rumen virome of cattle may be the key to unlocking the secretes of the optimally functioning rumen, and therefore, should be a consideration in the quest to develop the ultimate production animal.

3. CONCLUSIONS

We have completed the deepest sequencing of the rumen virome to date. The results indicate large taxonomic diversity between the rumen viromes of Australian Holstein dairy cattle, and that animals housed together may have more similar viromes than those housed separately. However, this taxonomic variation is not reflected in the functional characteristics of the rumen virome. In fact, the rumen virome appears to be functionally conserved between animals. This suggests that while rumen viral genome sequences diverge rapidly between hosts, functional characteristics are under significant evolutionary constraint.

The observed differences between the rumen virome and the rumen microbiome likely reflect the much narrower set of functions typically encoded in a viral genome compared to a microbial genome. As MPS becomes more widely used to investigate viromes, it will be interesting to observe how these species correlate with changes in the bacterial population, or even if population differences are associated or predictive of key traits such as feed conversion efficiency and methane yield.

4. METHODS

4.1 Animals

Animals used in this study have been previously described [6, 28]. All animals were adult lactating Holstein dairy cattle. Twelve of the 13 cows were house together, fed the same diet, had a shared location and were sampled on the same day, the remaining cow (2202) was a cannulated individual which had been five months earlier. Rumen fluid from animal 2202 was removed via the fistula, rumen fluid from all other animal was collected via stomach pump. All work involving animals was approved by the Department of Environment and Primary Industries Animal Ethics Committee.

4.2 Molecular techniques & sequence quality control

A cell pellet was removed from the rumen fluid via centrifugation and then viral DNA was extracted from the supernatant as per [29]. The putative phage DNA was sequenced on the Illumina HiSeq2000 on a 101 bp paired end run.

Poor quality sequence was removed from the dataset using dynamic trimming such that no read had an average Phred quality score below 20, and no read contained more than three bases with a Phred quality score below 15.

4.3 Assembly and annotation

Sequence from all animals was computationally pooled and assembled into contigs using Gossamer [30]. Because of the complexity of metagenome samples, several assemblies were performed at different K-mer lengths. These assemblies were then further combined in a single Gossamer assembly (K-mer = 51). The 14 contigs that were greater than 30 Kbp were inspected for even coverage and manually annotated. Annotation was performed using Integrative services for genomic analysis [31] and the BLASTp algorithm at NCBI to determine the presence of conserved motifs. The majority of genes could not be assigned a function, either because all hits with significant homology were either unannotated or predicted proteins, or because the open reading frame lacked homology to any sequence in the nr (non-redundant protein) database (E-value > 10^5).

To obtain independent rumen virome contigs the reads from [8] were also assembled using Gossamer [30] with K-mer = 60.

4.4 Identifying integrated phage

Reads from [15] (microbiome sequence from the same samples as the 12 co-habited viromes) were aligned to each of the 14 largest contigs annotated in Figure 1 using BWA [32]. The proportion of alignments was calculated as A/B, where A is the number of reads that aligned to that contig, and B is the number of reads that aligned to all 14 contigs. The same procedure was applied to the virome data. Differences between the proportion of reads that aligned from the virome and microbiome were assessed by a paired t-test with two tails. A Bonferroni correction was applied by multiplying each p-value by the number of tests (14). To ensure that the overrepresentation in the microbiome was not due to a single conserved gene, alignments from the microbiome on the two significant contigs were manually checked. Converge of the contigs was relatively even, suggesting that the entire sequence was over represented in the microbiome (compared to expected contamination proportions).

4.5 Richness

Overlapping reads were joined together using FLASH [33]. To estimate the number of species present in each of our samples, we used CatchAll [13], employing the minimo assembler within Circonspect [34] (the parameters changed from default were -u 20 -v 100 -r 1 -l 35 -s 100000) on 100000 joined reads.

4.6 Between sample comparison

To compare the abundance of each contig in each sample the sequence reads were mapped to the assembled contigs using BWA-backtrack [32], with no insertions/deletions permitted. To compare the abundance of KEGG pathways we used 100,000 joined reads per sample and preformed a BLASTx search of the nr (non redundant protein) database of reads. Of the reads, 16.4% had a hit to the database with extremely relaxed parameters (e-value < 0.02). We then used MEGAN [35, 36] (default parameters) to extract KEGG assignments form the BLAST output.

ACKNOWLEDGMENTS

The authors are very grateful to the Dairy Futures Cooperative Research Centre for funding this work. Funding for this work was also provided by the Department of Environment and Primary Industries Victoria (DEPI),

Dairy Australia, Meat and Livestock Australia, and the Australian Government Department of Agriculture, Fisheries and Forestry under its Australia's Farming Future Climate Change Research Program. Thank you also to the staff of DEPI Ellinbank who helped with sampling and the staff of DEPI Bundoora who helped with molecular techniques, Andrew Robinson (La Trobe University) for helping with the CatchAll analysis and Professor Ben Cocks (DEPI) for comments on the manuscript.

REFERENCES

1. Van Soest PJ: Nutritional ecology of the ruminant. 1994, Ithaca, NY: Cornell Univ. Press
2. Deng W, Xi D, Mao H, Wanapat M: The use of molecular techniques based on ribosomal RNA and DNA for rumen microbial ecosystem studies: a review. Mol Biol Rep. 2008, 35 (2): 265-274. 10.1007/s11033-007-9079-1.
3. Mann NH: The third age of phage. PLoS Biol. 2005, 3: 182-10.1371/journal.pbio.0030182.
4. Brulc JM, Antonopoulos DA, Berg Miller ME, Wilson MK, Yannarell AC, Dinsdale EA, Edwards RE, Frank ED, Emerson JB, Wacklin P, et al: Gene-centric metagenomics of the fiber-adherent bovine rumen microbiome reveals forage specific glycoside hydrolases. Proc Natl Acad Sci U S A. 2009, 106 (6): 1948-1953. 10.1073/pnas.0806191105.
5. Hess M, Sczyrba A, Egan R, Kim T-W, Chokhawala H, Schroth G, Luo S, Clark DS, Chen F, Zhang T, et al: Metagenomic discovery of biomass-degrading genes and genomes from cow rumen. Science. 2011, 331 (6016): 463-467.
6. Ross EM, Moate PJ, Bath CR, Davidson SE, Sawbridge TI, Guthridge KM, Cocks BG, Hayes BJ: High throughput whole rumen metagenome profiling using untargeted massively parallel sequencing. BMC Genet. 2012, 13: 53.
7. Li RW, Connor EE, Li C, Baldwin V, Ransom L, Sparks ME: Characterization of the rumen microbiota of pre-ruminant calves using metagenomic tools. Environ microbiol. 2012, 14 (1): 129-139.
8. Berg Miller ME, Yeoman CJ, Chia N, Tringe SG, Angly FE, Edwards RA, Flint HJ, Lamed R, Bayer EA, White BA: Phage–bacteria relationships and CRISPR elements revealed by a metagenomic survey of the rumen microbiome. Environ Microbiol. 2012, 14 (1): 207-227.
9. Swain RA, Nolan JV, Klieve AV: Natural variability and diurnal fluctuations within the bacteriophage population of the rumen. Appl Environ Microbiol. 1996, 62: 994-997.
10. Reyes A, Semenkovich NP, Whiteson K, Rohwer F, Gordon JI: Going viral: next-generation sequencing applied to phage populations in the human gut. Nat Rev Microbiol. 2012, 10 (9): 607-617.
11. Angly FE, Willner D, Prieto-Davó A, Edwards RA, Schmieder R, Vega-Thurber R, Antonopoulos DA, Barott K, Cottrell MT, Desnues C, et al: The GAAS Metagenomic Tool and Its Estimations of Viral and Microbial Average Genome Size in Four Major Biomes. PLoS Comput Biol. 2009, 5 (12): e1000593-10.1371/journal.pcbi.1000593.
12. Angly F, Rodriguez-Brito B, Bangor D, McNairnie P, Breitbart M, Salamon P, Felts B, Nulton J, Mahaffy J, Rohwer F: PHACCS, an online tool for estimating the structure and diversity of uncultured viral communities using metagenomic information. BMC bioinform. 2005, 6 (1): 41-10.1186/1471-2105-6-41.
13. Allen HK, Bunge J, Foster JA, Bayles DO, Stanton TB: Estimation of viral richness from shotgun metagenomes using a frequency count approach. Microbiome. 2013, 1: 5-10.1186/2049-2618-1-5.
14. Reyes A, Haynes M, Hanson N, Angly FE, Heath AC, Rohwer F, Gordon JI: Viruses in the faecal microbiota of monozygotic twins and their mothers. Nature. 2010, 466 (7304): 334-338.

15. Ross EM, Moate PJ, Marett L, Cocks BG, Hayes BJ: Investigating the effect of two methane mitigating diets on the rumen microbiome using massively parallel sequencing. J dairy sci. 2013, 96 (9): 6030-6046.
16. Letarov A, Kulikov E: The bacteriophages in human- and animal body-associated microbial communities. J Appl Microbiol. 2009, 107 (1): 1-13.
17. Coetzee B, Freeborough M-J, Maree HJ, Celton J-M, Rees DJG, Burger JT: Deep sequencing analysis of viruses infecting grapevines: Virome of a vineyard. Virology. 2010, 400 (2): 157-163.
18. Allen HK, Looft T, Bayles DO, Humphrey S, Levine UY, Alt D, Stanton TB: Antibiotics in Feed Induce Prophages in Swine Fecal Microbiomes. mBio. 2011, 2: 6.
19. Pride DT, Salzman J, Haynes M, Rohwer F, Davis-Long C, White RA, Loomer P, Armitage GC, Relman DA: Evidence of a robust resident bacteriophage population revealed through analysis of the human salivary virome. ISME Journal. 2012, 6 (5): 915-926. 10.1038/ismej.2011.169.
20. Rodriguez-Brito B, Li L, Wegley L, Furlan M, Angly F, Breitbart M, Buchanan J, Desnues C, Dinsdale E, Edwards R, et al: Viral and microbial community dynamics in four aquatic environments. ISME J. 2010, 4 (6): 739-751.
21. Willner D, Furlan M, Schmieder R, Grasis JA, Pride DT, Relman DA, Angly FE, McDole T, Mariella RP, Rohwer F, et al: Metagenomic detection of phage-encoded platelet-binding factors in the human oral cavity. Proc Natl Acad Sci. 2011, 108: 4547-4553.
22. Lamendella R, Santo Domingo JW, Ghosh S, Martinson J, Oerther DB: Comparative fecal metagenomics unveils unique functional capacity of the swine gut. BMC microbiol. 2011, 11 (1): 103-10.1186/1471-2180-11-103.
23. Sedghizadeh PP, Yooseph S, Fadrosh DW, Zeigler-Allen L, Thiagarajan M, Salek H, Farahnik F, Williamson SJ: Metagenomic investigation of microbes and viruses in patients with jaw osteonecrosis associated with bisphosphonate therapy. Oral Surg Oral Med Oral Pathol Oral Radiol. 2012, 114 (6): 764-770.
24. Hackmann TJ, Spain JN: Invited review: Ruminant ecology and evolution: Perspectives useful to ruminant livestock research and production. J dairy sci. 2010, 93 (4): 1320-1334.
25. Patra A: Enteric methane mitigation technologies for ruminant livestock: a synthesis of current research and future directions. Environ Monit Assess. 2012, 184 (4): 1929-1952.
26. Barr JJ, Auro R, Furlan M, Whiteson KL, Erb ML, Pogliano J, Stotland A, Wolkowicz R, Cutting AS, Doran KS, et al: Bacteriophage adhering to mucus provide a non–host-derived immunity. Proc Natl Acad Sci. 2013, 110: 10771-10776.
27. Lu TK, Collins JJ: Engineered bacteriophage targeting gene networks as adjuvants for antibiotic therapy. Proc Natl Acad Sci U S A. 2009, 106 (12): 4629-4634.
28. Moate PJ, Williams SRO, Ribaux BE, Wales WJ: The 5th Australasian Dairy Science Symposium. Feeding grape marc to dairy cows suppresses methane emissions. 2012, 454-455.
29. Petrovski S, Seviour RJ, Tillett D: Genome sequence and characterization of the Tsukamurella bacteriophage TPA2. Appl Environ Microbiol. 2011, 77 (4): 1389-1398.
30. Conway T, Wazny J, Bromage A, Zobel J, Beresford Smith B: Gossamer--a resource-efficient de novo assembler. Bioinform (Oxford England). 2012, 28 (14): 1937-1938.
31. Hemmerich C, Buechlein A, Podicheti R, Revanna KV, Dong Q: An Ergatis-based prokaryotic genome annotation web server. Bioinform (Oxford England). 2010, 26 (8): 1122-1124.
32. Li H, Durbin R: Fast and accurate short read alignment with Burrows–Wheeler transform. Bioinformatics (Oxford, England). 2009, 25 (14): 1754-1760.
33. Magoc T, Salzberg SL: FLASH: Fast Length Adjustment of Short Reads to Improve Genome Assemblies. Bioinform (Oxford England). 2011, 27: 2957-2963.
34. Angly FE, Felts B, Breitbart M, Salamon P, Edwards RA, Carlson C, Chan AM, Haynes M, Kelley S, Liu H: The marine viromes of four oceanic regions. PLoS biol. 2006, 4 (11): e368-10.1371/journal.pbio.0040368.
35. Huson DH, Auch AF, Qi J, Schuster SC: MEGAN analysis of metagenomic data. Genome res. 2007, 17 (3): 377-386.

36. Huson DH, Mitra S, Ruscheweyh H-J, Weber N, Schuster SC: Integrative analysis of environmental sequences using MEGAN4. Genome res. 2011, 21: 1552-1560.

CHAPTER 18

Characterisation of a mobilisable plasmid conferring florfenicol and chloramphenicol resistance in *Actinobacillus pleuropneumoniae*

Janine T Bossé[a,*], Yanwen Li[a,1], Tom G Atherton[a], Stephanie Walker[a], Susanna M Williamson[b], Jon Rogers[b], Roy R Chaudhuri[c], Lucy A Weinert[c], Matthew TG Holden[d], Duncan J Maskell[c], Alexander W Tucker[c], Brendan W Wren[e], Andrew N Rycroft[f], Paul R Langford[a]

[a]*Section of Paediatrics, Department of Medicine, Imperial College London, St. Mary's Campus, London, W2 1PG, UK*

[b]*Animal and Plant Health Agency (APHA) Bury St Edmunds, Rougham Hill, Bury St Edmunds, Suffolk, IP33 2RX, UK*

[c]*Department of Veterinary Medicine, University of Cambridge, Madingley Road, Cambridge, CB3 0ES, UK*

[d]*The Wellcome Trust Sanger Institute, Wellcome Trust Genome Campus, Hinxton, Cambridge CB10 1SA, UK*

[e]*Faculty of Infectious & Tropical Diseases, London School of Hygiene & Tropical Medicine, Keppel Street, London, WC1E 7HT, UK*

[f]*Department of Pathology and Pathogen Biology, The Royal Veterinary College, Hawkshead Campus, Hatfield, Hertfordshire, AL9 7TA, UK*

ABSTRACT

The complete nucleotide sequence of a 7.7 kb mobilisable plasmid (pM3446F), isolated from a florfenicol resistant isolate of *Actinobacillus pleuropneumoniae*, showed extended similarity to plasmids found in other members of the *Pasteurellaceae* containing the *floR* gene as well as replication and mobilisation genes. Mobilisation into other *Pasteurellaceae* species confirmed that this plasmid can be transferred horizontally.

1. INTRODUCTION

Actinobacillus pleuropneumoniae, one of the main pathogens contributing to swine respiratory disease throughout the world, has been used as an indicator organism in surveillance studies of antimicrobial resistance in bacteria of animal origin (de Jong et al., 2014; Hendriksen et al., 2008). Tetracyclines, beta-lactams, and trimethoprim/sulphonamides are the commonest frontline treatments for porcine pleuropnuemonia in Europe (European Medicines Agency, 2012), however, levels of resistance are increasing (de Jong et al., 2014; Vanni et al., 2012). To a lesser extent, florfenicol, a fluorinated thiamphenicol derivative licensed to treat respiratory diseases of pigs and cattle in Europe since 2000 and 1995, respectively (Kehrenberg et al., 2004), has been used. Most surveys of *A. pleuropneumoniae* antimicrobial susceptibility have shown that nearly all isolates are susceptible to florfenicol, except in Korea where resistance levels of 34% were recently reported (Gutiérrez-Martín et al., 2006; Kucerova et al., 2011; Priebe and Schwarz, 2003; Shin et al., 2005; Vanni et al., 2012; Yoo et al., 2014).

A chloramphenicol/florfenicol efflux pump, encoded by *floR* (Schwarz et al., 2004), has been detected by PCR in florfenicol-resistant *A. pleuropneumoniae* isolates (Kucerova et al., 2011; Yoo et al., 2014), and this resistance was transferrable to *Escherichia coli*(Yoo et al., 2014), however no plasmid was characterised.

Here we describe the isolation and characterisation of a 7.7 kb florfenicol resistance plasmid from a clinical isolate of *A. pleuropneumoniae*. Comparative sequence analysis of this plasmid with others reported as mediating florfenicol resistance in bovine isolates of *Pasteurella multocida* (Kehrenberg and Schwarz, 2005; Kehrenberg et al., 2008) and *Mannheimia haemolytica* (Katsuda et al., 2012) is also described.

1.1. Material and methods

1.1.1. Bacterial isolates and MIC measurements

A. pleuropneumoniae MIDG3446, a serovar 2 clinical isolate, was originally cultured from pneumonic lungs of a pig in Greece in 2010 and archived at the then Animal Health and Veterinary Laboratory Agency (now Animal and Plant Health Agency) in England. Minimum inhibitory concentrations (MIC) for florfenicol and chloramphenicol were determined by agar dilution using Chocolate Mueller-Hinton plates, with *A. pleuropneumoniae* ATCC 27090, *Histophilus somni* ATCC 70025 and *Staphylococcus aureus* ATCC 29213 used as quality control strains (CLSI, 2008).

1.1.2. Plasmid isolation and confirmation of floR by PCR

Plasmid DNA was isolated using a Qiaprep Spin Miniprep kit (Qiagen) according to the manufacturer's protocol. The presence of the *floR* gene was confirmed by PCR using primers floR_for (CGACGCCCGCTATGATCCAACTC) and floR_rev (CCCAAAAAGCCGGACTCGCGAAG). Initially, florfenicol resistance was transferred to *E. coli* Stellar cells (Clontech) by heat shock with the MIDG3446 plasmid extract, according to the manufacturer's protocol, with selection of transformants on LB agar supplemented with 10 μg/ml florfenicol. Subsequently, mobilisation into selected *Pasteurellaceae* strains was assessed using a mating protocol previously described (Bossé et al., 2009; Bossé et al., 2015). Recipient strains included nalidixic acid resistant derivatives of *M. haemolytica* (MIDG1579Nal^R), *P. multocida* (MIDG1570 Nal^R), and *Haemophilus parasuis* (MIDG3176Nal^R), as well as the NAD-independent *A. pleuropneumoniae* isolate, MIDG2331Δ*ureC*::*nadV*, previously described (Bossé et al., 2009; Bossé et al., 2014). Transconjugants were selected on Brain Heart Infusion agar (with or without 0.01% NAD or 20 μg/ml nalidixic acid, as appropriate) supplemented with 2 μg/ml florfenicol. Transformants and transconjugants were tested by PCR for the presence of *floR*, as above, and for the *nadV* gene, as previously described (Bossé et al., 2015), where appropriate. MICs for florfenicol and chloramphenicol were determined, as above, for recipient strains +/− plasmid.

1.2. Plasmid sequence

A plasmid carrying the *floR* gene was isolated from a selected *E. coli* transformant (above) and the complete nucleotide sequence was determined using a primer walking strategy. Sequence analysis was carried out using BLASTn and BLASTx. Alignments with other *floR* plasmids were done using ClustalW.

The sequence of the *A. pleuropneumoniae floR* plasmid (pM3446F) has been deposited to Genbank (accession number KP696484).

2. RESULTS AND DISCUSSION

A serovar 2 isolate of *A. pleuropneumoniae*, MIDG3446, was found to be resistant to florfenicol and chloramphenicol (Table 1). Analysis of DNA from MIDG3446 revealed multiple small plasmids, one of which when transformed into *E. coli* Stellar cells (Clontech), conferred resistance to florfenicol and chloramphenicol (Fig. 1; Table 1). Additionally, mobilisation from MIDG3446 was successful into *A. pleuropneumoniae* MIDG 2331Δ*ureC*::*nadV* and *M. haemolytica* MIDG1579Nal^R, but not *P. multocida* MIDG1570Nal^R nor *H. parasuis* MIDG3176Nal^R.

The location of the genes encoding the conjugation machinery was not determined, and no attempts were made to optimise mating conditions, nor were other isolates of these species tested. PCR confirmed the presence of *floR* in transformants and transconjugants (Fig. 1), and MICs for florfenicol and chloramphenicol, determined as above, for recipient strains +/− plasmid indicated transfer of resistance with the plasmid (Table 1).

(A) Comparison of plasmid extracts from donor strain MIDG3446 (lane 1), recipient strains *E. coli* Stellar, *M. heamolytica* MIDG1579 and *A. pleuropneumoniae* MIDG2331 Δ*ureC*::*nadV* (lanes 2-4), and respective florfenicol resistant transformant/transconjugants showing transfer of plasmid pM3446F into *E. coli* Stellar, *M. heamolytica* MIDG1579, and *A. pleuropneumoniae* MIDG2331 Δ*ureC*::*nadV* (lanes 5-7)

(B) Multiplex PCR amplification of *floR* (510-bp amplicon) and *nadV* (1.5 kb amplicon) from *A. pleuropneumoniae* MIDG3446 (lane 1), *E. coli* Stellar (lane 2), *M. heamolytica* MIDG1579 (lane 3), *A. pleuropneumoniae* MIDG2331 Δ*ureC*::*nadV* (lane 4), *E. coli* Stellar + pM3446F (lane 5), *M. heamolytica* MIDG1579 + pM3446F (lane 6), *A. pleuropneumoniae* MIDG2331 Δ*ureC*::*nadV* + pM3446F (lane 7).

The complete 7,709-bp sequence of pM3446F, isolated from a selected *E. coli* transformant, was determined using a primer walking strategy. Sequence analysis using BLASTn and BLASTx (http://blast.ncbi.nlm.nih.gov/Blast.cgi) revealed extended similarity to pMh1405 (AB621552), a 7,674-bp plasmid isolated in 2009 from an *M. haemolytica* strain recovered from a calf in Japan (Katsuda et al., 2012), and to a lesser extent to pCCK381 (AJ871969), a 10,874-bp plasmid isolated in 2005 from a *P. multocida* strain recovered from a calf in Thirsk, England (Kehrenberg and Schwarz, 2005), and subsequently from bovine and porcine isolates of *P. multocida* in Germany (Kehrenberg et al., 2008).

Table 1. Antimicrobial susceptibility for strains with and without plasmids.

Strain	Species	MIC (μg/ml)		Source or Reference
		Florfenicol	Chloramphenicol	
MIDG3446	Ap[a]	8	8	This study
MIDG2331 Δ*ureC*::*nadV*	Ap	2	2	(Bossé et al., 2014)
MIDG2331 Δ*ureC*::*nadV* + pM3446F	Ap	8	8	This study
MIDG1579	Mh[b]	1	1	(Bossé et al., 2009)
MIDG1579 + pM3446F	Mh	16	16	This study
Stellar	Ec[c]	2	4	Clontech
Stellar + pM3446F	Ec	16	32	This study

[a] Ap = *A. pleuropneumoniae*.
[b] Mh = *M. haemolytica*.
[c] Ec = *E. coli*.

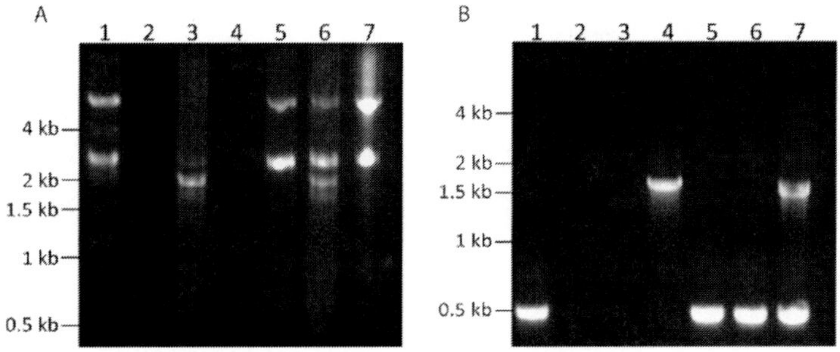

Figure. 1. Transfer of florfenicol resistance plasmid from *A. pleuropneumoniae* MIDG3446 by transformation into *E. coli* Stellar, or mating into *M. heamolytica* MIDG1579 and *A. pleuropneumoniae* MIDG2331 Δ*ureC*::*nadV*.

All three *Pasteurellaceae*-derived plasmids (Fig. 2) have virtually identical replication (*repA-C*) and mobilisation genes (*mobA-C*), as well as *oriV* and *oriT* sequences, as those comprising pDN1 (Fig. 2), a 5,112-bp RSF1010-like IncQ broad-host-range plasmid (Y19120) isolated from *Dichelobacter nodosus*, the causative agent of ovine footrot (Kehrenberg and Schwarz, 2005; Rawlings and Tietze, 2001; Whittle et al., 2000). However, whereas pDN1 does not encode resistance genes, these *Pasteurellaceae* plasmids contain differing fragments of what was initially described as Tn*floR*, a transposable element encoding *floR*, a *lysR* transcriptional regulator, and the *tnpA* transposase (Doublet et al., 2005), and subsequently shown to be an IS*CR2-floR* element (Toleman et al., 2006). Similar sequences have been found in various plasmids and chromosomes of different bacterial species, suggesting recombination and/or deletion events following integration have given rise to different truncated forms of the genes flanking *floR* (Doublet et al., 2005; Schwarz et al., 2004; Toleman et al., 2006).

The sequences flanking *floR* in these *Pasteurellaceae* plasmids show greatest similarity to those found in pAQU1 (AB571865), a 200 kb conjugative plasmid isolated from *Photobacterium damselae*, subsp. *damselae* in Japan (Nonaka et al., 2012). In Figure 2, only the sequence around the IS*CR2-floR* element of pAQU1 is shown (from bases 138014 to 143017 in the Genbank annotated sequence), with the Δ*tnp* gene upstream of *floR* almost identical to the last 720 bases of the *tnp* gene downstream. In pCCK381, in addition to the partial IS*CR2-floR* (99% identity with pAQU1 bases 138160–141381), the 2062 bases comprising the end of the *lysR* gene to the start of the *oriV* are 99% identical to bases 146509–148591 of pAQU1. Whereas, upstream of the Δ*tnp* gene in pCCK381, there are sequences 99-100% identity with other regions (bases 148616–148778 and bases 148992–149190) of pAQU1. In pMh1405, the 2.5 kb upstream of *oriV* shares 99% identity with bases 139426–141904 of pAQU1 (a different deletion fragment of IS*CR2-floR* than that seen in pCCK381); and in pM3446F no *tnp* sequences remain (with bases 622–2522 of pM3446F being 99% identical to bases 139426–141904 of pAQU1). An orf downstream of *floR* in pM3446F shows partial identity with *lysR*, with the first 223/348 bp being 100% identical to bases 141107–141329, and bases 214–334 of the orf having 98% identity with bases 148514–148633, of pAQU1. In addition, bases 25– 597 of pM3446F are 99% identical to bases 148616–149190 of pAQU1, parts of which are conserved in the sequence upstream of the Δ*tnp* gene in pCCK381. These data suggest that these *Pasteurellaceae* plasmids share a common origin, with sequences from pAQU1 (or a related plasmid) having integrated into pDN1, and subsequent deletions/rearrangements of the inserted sequences.

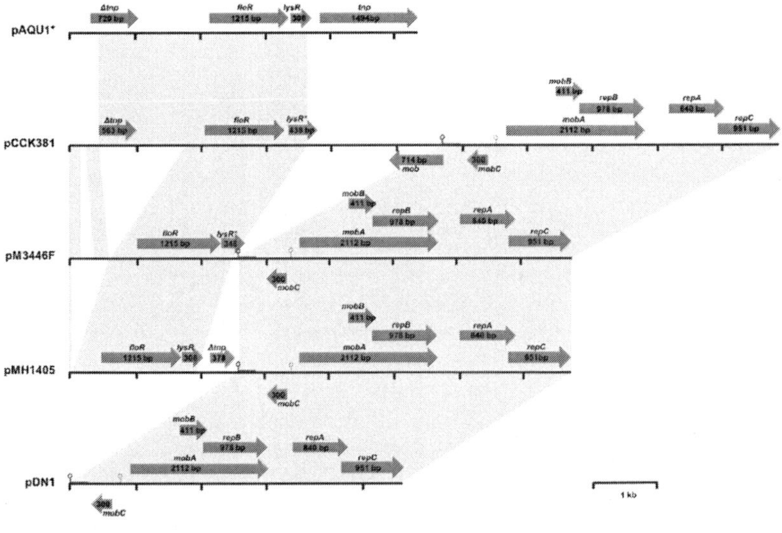

Figure. 2. Schematic comparison of florfenicol resistance plasmid pM3446F from *A. pleuropneumoniae* with plasmids pAQU1 (from *Photobacterium damselae*, subsp. *damselae*), pCCK381 and pMH1405 (from related *Pasteurellaceae*), and pDN1 (from *Dichelobacter nodosus*). Note: only 5.3 kb (from bases 138014–143017 of 204052 total) of pAQU1 is shown. Reading frames are indicated by arrows, with arrowheads showing direction of transcription (*floR*: florfenicol resistance; *lysR*: transcriptional regulator; *lysR**: partial *lysR*; *tnp*: transposase; Δ*tnp*: truncated transposase; *mob*, *mobA*, *mobB*, *mobC*: plasmid mobilisation; *repA*, *repB*, *repC*: plasmid replication). The predicted origins of replication (*oriV*;) and transfer (*oriT*;) are shown. Grey blocks between sequences indicate ≥ 98% nucleotide sequence identity. A distance scale in kb is shown.

Two further *Pasteurellaceae floR* plasmids have been identified: pCCK1900 (NC_011378), a 10,226-bp plasmid from a porcine isolate of *P. multocida* in Germany (Kehrenberg et al., 2008) that in encodes resistance to florfenicol (*floR*), sulphonamide (*sul2*), and streptomycin (*strA, strB*); pCCK13698 (NC_007800), a 14,969-bp plasmid from a bovine isolate of *Bibersteinia trehalosi* in France (Kehrenberg et al., 2006) that encodes resistance to florfenicol (*floR*), chloramphenicol (*catA3*) and sulphonamide (*sul2*). The sequence of pCCK1900 shows simple integration of the *floR* and *lysR* genes into an RSF1010 backbone (Kehrenberg et al., 2008). Although pCCK13698 shares some *rep* and *mob* gene sequences, as well as the *sul2* gene, with RSF1010, it shows more extensive rearrangements with sequences from other plasmids including pHS-rec from *H. parasuis* (Lancashire et al., 2005) and pMVSCS1 from *Mannheimia varigena* (Kehrenberg and Schwarz, 2002), and the *floR* and *lysR* genes from IS*CR2-floR* (Kehrenberg et al., 2006). Both pCCK1900 and pCCK13698 appear to have arisen separately from the pDN1-like plasmids described above.

3. CONCLUSION

In summary, to our knowledge, this is the first report of a complete sequence of a mobilisable florfenicol resistance plasmid from *A. pleuropneumoniae*. Structural analysis of pM3446F revealed extensive similarity to two florfenicol resistance plasmids found in other members of the *Pasteurellaceae*, and mating experiments confirmed the ability to mobilise between species. This highlights the importance of continued surveillance of florfenicol susceptibility in *A. pleuropneumoniae* and other Gram-negative pathogens that may co-exist within the respiratory tract of pigs.

3.1. Funding

This work was supported by a Longer and Larger (LoLa) grant from the Biotechnology and Biological Sciences Research Council (grant numbers BB/G020744/1, BB/G019177/1, BB/G019274/1 and BB/G018553/1), the UK Department for Environment, Food and Rural Affairs, and Zoetis (formerly Pfizer Animal Health) awarded to the Bacterial Respiratory Diseases of Pigs-1 Technology (BRaDP1T) consortium. MTGH was supported by Wellcome Trust grant 98051. The MIC work was funded from the former AHVLA's Research and Development Internal Investment Fund RD0030c. The funders had no role in study design, data collection and analysis, decision to publish, or preparation of the manuscript.

ACKNOWLEDGEMENTS

The BRaDP1T Consortium comprises: Duncan J. Maskell, Alexander W. (Dan) Tucker, Sarah E. Peters, Lucy A. Weinert, Jinhong (Tracy) Wang, Shi-Lu Luan, Roy R. Chaudhuri (University of Cambridge; present address for R. Chaudhuri is: Department of Molecular Biology and Biotechnology, University of Sheffield, Firth Court, Western Bank, Sheffield, S10 2TN, UK), Andrew N. Rycroft, Gareth A. Maglennon, Dominic Matthews (Royal Veterinary College); Brendan W. Wren, Jon Cuccui, Vanessa S. Terra (London School of Hygiene and Tropical Medicine); and Paul R. Langford, Janine T. Bossé, Yanwen Li (Imperial College London). We wish to thank Susanna Williamson and Chris Teale from the APHA for their advice and input.

REFERENCES

1. Bossé, J.T., Durham, A.L., Rycroft, A.N., Kroll, J.S., Langford, P.R., 2009. New plasmid tools for genetic analysis of Actinobacillus pleuropneumoniae and other Pasteurellaceae. Appl. Environ. Microbiol. 75, 6124–6131. doi:http://dx.doi.org/ 10.1128/AEM.00809-09.
2. Bossé, J.T., Soares-Bazzolli, D.M., Li, Y., Wren, B.W., Tucker, A.W., Maskell, D.J., Rycroft, A.N., Langford, P.R., BRaDP1T consortium, 2014. The generation of successive unmarked mutations and chromosomal insertion of heterologous genes in Actinobacillus pleuropneumoniae using natural transformation. PLoS ONE 9, e111252. doi:http://dx.doi.org/10.1371/journal.pone.0111252.
3. Bossé, J.T., Li, Y., Walker, S., Atherton, T., Fernandez Crespo, R., Williamson, S.M., Rogers, J., Chaudhuri, R.R., Weinert, L.A., Oshota, O., Holden, M.T.G., Maskell, D.J., Tucker, A.W., Wren, B.W., Rycroft, A.N., Langford, P.R., 2015. BRaDP1T consortium, 2015. Identification of dfrA 14 in two distinct plasmids conferring trimethoprim resistance in Actinobacillus pleuropneumoniae. J. Antimicrob. Chemother. doi:http://dx.doi.org/10.1093/ jac/ dkv121.
4. de Jong, A., Thomas, V., Simjee, S., Moyaert, H., Garch, El, F., Maher, K., Morrissey, I., Butty, P., Klein, U., Marion, H., Rigaut, D., Vallé, M., 2014. Antimicrobial susceptibility monitoring of respiratory tract pathogens isolated from diseased cattle and pigs across Europe: the vetPath study. Vet. Microbiol. 172, 202–215.
5. Clinical Laboratory Standards Institute, C.L.S.I., 2008. Performance Standards for Antimicrobial Disk and Dilution Susceptibility Tests for Bacteria Isolated From Animals - Third Edition: Approved Standard M31-A3. CLSI, Wayne, PA.
6. Doublet, B., Schwarz, S., Kehrenberg, C., Cloeckaert, A., 2005. Florfenicol resistance gene floR is part of a novel transposon. Antimicrob. Agents Chemother. 49, 2106–2108.
7. European Medicines Agency, 2012. Sales of veterinary antimicrobial agents in 19 EU/ EEA countries in 2010. (EMA/88728/2012).
8. Gutiérrez-Martín, C.B., del Blanco, N.G., Blanco, M., Navas, J., Rodríguez-Ferri, E.F., 2006. Changes in antimicrobial susceptibility of Actinobacillus pleuropneumoniae isolated from pigs in Spain during the last decade. Vet. Microbiol. 115, 218–222.
9. Hendriksen, R.S., Mevius, D.J., Schroeter, A., Teale, C., Jouy, E., Butaye, P., Franco, A., Utinane, A., Amado, A., Moreno, M., Greko, C., Stärk, K.D., Berghold, C., Myllyniemi, A.-L., Hoszowski, A., Sunde, M., Aarestrup, F.M., 2008. Occurrence of antimicrobial resistance among bacterial pathogens and indicator bacteria in pigs in different European countries from year 2002–2004: the ARBAO-II study. Acta. Vet. Scand. 50, 19.

10. Katsuda, K., Kohmoto, M., Mikami, O., Tamamura, Y., Uchida, I., 2012. Plasmid-mediated florfenicol resistance in Mannheimia haemolytica isolated from cattle. Vet. Microbiol. 155, 444–447.
11. Kehrenberg, C., Schwarz, S., 2005. Plasmid-borne florfenicol resistance in Pasteurella multocida. J. Antimicrob. Chemother. 55, 773–775.
12. Kehrenberg, C., Meunier, D., Targant, H., Cloeckaert, A., Schwarz, S., Madec, J.-Y., 2006. Plasmid-mediated florfenicol resistance in Pasteurella trehalosi. J. Antimicrob. Chemother. 58, 13–17.
13. Kehrenberg, C., Mumme, J., Wallmann, J., Verspohl, J., Tegeler, R., Kühn, T., Schwarz, S., 2004. Monitoring of florfenicol susceptibility among bovine and porcine respiratory tract pathogens collected in Germany during the years 2002 and 2003. J. Antimicrob. Chemother. 54, 572–574.
14. Kehrenberg, C., Schwarz, S., 2002. Nucleotide sequence and organization of plasmid pMVSCS1 from Mannheimia varigena: identification of a multiresistance gene cluster. J. Antimicrob. Chemother. 49, 383–386.
15. Kehrenberg, C., Wallmann, J., Schwarz, S., 2008. Molecular analysis of florfenicol-resistant Pasteurella multocida isolates in Germany. J. Antimicrob. Chemother. 62, 951–955. doi:http://dx.doi.org/10.1093/jac/dkn359.
16. Kucerova, Z., Hradecka, H., Nechvatalova, K., Nedbalcova, K., 2011. Antimicrobial susceptibility of Actinobacillus pleuropneumoniae isolates from clinical outbreaks of porcine respiratory diseases. Vet. Microbiol. 150, 203–206.
17. Lancashire, J.F., Terry, T.D., Blackall, P.J., Jennings, M.P., 2005. Plasmid-encoded Tet B tetracycline resistance in Haemophilus parasuis. Antimicrob. Agents Chemother. 49, 1927–1931.
18. Nonaka, L., Maruyama, F., Miyamoto, M., Miyakoshi, M., Kurokawa, K., Masuda, M., 2012. Novel conjugative transferable multiple drug resistance plasmid pAQU1 from Photobacterium damselae subsp. damselae isolated from marine aquaculture environment. Microbes Environ. 27, 263–272.
19. Priebe, S., Schwarz, S., 2003. In vitro activities of florfenicol against bovine and porcine respiratory tract pathogens. Antimicrob. Agents Chemother. 47, 2703–2705.
20. Rawlings, D.E., Tietze, E., 2001. Comparative biology of IncQ and IncQ-like plasmids. Microbiol. Mol. Biol. Rev. 65, 481–496.
21. Schwarz, S., Kehrenberg, C., Doublet, B., Cloeckaert, A., 2004. Molecular basis of bacterial resistance to chloramphenicol and florfenicol. FEMS Microbiol. Rev. 28, 519–542.
22. Shin, S.J., Kang, S.G., Nabin, Kang, R., Yoo, M.L., HS, 2005. Evaluation of the antimicrobial activity of florfenicol against bacteria isolated from bovine and porcine respiratory disease. Vet. Microbiol. 106, 73–77.
23. Toleman, M.A., Bennet, P.M., Walsh, T.R., 2006. ISCR Elements: novel gene-capturing systems of the 21st century. Micriobiol. Mol. Biol. Rev. 70, 296–316.
24. Vanni, M., Merenda, M., Barigazzi, G., Garbarino, C., Luppi, A., Tognetti, R., Intorre, L., 2012. Antimicrobial resistance of Actinobacillus pleuropneumoniae isolated from swine. Vet. Microbiol. 156, 172–177.
25. Whittle, G., Katz, M.E., Clayton, E.H., Cheetham, B.F., 2000. Identification and characterization of a native Dichelobacter nodosus plasmid, pDN1. Plasmid 43, 230–234.
26. Yoo, A.N., Cha, S.B., Shin, M.K., Won, H.K., Kim, E.H., Choi, H.W., Yoo, H.S., 2014. Serotypes and antimicrobial resistance patterns of the recent Korean Actinobacillus pleuropneumoniae isolates. Vet. Rec. 174, 223.

CHAPTER 19

The role of the environment in transmission of Dichelobacter nodosus between ewes and their lambs

Mohd Muzafar [a], Leo A. Calvo-Bado [a], Laura E. Green [a], Edward M. Smith [a], Claire L. Russell [b,1], Rose Grogono-Thomas [b], Elizabeth M.H. Wellington [a,*]

[a]*School of Life Sciences, University of Warwick, Gibbet Hill Road, Coventry CV4 7AL, UK*

[b]*Department of Clinical Veterinary Sciences, University of Bristol, Langford House, Langford BS40 5DU, UK*

ABSTRACT

Dichelobacter nodosus (D. nodosus) is the essential causative agent of footrot in sheep. The current study investigated when D. nodosus was detectable on newborn lambs and possible routes of transmission. Specific qPCR was used to detect and quantify the load of D. nodosus in foot swabs of lambs at birth and 5–13 h post-partum, and their mothers 5–13 h post-partum; and in samples of bedding, pasture, soil and faeces. D. nodosus was not detected on the feet of newborn lambs swabbed at birth, but was detected 5–13 h after birth, once they had stood on bedding containing naturally occurring D. nodosus. Multiple genotypes identified by cloning and sequencing a marker gene, pgrA, and by multi locus variable number tandem repeat analysis (MLVA) of community DNA from swabs on individual feet indicated a mixed population of D. nodosus was present on the feet of both ewes and lambs. There was high variation in pgrA tandem repeat number (between 3 and 21 repeats), and multiple MLVA types. The overall similarity index between the populations on ewes and lambs was 0.45, indicating moderate overlap. Mother offspring pairs shared some alleles but not all, suggesting lambs were infected from sources(s) other than just their mother's feet. We hypothesise that D. nodosus is transferred to the feet of lambs via bedding containing naturally occurring populations of D. nodosus, probably as a result of transfer from the feet of the group of housed ewes. The results support the hypothesis that the environment plays a key role in the transmission of D. nodosus between ewes and lambs.

Keywords: D. nodosus, Footrot, Pathogen transmission, Disease reservoirs

1. INTRODUCTION

Footrot is an economically important disease of sheep. The aerotolerant anaerobe Dichelobacter nodosus (D. nodosus) is the essential causative agent (Beveridge, 1941) and Fusobacterium necrophorum has been suggested as a secondary bacterium after the development of disease (Beveridge, 1941; Witcomb et al., 2014). The disease is present worldwide and accounts for annual losses of between £24 and £84 million to the UK sheep industry alone (Nieuwhof and Bishop, 2005; Wassink et al., 2010). The severity of ovine footrot can vary from mild interdigital dermatitis (synonymous with benign footrot in Australian research) to virulent footrot causing severe under-running of the hoof horn with separation from the underlying tissue (Stewart, 1989). D. nodosus can be detected on the feet of sheep with no sign of disease (Calvo-Bado et al., 2011b; Witcomb et al., 2014) but the load is higher both before and during episodes of interdigital dermatitis and virulent footrot than on healthy feet (Witcomb et al., 2014).

Temporal clustering of footrot between mothers and offspring was observed in a state transition study of factors associated with development of, and recovery from, footrot. Given that families cluster spatially this suggests spatiotemporal transmission of D. nodosus between family members (Kaler et al., 2010). D. nodosus has been isolated from pasture and barns where sheep are kept, indicating that contamination of the environment occurs (Witcomb, 2012). Contaminated holding areas have also been shown to cause disease in sheep put into such environments up to 2 weeks from initial seeding (Beveridge, 1941; Whittington, 1995). Recent work has indicated that D. nodosus can survive up to 14 days at 5 °C in soil, and at least 24 days when hoof material was present (Cederlof et al., 2013) and under certain conditions, D. nodosus has survived for at least 40 days in soil microcosms (unpublished data), however, further work is required to determine if survival is at a dose that could cause disease in sheep.

Multiple strains of D. nodosus detected by serogroup typing have been reported to co-exist in individual feet during subclinical and clinical infections (Claxton et al., 1983; Hindmarsh and Fraser, 1985; Jelinek et al., 2000; Moore et al., 2005). Molecular detection of strain differences is now possible using typing the pgr locus and by MLVA of D. nodosus (Calvo-Bado et al., 2011a; Russell et al., 2014).

The aims of this study were to investigate whether D. nodosus was present on the feet of newborn lambs at or after birth and the potential role played by the environment in pathogen transmission.

2. MATERIALS AND METHODS

2.1. Selection of animals

In April 2011 10 ewes with no clinical signs of disease and one lamb per ewe were convenience selected from a flock of 99 Mule and Suffolk crossbred ewes. Ewes were housed on the 28th March 2011, and samples collected on the 1st–6th April 2011 (Supplementary Table 1). Lambs were born in a large communal straw bedded pen, ewes and their lambs were moved to individual pens once the ewe had given birth to all her lambs. Sampled lambs were marked with tape so they could be identified for subsequent sampling.

2.2. Collection of environmental and foot swab samples

Environmental samples were taken in March prior to lambing and included swab samples of 30 fresh hoof prints in soil, four soil samples from the area around water containers, 10 samples of faecal material on the ground and compacted in the interdigital space and three straw samples collected from the storage area. In April, 10 straw bedding samples were collected from the communal pen where pregnant ewes were housed. All samples were stored at 4 °C for transportation and at −80 °C until analysed. All four feet of each lamb was swabbed using sterile cotton swabs (EUROTUBO collection swab; Delta lab, Rubi, Spain) directly after birth and before the lamb touched the ground. The lamb and its dam were sampled 5–13 h later once the lamb had stood and been transferred, with its mother, to an individual pen. Swabs were stored at 4 °C for transportation and at −80 °C on arrival at the laboratory.

2.3. Detection limit assay by direct PCR and nested PCR from swabs

The D. nodosus strain VCS1703A was used as a positive control for all PCR reactions. To determine the PCR detection limits, cells were harvested from a 5 d culture grown on 2% hoof agar, and 10-fold serial dilutions (10^{-1} to 10^{-10}) were made in triplicate in sterile phosphate buffered saline (PBS). The numbers of cells in the initial concentration and all dilutions were counted using a haemocytometer. Sterile swabs were inoculated with 500 μl of each dilution, and frozen at −20 °C to produce swabs containing a known bacterial load. Microbial DNA was extracted from swabs as described below and the DNA used to determine assay detection limits.

2.4. DNA extraction from swabs

Total genomic DNA was extracted using the NucleoSpin Tissue Kit (Macherey-Nagel, GmbH and Co, Düren, Germany) with modifications. Swabs were thawed at 4 °C and 400 μl of buffer T1 was added followed by 40 μl of proteinase K. The samples were vortexed twice for 5 s and incubated for 10 min at 56 °C. The mixtures were transferred to microcentrifuge tubes and 400 μl of buffer B3 was added. The samples were vortexed twice for 5 s and incubated for 5 min at 70 °C then allowed to cool before adding 400 μl of 100% ethanol. The samples were again vortexed twice and the supernatant transferred to a NucleoSpin Tissue column and centrifuged at 11,000 × g for 1 min. The flow-through was discarded, the membrane was washed with 500 μl of buffer B5 and centrifuged at 11,000 × g for 1 min. The flow-through was again discarded, the column was washed with 600 μl of buffer B5 and centrifuged at 11,000 × g for 1 min. The flow-through was again discarded and the membrane dried by centrifugation at 11,000 × g for 1 min to remove residual ethanol. The DNA was eluted into 40 μl of elution buffer, warmed to 70 °C and centrifuged at 11,000 × g for 1 min and the resultant DNA was stored at −20 °C.

2.5. DNA extraction from soil and faeces

DNA was extracted from soil and faecal samples (one gram each) using the Fast DNA Spin Kit for soil (QBiogene, Carlsbad, CA, USA) according to the manufacturer's instructions, and eluted in 70 μl DES (DNase/Pyrogen Free Water). Sterile soil (autoclaved twice at 121 °C for 15 min) was used as negative controls for each set of extractions. The resultant DNA was stored at −20 °C.

2.6. DNA extraction from bedding

One gram of each of 10 bedding samples was thawed, and suspended in 40 ml of transport buffer (sterile PBS containing 20 mM Na2 EDTA; pH 8.0). The samples were shaken for 1 h at 37 °C followed by centrifugation for 15 min at 13,523 × g at 4 °C. The supernatant was removed and the pellet resuspended in 2 ml sterile PBS. DNA was extracted from 200 µl of this solution using the NucleoSpin Blood kit (Macherey-Nagel, GmbH and Co, Düren, Germany) according to the manufacturer's recommendations, and DNA was stored at −20 °C.

2.7. End point and nested PCR

PCR amplifications were performed using an Eppendorf vapo.protect Mastercycler (Eppendorf, Hamburg, Germany). Each 50 µl reaction contained 25 µl Promega PCR master mix (Promega, Southampton, UK), 1.0 µl each of primers Cc and Ac [10 mM] (La Fontaine et al., 1993) (Table 1), 2.5 µl of dimethyl sulfoxide (Fisher Scientific, Leicestershire, UK), 2 µl bovine serum albumin [10 mg ml^{-1}] (Sigma–Aldrich Ltd., Poole, Dorset, UK), 16.5 µl of nuclease free water and 2 µl of template DNA. For direct detection of D. nodosus, PCR was performed using Cc and Ac primers (La Fontaine et al., 1993) (Table 1) under the following conditions: 1 cycle of 95 °C for 2 min, 40 cycles of 95 °C for 1 min, 60 °C for 45 s and 72 °C for 2 min and a final extension step of 72 °C for 5 min.

Samples that were negative using this approach, were tested further using nested PCR. In the first round 16S rRNA universal primers 27F and 1525R (Table 1) (Baker et al., 2003; Lane, 1991) were used in the conditions described above but with an annealing temperature of 55 °C, 1 µl of this product was used in the second round of PCR as described above. The PCR products were visualised under UV light.

2.8. Quantitative PCR of D. nodosus

The load of D. nodosus was determined using the Applied Biosystems 7500 Fast real-time detection system (Applied Biosystems, Warrington, UK). The qPCR targeted the rpoD gene (RNA polymerase sigma 70 factor, single copy number in D. nodosus genome) as described previously (Calvo-Bado et al., 2011b).

Table 1. All primers and probes used in the study.

Primer (5′–3′)	Sequence	Expected size in VCS1703A (BP)	Reference
Cc	TCGGTACCGAGTATTTCTACCCAACACCT	783	La Fontaine et al. (1993)
Ac	CGGGGTTATGTAGCTTGC	783	La Fontaine et al. (1993)
27F	AGAGTTTGATCMTGGCTCAG	1500	Lane (1991); Baker et al. (2003)
1525R	AAGGAGGTGWTCCARCC	1500	Lane (1991); Baker et al. (2003)
pgrAF1	CCTGCACCATGCTTGTTAAA	290	Calvo-Bado et al. (2011a)
pgrAR1	GCTGTTGGTGGTTTGGCTAT	290	Calvo-Bado et al. (2011a)
M13F	GTAAAACGACGGCCAG	N/A	Supplied in the cloning kit
M13R	CAGGAAACAGCTATGAC	N/A	Supplied in the cloning kit
DNTR02F	(FAM)-GATCCATCGTTTCATCGTCA	549	Russell et al. (2014)
DNTR02R	CGCACTTTAGCCGTTATGTTT	549	Russell et al. (2014)
DNTR09F	(VIC)-GGCGTAAACGAAATGCCTAA	987	Russell et al. (2014)
DNTR09R	ATCGGCGGAAGATTGTCTC	987	Russell et al. (2014)
DNTR10F	(NED)-CCGTCTATCCACCCGATTTA	626	Russell et al. (2014)
DNTR10R	TTGAACCGCGTCACTATCAG	626	Russell et al. (2014)
DNTR19F	(PET)-CCCGTCGAATCACTCCAG	854	Russell et al. (2014)
DNTR19R	GGTAGCGCCGAAGAAAGA	854	Russell et al. (2014)
rpoDF	GCTCCCATTTCGCGCATAT	61	Calvo-Bado et al. (2011b)
rpoDR	CTGATGCAGAAGTCGGTAGAACA	61	Calvo-Bado et al. (2011b)
rpoD Taqman probe	(6FAM)-CATTCTTACCGGKCG-(BBQ)	61	Calvo-Bado et al. (2011b)
pgrAF	CATGAATGATAATATTTACCTTTTCGTT	298	
pgrAR	AAGATTGATGATGCTCCAGAAGAAG	298	
pgrA Taqman probe	(6FAM)-CCTGCACCATGCTTGTTAAACTCT AATTTT-(BBQ)	298	
	AATTTT-(BBQ)		
pgrBF	AAAGGTGATCTCAACTGTATCGTCAT	N/A	
pgrBR	AATYARCARMGCCARAATTAGAGCTTAAT	N/A	
pgrB Taqman probe	(6FAM)-TTTACCCGCACCGTKCT-(BBQ)	N/A	

FAM – Carboxyfluorescein, BBQ (Black Berry Quencher). BP is the size of fragment in base pairs.

All PCR reactions were performed in triplicate and each contained 12.5 µl TaqMan Universal Master Mix (Applied Biosystems, Warrington, UK), 2.25 µl each of *rpoDF* and rpoDR [10 pmol μl^{-1}], 0.625 ml *rpoD* probe [10 pmol μl^{-1}] (Table 1), 1.25 µl bovine serum albumin [10 mg ml^{-1}], 5.375 µl nuclease free water and 1 µl of template DNA. DNA dilutions of 1:10 were also used to investigate potential inhibitors of the reaction. In addition, known concentrations of target DNA were added to negative samples as internal controls. A non-template control (nuclease free water) was included in triplicate in all PCR reactions. The reaction was carried out under the following conditions: one cycle at 50 °C for 2 min, one cycle at 95 °C for 10 min, followed by 40 cycles at 95 °C for 15 s and 55 °C for 1 min. The rpoD copy number was estimated based on the standard curve obtained from analysis of 10-fold serial dilutions of DNA extracted from D. nodosus strain VCS1703A.

2.9. Quantitative PCR assay for pgrA and pgrB

A fluorescent PCR was designed to increase the sensitivity of detection of *pgrA* and B (Table 1). The specificity of the *pgrA* and *B* primers was tested against genomic DNA isolated from 11 bacterial species and from a range of diverse environmental samples. D. nodosus DNA from VCS1703A was used as a positive control and genomic DNA from Escherichia coli and sterile water as negative and no target controls respectively (Supplementary Table 2). The cycling conditions were modified slightly from those described above by reducing the number of cycles to 38 and 36 for *pgrA* and *pgrB* respectively.

2.10. Cloning and sequencing of pgrA amplicons

The pgrA gene was amplified using the primers pgrAF1 and pgrAR1 (Table 1; (Calvo-Bado et al., 2011a)) from the DNA extracted from the foot swabs of five ewes (n = 14 feet) and their lambs (n = 10 feet) and cloned using the TOPO TA Cloning Kit (Invitrogen Ltd., Paisley, UK). The cloning reactions were set up following the manufacturer's recommendations. Transformations were carried out using chemically competent (TOP10) E. coli cells (Invitrogen Ltd., Paisley, UK) with 50 µg ml^{-1} kanamycin. One-hundred microliters of the resulting solution was cultured on LB plates containing 50 µg ml^{-1} kanamycin. Fifty colonies per sample were inoculated into individual wells of a 96-well plate, each containing 50 µl sterile water. The samples were heated to 75 °C for 10 min, and 1 µl of this solution was used as a template for PCR, resulting in analysis of 1200 transformants. The PCR products were run on a 1% high resolution agarose gel and visualised under UV light. The clones that showed variation in size within each foot were inoculated into LB media to provide sufficient biomass for plasmid

DNA extraction using the Qiagen MiniPrep kit (Qiagen, West Sussex, UK). The plasmid DNA was digested using EcoR1 and sequenced by GATC Biotech (London, UK) using the supplied M13f and M13r primers (Invitrogen, Paisley, UK; Table 1). All sequences were deposited in GenBank database under accession numbers KR105403–KR105413.

2.11. Multi Locus VNTR analysis

The four D. nodosus tandem repeat (DNTR) loci were amplified individually as described previously (Russell et al., 2014) (Table 1) from the DNA isolated from the feet of all ewes and lambs. Primer specificity was tested against genomic DNA isolated from 11 bacterial species and from a range of diverse environmental samples (Supplementary Table 2). D. nodosus strain VCS1703A was included as a positive control for all PCR reactions. The amplified products from the community DNA were pooled in the ratio of 1:1:1:1 and submitted for fragment analysis to the University of Dundee (DNA Sequencing & Services, Dundee, Scotland). GeneScan 1200 LIZ dye (Applied Biosystems, Warrington, UK) was used as a size standard in fragment analysis. The data obtained were processed using Peak Scanner Software (Applied Biosystems, Warrington, UK) and analysed using T-REX (Culman et al., 2009) with a minimum fragment length cut off values of 300 bp and 500 bp for DNTR10 and DNTR19 respectively, peak height baseline threshold of 40 and bin range of 4 bp.

2.12. Statistical analysis

To determine the strain overlap between all ewes and lambs defined by MLVA, the coincidence index of overlap was calculated using the formula: $C = 2B/(E + L)$, where C = coincidence index of overlap, B = occurrence of the same allele in ewes and lambs, E = occurrence of the allele in ewes, L = occurrence of the allele in lambs. Results can range from $C = 0$, no overlap between ewes and lambs to $C = 1$, identical strains occur in both ewes and lambs (Dice, 1945).

The *rpoD* copy number in ewes and lambs was not normally distributed. Therefore, a Mann–Whitney U test (Mann and Whitney, 1947) was used to test for differences in copy number between ewes and lambs.

3. RESULTS

3.1. D. nodosus copy number detection limit and its persistence in the environment

End point PCR and qPCR data of the inoculated swabs suggested that the minimum detection level was 10^4 cells and 10^2*rpoD* genome equivalents (i.e. 10^2 cells) per swab respectively assuming 100% DNA recovery. Below this concentration, detection was not reproducible. D. nodosus was detected by end point PCR in 5/10 faecal samples/balls from the interdigital space, 2/10 straw bedding samples collected after ewes were housed, 5/30 fresh hoof prints and 2/4 soil samples taken from the areas surrounding water containers. However qPCR analysis revealed that D. nodosus was present at loads of 10^2–10^4*rpoD* genome equivalents per gram in all the used straw bedding samples and 10^3–10^4*rpoD* genome equivalents per gram in all the faecal samples. Quantitative PCR also confirmed that D. nodosus was not detectable in the three stored straw samples (Fig. 1).

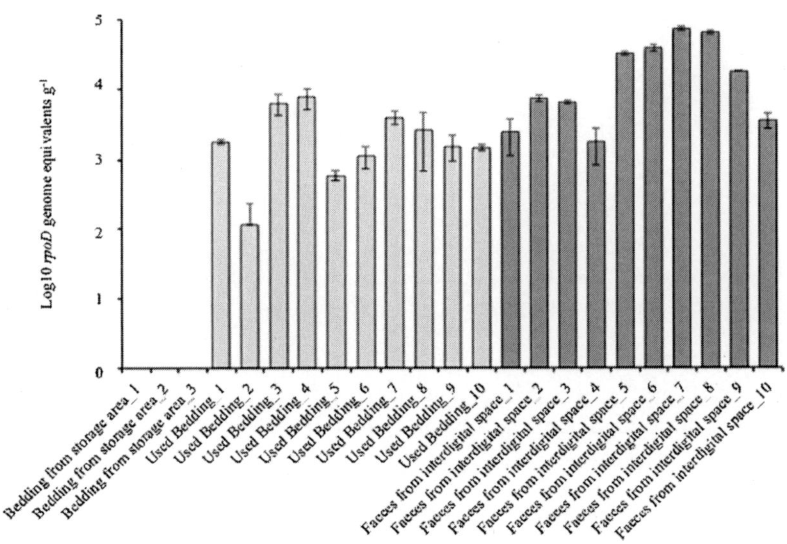

Figure . 1. Presence of D. nodosus in the environmental samples. Absolute quantification of rpoD gene in bedding samples from the storage area, used bedding samples and faeces compacted within the interdigital space. Each bar is the average of triplicate analyses, error bars represent ± standard deviation.

3.2. Quantification of D. nodosus in ewes and lambs

D. nodosus was not detected in lambs' feet at birth but was detected in all lambs and ewes 5–13 h later after their feet had touched the floor initially in a large communal pen, and subsequently in an individual pen. D. nodosus was detected on 39/40 (97.5%) of ewes' feet and 39/40 (97.5%) of lambs' feet (Fig. 2). Whilst overall the population loads were significantly higher in ewes than lambs (Mann–Whitney U test; p-value < 0.001); analysis of ewe/lamb pairs suggested only ewes 2, 3 and 5 had a higher load than their lambs. The D. nodosus load ranged from 10^3 to 10^5*rpoD* genome equivalents per swab in lambs and 10^2–10^7*rpoD* genome equivalents per swab in ewes (Supplementary Fig. 1).

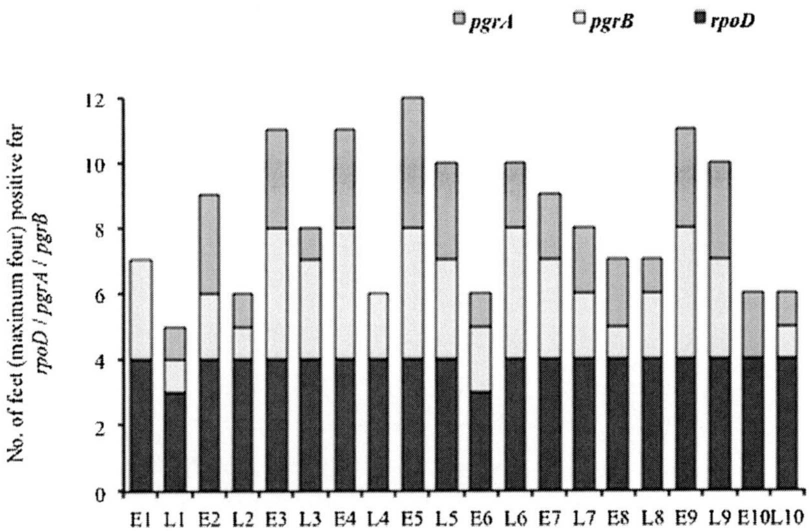

Figure . 2. Detection of *pgr* variants in the community DNA. Presence of *pgrA, pgrB* and *rpoD* on the feet of 10 ewes and 10 lambs. *pgrA/B* was absent/below the detection limit in the samples where no data is shown. (E = Ewe; L = Lamb).

3.3. Detection of pgr variants in the community DNA

pgrA was detected on 23/40 (57.5%) of ewes' feet and 15/40 (37.5%) of lambs' feet, whereas *pgrB* was detected on 27/40 (67.5%) of ewes' feet and

22/40 (55%) of lambs' feet (Fig. 2). Both variants *pgrA* and *pgrB* were detected on eight ewes and nine lambs.

Forty-two *pgrA* clones were sequenced from 14 foot swabs from five ewes and 10 foot swabs from five lambs. This resulted in the detection of 11 variants containing 3–21 tandem repeats in the R1 region, with 2–6 variants per animal (Table 2). Multiple *pgrA* variants with varying numbers of tandem repeats were observed in a single foot swab (data not shown).

Table 2. Distribution of pgrA R1 tandem repeats in five pairs of ewes and lambs (14 ewe and 10 lamb feet).

Ewe/Lamb ID	Number of clones sequenced	Number of *pgrA* tandem repeats in the R1 region
E 1	6	3, 4, 5, 11, 13, 16
L 1	4	4, 11, 12, 16
E 2	3	4, 11, 15
L 2	5	3, 4, 6, 11, 16
E 3	4	6, 15, 16, 21
L 3	2	16, 20
E 4	4	4, 6, 11, 13
L 4	5	4, 6, 11, 12, 16
E 5	4	5, 12, 15, 16
L 5	5	4, 11, 12, 15, 16

3.4. Molecular typing MLVA from the community DNA

Alleles at D. nodosus tandem repeat DNTR19 were detected on the feet of six ewes and their lambs whereas for DNTR10 alleles were only detected on two ewe/lamb pairs (Table 3) (Supplementary Fig. 2). The fluorescent data for DNTR09 was below the peak height threshold level (40 fluorescence units) and DNTR02 primers demonstrated some non-specific binding, so were excluded from the analysis. For both loci (DNTR10 and DNTR19), one or two alleles occurred in lambs, but there was greater diversity in ewes, with up to six alleles detected (Table 3).

As detectable diversity in ewes increased, so did the likelihood of detecting the same strain on its offspring. The overall coincidence index of overlap between ewes and lambs was 0.45.

Table 3. DNTR19 and DNTR10 allelic distribution between six and two pairs of ewes and lambs.

ID	DNTR19							DNTR10					
	3+	4	5	6	7	8	Total	3	4	7	9	10	Total
E 1	1	1	1	1	1	1	6	1	1	1	1	1	5
L 1	0	0	0	1	0	0	1	1	0	0	0	0	1
E 2	1	1	0	1	0	0	3	1	0	0	1	0	2
L 2	1	0	0	0	0	0	1	1	0	0	0	0	1
E 3	0	1	0	0	0	0	1						
L 3	1	0	0	0	0	0	1						
E 4	0	1	1	0	0	0	2						
L 4	0	0	0	1	0	0	1						
E 5	1	1	1	1	1	0	5						
L 5	0	0	0	0	1	0	1						
E 6	0	1	0	0	0	0	1						
L 6	0	0	0	0	0	1	1						

+ The numbers given in the table heading are the number of tandem

4. DISCUSSION

In the current study we demonstrated, for the first time, that lambs are born free from D. nodosus (given the sensitivity of the tests used) but that we were able to detect high levels of D. nodosus on foot swabs within a few hours of birth. The most likely route of transmission is from standing on contaminated bedding in communal pens, however, recent work (Witcomb, 2012) indicated that D. nodosus can be detected in the mouths of ewes, so transmission might also have occurred as the lamb was being cleaned by the ewe after birth. Quantitative PCR with specific primers is widely used to determine bacterial load from swabs (Fredricks et al., 2009; Jaton et al., 2006; Lund et al., 2004) especially when studying disease development over time (Srinivasan et al., 2010). Here, as elsewhere (Witcomb et al., 2014) qPCR of *rpoD* was used to determine D. nodosus load on the feet of ewes and lambs. It was striking that such a large D. nodosus population was present on the feet of lambs within a few hours of birth. This suggests that D. nodosus might be an early coloniser of lambs' feet, although, it could be that naïve feet are simply colonized by the first bacterial species they encounter.

Several serogroups of D. nodosus on individual feet have been reported elsewhere (Claxton et al., 1983; Hindmarsh and Fraser, 1985; Jelinek et al., 2000; Moore et al., 2005) but not using the typing methods used in the current study where both pgr data and MLVA provided evidence for the occurrence of multiple strains in most animals studied. Not all the D. nodosus strains detected on the feet of lambs were on their mother's feet, indicating that the strains detected on lambs' feet originated from sources additional to their mothers' feet, most likely contaminated bedding in the communal pen. In addition, not all the strains detected in ewes were present in their offspring. It is possible that the specific variants on the ewes were not detected by chance, or were present below the minimum detection level or that only some of the strains were transferred via bedding.

Given the survival time of D. nodosus off host and the D. nodosus-negative stored straw samples, the most probable source of D. nodosus was the population of ewes in the communal pen. Methods to reduce the load of D. nodosus in ewes and the environment would probably have the biggest impact on transmission to newborn lambs. Management strategies that have been linked to a reduction in footrot prevalence and incidence include rapid appropriate treatment of diseased sheep (Kaler et al., 2010; Wassink et al., 2010), segregation of diseased sheep and footbathing healthy sheep (Wassink et al., 2004).

The identification of transmission routes and understanding the role of the environment is critical for the control of footrot. Previous studies have highlighted that diseased sheep are a reservoir of infection (Green et al., 2007; Kaler et al., 2010; Smith et al., 2014; Whittington, 1995), although transmission occurs indirectly via contaminated pasture or floors (Beveridge, 1941; Whittington, 1995). The current study has demonstrated that the environment potentially forms at least a temporary reservoir of infection for lambs because D. nodosus was detected in the straw bedding in the communal pens, and lambs have strains of D. nodosus on their feet not detected on their mother's feet.

The fact that lambs were D. nodosus-negative at birth suggests that it might be possible to produce D. nodosus-free lambs without the need for a caesarean birth. This finding is of interest to those performing challenge studies on pathogen-free individuals. At the practical, farm level this information might also be useful in countries that have eradicated footrot. In countries where footrot is endemic, even if lambs could be kept D. nodosus-free around lambing time, it is highly unlikely that this status could be maintained because of the high levels of infection in ewes. Reducing initial exposure (as suggested above) might be beneficial to subsequent disease severity, however, it might be detrimental if later age at first exposure increases disease severity.

5. CONCLUSIONS

We have provided evidence that lambs are born D. nodosus-negative, but within hours of birth several strains of D. nodosus are detectable on their feet. The strains detected were a combination of those present on their mother's feet and on the feet of other ewes. Straw bedding in the communal pen was D. nodosus-positive and the most likely source of D. nodosus for newborn lambs.

ACKNOWLEDGEMENTS

This work was supported in part by the Combating Endemic Diseases of Farmed Animals for Sustainability (CEDFAS) initiative, Grant No. BBE01870X1 from the Biotechnology and Biological Sciences Research Council (BBSRC). MM was in receipt of a Warwick Chancellor's International Scholarship Ref: 1150257/School of Life Sciences.

REFERENCES

1. Baker, G.C., Smith, J.J., Cowan, D.A., 2003. Review and re-analysis of domain-specific 16S primers. J. Microbiol. Meth. 55, 541–555.
2. Beveridge, 1941. A transmissible disease due to infection with Fusiformis nodusus (n.sp.): studies on its cause, epidemiology, and control, Bulletin of the Council of Scientific and Industrial Research Australia (Australia), , pp. 1–58.
3. Calvo-Bado, L.A., Green, L.E., Medley, G.F., Ul-Hassan, A., Grogono-Thomas, R., Buller, N., Kaler, J., Russell, C.L., Kennan, R.M., Rood, J.I., Wellington, E.M., 2011a. Detection and diversity of a putative novel heterogeneous polymorphic proline-glycine repeat (Pgr) protein in the footrot pathogen Dichelobacter nodosus. Vet. Microbiol. 147, 358–366.
4. Calvo-Bado, L.A., Oakley, B.B., Dowd, S.E., Green, L.E., Medley, G.F., Ul- Hassan, A., Bateman, V., Gaze, W., Witcomb, L., Grogono-Thomas, R., Kaler, J., Russell, C.L., Wellington, E.M., 2011b. Ovine pedomics: the first study of the ovine foot 16S rRNA-based microbiome. ISME J. 5, 1426–1437.
5. Cederlof, S.E., Hansen, T., Klaas, I.C., Angen, O., 2013. An evaluation of the ability of Dichelobacter nodosus to survive in soil. Acta Vet. Scand. 55, 4.
6. Claxton, P.D., Ribeiro, L.A., Egerton, J.R., 1983. Classification of Bacteroides nodosus by agglutination tests. Aus. Vet. J. 60, 331–334.
7. Culman, S., Bukowski, R., Gauch, H., Cadillo-Quiroz, H., Buckley, D., 2009. T-REX: software for the processing and analysis of T-RFLP data. BMC Bioinformatics 10, 171.
8. Dice, L.R., 1945. Measures of the amount of ecologic association between species. Ecology 26, 297–302.
9. Fredricks, D.N., Fiedler, T.L., Thomas, K.K., Mitchell, C.M., Marrazzo, J.M., 2009. Changes in vaginal bacterial concentrations with intravaginal metronidazole therapy for bacterial vaginosis as assessed by quantitative PCR. J. Clin. Microbiol. 47, 721–726.
10. Green, L.E., Wassink, G.J., Grogono-Thomas, R., Moore, L.J., Medley, G.F., 2007. Looking after the individual to reduce disease in the flock: a binomial mixed effects model

investigating the impact of individual sheep management of footrot and interdigital dermatitis in a prospective longitudinal study on one farm. Prev. Vet. Med. 78, 172–178.
11. Hindmarsh, F., Fraser, J., 1985. Serogroups of Bacteroides nodosus isolated from ovine footrot in Britain. Vet. Rec. 116, 187–188.
12. Jaton, K., Bille, J., Greub, G., 2006. A novel real-time PCR to detect Chlamydia trachomatis in first-void urine or genital swabs. J. Med. Microbiol. 55, 1667–1674.
13. Jelinek, P.D., Depiazzi, L.J., Galvin, D.A., Spicer, I.T., Palmer, M.A., Pitman, D.R., 2000. Occurrence of different strains of Dichelobacter nodosus in new clinical lesions in sheep exposed to footrot associated with multi-strain infections. Aust. Vet. J. 78, 273–276.
14. Kaler, J., Medley, G.F., Grogono-Thomas, R., Wellington, E.M.H., Calvo- Bado, L.A., Wassink, G.J., King, E.M., Moore, L.J., Russell, C., Green, L.E., 2010. Factors associated with changes of state of foot conformation and lameness in a flock of sheep. Prev. Vet. Med. 97, 237–244.
15. La Fontaine, S., Egerton, J.R., Rood, J.I., 1993. Detection of Dichelobacter nodosus using species-specific oligonucleotides as PCR primers. Vet. Microbiol. 35, 101–117.
16. Lane, D.J., 1991. 16S'23S rRNA sequencing. In: Stackebrandt, E., Goodfellow, M. (Eds.), Nucleic Acid Techniques in Bacterial Systematics. Wiley, Chichester, pp. 115–175.
17. Lund, M., Nordentoft, S., Pedersen, K., Madsen, M., 2004. Detection of Campylobacter spp. in chicken fecal samples by real-time PCR. J. Clin. Microbiol. 42, 5125–5132.
18. Mann, H.B., Whitney, D.R., 1947. On a Test of Whether one of Two Random Variables is Stochastically Larger than the Other, pp. 50–60.
19. Moore, L.J., Wassink, G.J., Green, L.E., Grogono-Thomas, R., 2005. The detection and characterisation of Dichelobacter nodosus from cases of ovine footrot in England and Wales. Vet. Microbiol. 108, 57–67.
20. Nieuwhof, G.J., Bishop, S.C., 2005. Costs of the major endemic diseases of sheep in Great Britain and the potential benefits of reduction in disease impact. Anim. Sci. 81, 23–29.
21. Russell, C.L., Smith, E.M., Calvo-Bado, L.A., Green, L.E., Wellington, E.M., Medley, G.F., Moore, L.J., Grogono-Thomas, R., 2014. Multiple locus VNTR analysis highlights that geographical clustering and distribution of Dichelobacter nodosus, the causal agent of footrot in sheep, correlates with inter-country movements. Infect. Genet. Evol. 22, 273–279.
22. Smith, E.M., Green, O.D., Calvo-Bado, L.A., Witcomb, L.A., Grogono-Thomas, R., Russell, C.L., Brown, J.C., Medley, G.F., KilBride, A.L., Wellington,E.M., Green, L.E., 2014. Dynamics and impact of footrot and climate on hoof horn length in 50 ewes from one farm over a period of 10 months. Vet. J. 201, 295–301.
23. Srinivasan, S., Liu, C., Mitchell, C.M., Fiedler, T.L., Thomas, K.K., Agnew, K.J., Marrazzo, J.M., Fredricks, D.N., 2010. Temporal variability of human vaginal bacteria and relationship with bacterial vaginosis. PloS One 5, e10197.
24. Stewart, D.J., 1989. Footrot of sheep. In: Egerton, J.R., Yong, W.K., Riffin, G.G. (Eds.), Footrot and Foot Abscess of Ruminants. CRC Press, Boca Raton, Florida, USA.
25. Wassink, G.J., Grogono-Thomas, R., Moore, L.J., Green, L.E., 2004. Risk factors associated with the prevalence of interdigital dermatitis in sheep from 1999 to 2000. Vet. Rec. 154, 551–555.
26. Wassink, G.J., King, E.M., Grogono-Thomas, R., Brown, J.C., Moore, L.J., Green, L.E., 2010. A within farm clinical trial to compare two treatments (parenteral antibacterials and hoof trimming) for sheep lame with footrot. Prev. Vet. Med. 96, 93–103.
27. Whittington, R.J., 1995. Observations on the indirect transmission of virulent ovine footrot in sheep yards and its spread in sheep on unimproved pasture. Aust. Vet. J. 72, 132–134.
28. Witcomb, L.A., 2012. The In Situ Analysis of the Microbial Community Associated with Footrot of Sheep. University of Warwick Ph.D. thesis.
29. Witcomb, L.A., Green, L.E., Kaler, J., Ul-Hassan, A., Calvo-Bado, L.A., Medley, G.F., Grogono-Thomas, R., Wellington, E.M., 2014. A longitudinal study of the role of Dichelobacter nodosus and Fusobacterium necrophorum load in initiation and severity of footrot in sheep. Prev. Vet. Med. 115, 48–55.

CHAPTER 20

Selective medium for culture of *Mycoplasma hyopneumoniae*

Beth S. Cook[1], Jessica G. Beddow, Lucía Manso-Silván[2], Gareth A. Maglennon, Andrew N. Rycroft*

Department of Pathology & Pathogen Biology, Royal Veterinary College, Hawkshead Lane, North Mymms, AL9 7TA, UK

ABSTRACT

The fastidious porcine respiratory pathogen *Mycoplasma hyopneumoniae* has proven difficult to culture since it was first isolated in 1965. A reliable solid medium has been particularly challenging. Moreover, clinical and pathological samples often contain the fast-growing *M. hyorhinis* which contaminates and overgrows *M. hyopneumoniae* in primary culture. The aim of this study was to optimise the culture medium for recovery of *M. hyopneumoniae* and to devise a medium for selection of *M. hyopneumoniae* from clinical samples also containing *M. hyorhinis*. The solid medium devised by Niels Friis was improved by use of Purified agar and incorporation of DEAE-dextran. Addition of glucose or neutralization of acidity in liquid medium with NaOH did not improve the final yield of viable organisms or alter the timing of peak viability. Analysis of the relative susceptibility of *M. hyopneumoniae* and *M. hyorhinis* strains to four antimicrobials showed that *M. hyopneumoniae* is less susceptible than *M. hyorhinis* to kanamycin. This was consistent in all UK and Danish strains tested. A concentration of 2 μg/ml of kanamycin selectively inhibited the growth of all *M. hyorhinis* tested, while *M. hyopneumoniae* was able to grow.

This forms the basis of an effective selective culture medium for *M. hyopneumoniae*.

Keywords: *Mycoplasma hyopneumoniae,* Selective culture, Enzootic pneumonia, Pig

1. INTRODUCTION

Mycoplasma hyopneumoniae is the primary cause of enzootic pneumonia (EP) in pigs. The disease is of global significance both in reducing growth efficiency and in promoting susceptibility to concurrent bacterial and viral infections (Thacker et al., 1999). Diagnosis of EP is usually by a combination of serological tests (Feld et al., 1992; Djordjevic et al., 1994), PCR of nasal secretion or lung tissue at slaughter (Mattsson et al., 1995; Baumeister et al., 1998; Stemke, 1997), real-time PCR (Dubosson et al., 2004; Marois et al., 2010) and isolation in culture (Kobisch and Friis, 1996). However, *M. hyopneumoniae* is notoriously fastidious (Goodwin and Hurrell, 1970) and culture remains challenging and time consuming. The most widely used liquid medium for culture of *M. hyopneumoniae* was developed by Niels Friis (1975). Homogenised lung tissue was serially diluted in tubes of the medium and incubation led to a gradual colour change of the phenol red indicator from pink to yellow, without turbidity, over a period of 3–10 days growth. However, individual mycoplasmas could only be purified by terminal dilution in the liquid medium.

A commercial medium is available from Mycoplasma Experience® Ltd. U.K. (ME). Both liquid and solid media are available. These support the growth of *M. hyopneumoniae* but the constituents of this commercial medium are not publicly available. Solidification of Friis medium with Agar did not allow growth of colonies of *M. hyopneumoniae*, suggesting that the agar was inhibitory to growth or sequestered an essential nutrient. The use of a better defined solid medium would allow greater flexibility for research, such as transformation of *M. hyopneumoniae* (Maglennon et al., 2013a).

When used for diagnostic purposes, both Friis medium and ME media are easily overgrown by the faster-growing *M. hyorhinis* which is sometimes present in pig lung lesions alongside *M. hyopneumoniae* and frequently in apparently normal tissue (Kobisch and Friis, 1996). To suppress the growth of *M. hyorhinis*, Friis recommended the use of 5% hyperimmune anti-*M. hyorhinis* rabbit serum and 500 μg/ml cycloserine (Friis, 1971). However, such serum is not always available, is expensive, and batches of serum vary in their capacity to suppress *M. hyorhinis*.

The purpose of this study was therefore to improve the culture medium for transformation and mutagenesis of *M. hyopneumoniae*, allowing single colonies to be readily distinguished and selected, and to devise an improved selection medium for the suppression of *M. hyorhinis*.

2. MATERIALS & METHODS

2.1. Mycoplasma strains

M. hyopneumoniae strains 277/94 and 325/95 were the gift of Niels Friis, Danish Veterinary Institute, Copenhagen. Other strains of *M. hyopneumoniae* were Danish field isolates from lesions of EP, the gift of Dr Branko Kokotovic, Danish Veterinary Institute, or UK field isolates from slaughterhouse lesions of EP. *M. hyorhinis* strains were UK field isolates obtained from pig lungs with or without gross lesions at post-mortem. The identity of all organisms was confirmed using species-specific PCR amplifying a region of the conserved hypothetical protein mhp165 from *M. hyopneumoniae* (Baumeister et al., 1998) and the highly conserved 16S rRNA region of *M. hyorhinis* (Lin et al., 2006). The identity of all *M. hyopneumoniae* strains was subsequently confirmed by whole genome sequencing and genome-wide analysis (J. Welch – personal communication).

2.2. Liquid culture medium

Friis medium was prepared largely as described by Kobisch and Friis (1996) with the following modifications: bacitracin and meticillin were replaced with azlocillin and flucloxacillin (final concentration 50 µg/ml). To make 500 ml of Friis medium, 1.5 g Brain Heart Infusion (BHI) (Difco) and 1.6 g PPLO (Difco) was dissolved in 365 ml water and sterilised by autoclaving. To this were added 18 ml of yeast extract (prepared from dried bakers yeast), 12.5 ml sterile solution A (160 g/l NaCl, 4 g/l; 8 g/l KCl, 2 g/l $MgSO_4 \cdot 7H2O$, 2 g/l $MgCl_2.6H_2O$, 3.7 g/l $CaCl_2 \cdot 2H_2O$), 12.5 ml sterile solution B (3.0 g/l $Na_2HPO_4.12H_2O$, 1.2 g/l KH_2PO_4), 50 ml pig serum (Invitrogen) heat-treated at 56 °C for 20 min, 50 ml heat-treated horse serum (Invitrogen), 1 ml phenol red solution (0.6% in 0.1 M NaOH), 25 µl of each of 100 mg/ml azlocillin and flucloxacillin and adjusted to pH 7.4 with 1.0 M NaOH.

2.3. Solid culture medium

Concentrated (2.8×) Friis medium was prepared by the addition of 88 ml water to 4.3 g BHI, 4.6 g PPLO sterilised by autoclaving, 51.4 ml yeast extract, 35.7 ml sterile solution A, 35.7 ml sterile solution B, 143 ml heat-treated horse serum, 143 ml heat-treated pig serum, 2.8 ml 0.6% phenol red solution, azlocillin and flucloxacillin (final concentration 140 μg/ml), pH 7.4. Friis agar was prepared by mixing concentrated Friis medium (2.8×), mixed at a ratio of 35: 65 with either 1.8% agar No. 1 (Oxoid, L11), Purified agar (Oxoid, L28) or agarose (Invitrogen). Where indicated, DEAE-dextran (Sigma-Aldrich, Gillingham, UK) was added to agar at 0.1 mg/ml. Mycoplasma Experience® solid medium was prepared according to the manufacturer's instructions (Mycoplasma Experience® Ltd, Reigate, UK). Cultures were incubated in a humidified incubator at 37 °C in 5% CO_2.

Charcoal-treated agar was prepared by incubating a 1.3% solution of Bacteriological agar with 2% w/v of activated charcoal (Oxoid, Basingstoke, UK) for 10 min at 55° C, the charcoal was then removed using a sieve. Following sterilisation, each agar type was mixed with medium supplement (Mycoplasma Experience Ltd, UK) or concentrated Friis medium at a ratio of 35: 65 (supplement: agar) and plates poured.

2.4. Viability determination

Ten-fold serial dilutions in Hanks' Balanced salts Solution (HBSS; Life Technologies) were spotted (10 μl in triplicate) on to solid medium. Cultures were incubated for 7 days in 5% CO_2 at 37 °C. Colonies were counted, from those dilutions showing the largest number of separate colonies, using a dissecting microscope; the triplicate counts were combined and the viability calculated. Growth was also estimated from the metabolic activity recorded as a colour shift of the phenol red pH indicator from pH 7.4 to 6.8 as described by Friis (1975).

2.5. Antibiotic susceptibility testing

Solutions of kanamycin (Sigma), tetracycline (Sigma) and puromycin (Fisher) were prepared in de-ionised water. Chloramphenicol (Sigma) was dissolved in ethanol at 10 mg/ml final concentration. All were sterilised through a 0.2 μm filter (Nalgene, UK). A final antibiotic concentration range was achieved by two-fold serial dilutions of each antibiotic in sterile water or ethanol as appropriate. Antibiotics were incorporated into solid medium over a range of concentrations 20–0.01 μg/ml (kanamycin), 4.0–0.03 μg/ml (tetracycline), 20–0.06 μg/ml (chloramphenicol and puromycin)

M. hyopneumoniae and *M. hyorhinis* inocula were prepared in 1 x sterile phosphate buffered saline (BR0014G, Oxoid) as a suspension containing approximately 10^3 CFU/ml. A sample of 10 µl of each strain was inoculated onto each plate in triplicate. Plates were incubated at 37° C with 5% CO_2 and observed for growth of colonies between day 2 and day 10. The MIC was determined as the highest dilution of antibiotic giving an absence of growth as determined by the formation of mycoplasma colonies.

3. RESULTS

3.1. The relationship of viability and colour shift

The viability of *M. hyopneumoniae* to the commonly used colour shift in liquid medium was determined (Fig. 1). Cultures reached a maximum viability of $10^{8.9}$at approximately 2.5 days incubation. This corresponded to a colour shift between 2 (pH 7.2) and 3 (pH 7.1). After this point there was an exponential decline reaching 10^6 CFU/ml after 8 days (beyond colour shift 4).

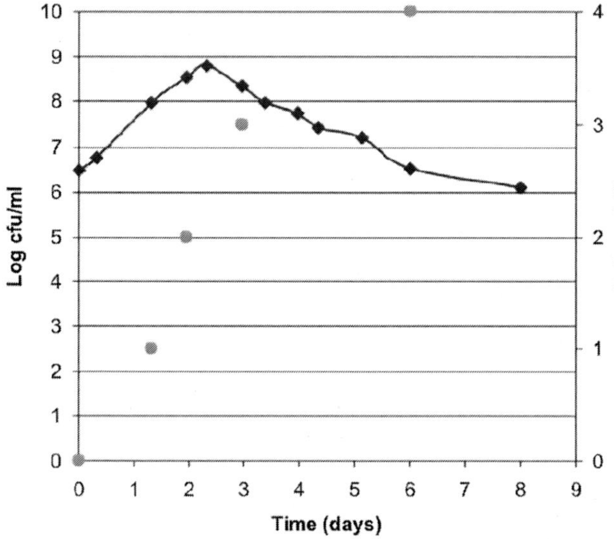

Figure. 1. Viability and colour shift of *M. hyopneumoniae* cultured in Friis broth. Viable cell counts were performed every 12 h and the colour shift recorded. Results are from two independent determinations of the growth curve.

3.2. The effect of glucose and NaOH

The concentration of glucose in Friis medium is approximately 2 mM: 1 mM is derived from the brain heart infusion broth and another 1 mM from the diluted serum. To investigate whether glucose is limiting the growth, further glucose was supplemented into the liquid medium to give 4 mM (×2) and 8 mM (×4) glucose, and viable counts performed through the growth cycle. The growth curve was unchanged by the increased glucose concentrations (Fig. 2). Friis suggested that the decline in viability when yellow colour shift was reached was due to the acidification of the medium. If so, this might be prevented by addition of alkali to the medium as the colour changes at the growth peak (60 h). Addition of alkali to neutralize the pH caused a marginal improvement in survival at 4–6 days although this difference was not marked (Fig. 2).

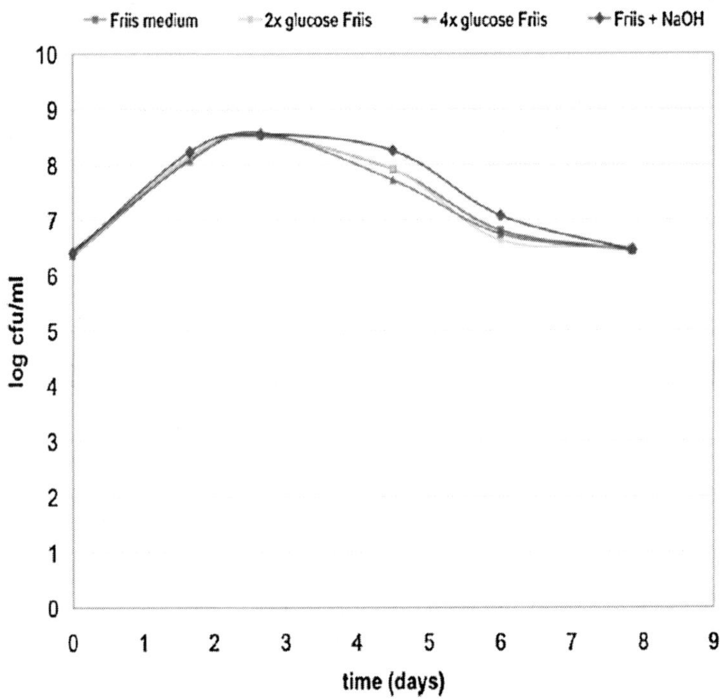

Figure. 2. Effect of glucose and alkali on the growth curve of *M. hyopneumoniae*. Viable counts were performed at intervals. Friis broth contained either ×2 or ×4 the normal concentration of glucose. The culture (NaOH) was neutralized to pH 7.3 at 2.5 days post inoculation.

Table 1. Growth of *M. hyopneumoniae* and *M. hyorhinis* on different solid media following 10 and 5 days incubation at 37 °C.

Sample	Liquid medium	Test agar	*M. hyopneumoniae*	*M. hyorhinis*
1	ME supplement	ME agar	+++	+++
2		1.3% Agar No. 1.	NG	NG
3		Charcoal-treated No. 1	NG	NG
4		1.3% agarose	NG	NG
5		1.3% Purified agar	NG	NG
6	Modified Friis (x2.8)	ME agar	+++	+++
7		1.3% Agar No. 1.	NG	+++
8		Charcoal-treated No. 1	NG	+++
9		1.3% agarose	+++	+++
10		1.3% Purified agar	++	+++

3.3. Solidifying Friis medium

In order to use Friis medium as the base for a solid medium to culture *M. hyopneumoniae*, different agar preparations were mixed with Friis base and or Mycoplasma Experience® (ME) supplement and investigated for the ability to support the growth of *M. hyopneumoniae* colonies. The results are shown in Table 1.

Friis medium solidified with the commercially supplied ME agar grew colonies of *M. hyopneumoniae* 277/94 & 325/95. When Agar no.1, used widely in the solidification of bacteriological culture medium, was mixed with either Friis medium or ME supplement there was no growth of *M. hyopneumoniae*. Pre-treatment of the agar with activated charcoal, in an attempt to remove any growth-inhibiting components, did not alter the ability of medium made with either supplement to support *M. hyopneumoniae.* However, when either Purified agar or agarose were used as the solidifying agent, Friis medium, but not the ME supplement, grew colonies of *M. hyopneumoniae* (Table 1).

ME solid medium supported the growth of *M. hyorhinis*. When charcoal-treated agar, agarose or Purified agar were used, ME supplement did not grow *M. hyorhinis* colonies while Friis medium supported *M. hyorhinis* when solidified with each of the different agars.

Based on the findings of Tauraso (1967) and as noted by Friis (1975) & Kobisch and Friis (1996), DEAE-dextran was added to the agar while molten and the growth of two strains of *M. hyopneumoniae* and *M. hyorhinis* were compared. Results are shown in Table 1. The inclusion of DEAE-dextran into Purified agar further enhanced the growth of colonies for strains of *M. hyopneumoniae* (strains 277/94 and 325/95) and appeared to be superior to the medium commercially available (Fig. 3).

3.4. Selective culture of *M. hyopneumoniae*

The susceptibility of *M. hyopneumoniae* 277/94 and 325/95, and *M. hyorhinis* Mhr1/09 was measured on solid medium containing four antimicrobials over a range of concentrations. The results are shown in Table 2. The MIC of kanamycin for *M. hyopneumoniae* isolates 325/95 and 277/94 was shown to be 8.0 μg/ml while *M. hyorhinis* appeared to be susceptible to a lower concentration, with an MIC of 1.0 μg/ml. Both strains of *M. hyopneumoniae* grew on medium containing 6.5 μg/ml. A mixed culture (equal numbers) of *M. hyopneumoniae* strain 277/94 and *M. hyorhinis* Mhr1/09 was used to assess the ability of kanamycin to selectively suppress the growth of *M. hyorhinis* on solid medium. After 8 days incubation in the absence of kanamycin, *M. hyorhinis* colonies outgrew

M. hyopneumoniae as recognised by colony morphology (Fig. 4A & B). However, with addition of 2 µg/ml kanamycin only *M. hyopneumoniae* colonies appeared to grow (Fig. 4C & D).

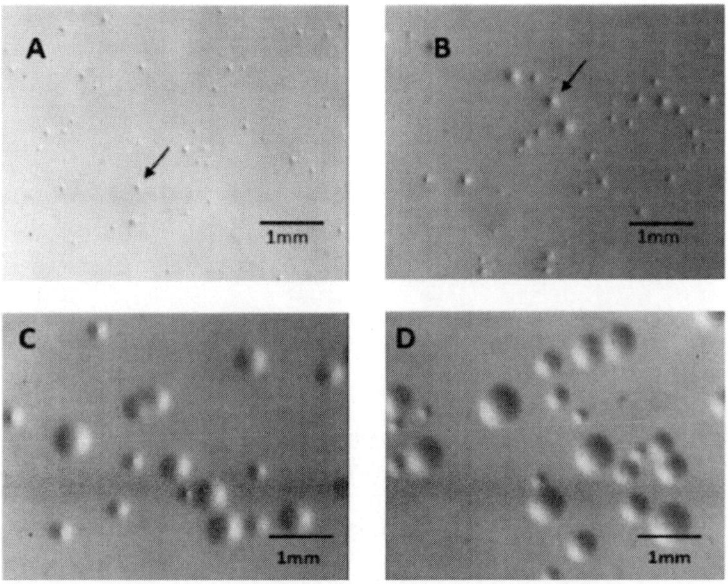

Figure. 3. Colonies of *M. hyopneumoniae* 325 on different solid media preparations. Colonies were incubated for 7 days at 37 °C in a humidified box and photographed using a dissecting microscope. **A,** ME supplement solidified with ME agar; **B,** modified Friis medium solidified with 0.85% Purified agar; **C,** modified Friis medium solidified with 0.8% Purified agar and 0.1 mg/ml DEAE-dextran; **D,** modified Friis medium solidified with 0.9% agarose. Arrows indicate small *M. hyopneumoniae* colonies..

Table 2. MICs of antimicrobial agents for *M. hyopneumoniae* strain 277/94 and 325/95, and *M. hyorhinis* Mhr1/09 for kanamycin, tetracycline, chloramphenicol and puromycin determined on solid medium.

Antimicrobial agent	**MIC µg/ml**		
	277/94	**325/95**	**Mhr1/09**
Kanamycin	8.0	8.0	1.0
Tetracycline	0.125	0.125	0.06
Chloramphenicol	2.5	2.5	2.5
Puromycin	1.25	1.25	2.5

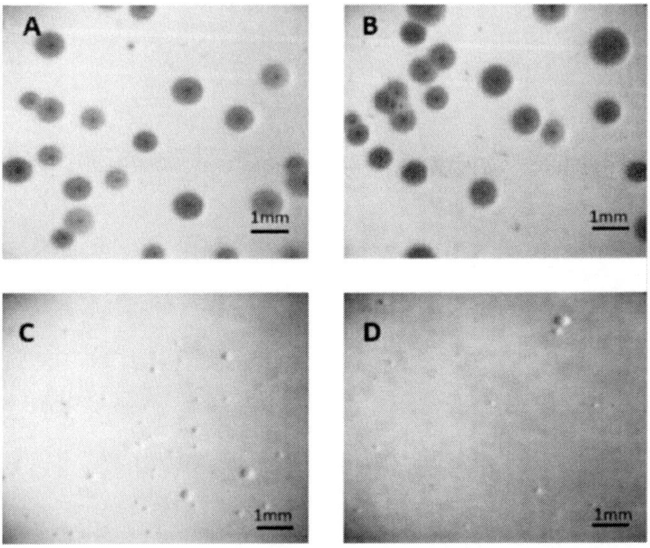

Figure. 4. Suppression of *M. hyorhinis* growth following kanamycin selection on solid medium. Growth of a mixed mycoplasma culture containing 1:1 equal volumes of *M. hyopneumoniae* and *M. hyorhinis* mid-log phase culture, following 9 days incubation on Friis solid medium with and without 2 µg/ml kanamycin at 37 °C in a humidified box. **A,***M. hyopneumoniae* 277 and *M. hyorhinis*; **B,***M. hyopneumoniae* 325 and *M. hyorhinis*; **C,***M. hyopneumoniae* 277 and *M. hyorhinis* with 2 µg/ml kanamycin; **D,***M. hyopneumoniae* 277 and *M. hyorhinis* with 2 µg/ml kanamycin..

The relative sensitivity of Mhr1/09 to kanamycin prompted the testing of more strains. A further 19 *M. hyopneumoniae* and 12 *M. hyorhinis* field isolates from the UK and Denmark, whose identity was confirmed by both PCR and growth characteristics, were tested for their susceptibility to kanamycin on MH selective medium (Table 3). All *M. hyopneumoniae* isolates grew in the presence of 2 µg/ml kanamycin. The isolates appeared to be unaffected by the incorporation of kanamycin into the medium with the exception of isolate Mp258/94 in which a small amount of growth suppression was noted in terms of colony size. The growth of all 12 *M. hyorhinis* isolates was completely inhibited by the addition of kanamycin in solid medium. Samples of pneumonic lung lesions were obtained from the abattoir over a period of 12 months. Cultures of these grown on MH selective medium were free from *M. hyorhinis* while approximately 60% of samples grown on non-selective Friis liquid and solid medium contained *M. hyorhinis* and became rapidly overgrown (Table 3).

Table 3. Growth of *M. hyopneumoniae* and *M. hyorhinis* isolates on solid medium with and without 2 μg/ml kanamycin.

***M. hyopneumoniae* isolate**	**Source**	**Growth on solid medium**	
		No kanamycin	**2 μg/ml kanamycin**
MH001	UK field isolates from lung lesions 2009–2010	+++	++
MH002		+++	+++
MH003		+++	+++
MH005		+++	+++
MH008		+++	+++
MH010		+++	+++
MH011		+++	+++
MH015		+++	++
MH016		+++	+++
MH031		+++	+++
MH033		+++	+++
MS2	Danish field isolates from lung lesions 1969	+++	+++
MS3		+++	+++
MS8		+++	+++
MS13		+++	+++
MS15		+++	+++
MS25		+++	+++
MS29		+++	+++
Mp258/94	Danish field isolates from lung lesions 1994–2007.	+++	+
Mp277/94		++	++
Mp473/95		+	+
Mp738/95		+++	+++
Mp325/95		++	++
Mp79/06		+++	++
Mp87/06		++	+

Mp92/06		+++	+++
Mp96//06		+++	+++
Mp85/07		+	+
Mp261/07		+++	+++
MHR01	UK field isolates from lung lesions 2009–2010.	+++	NG
MHR02		+++	NG
MHR04		+++	NG
MHR05		+++	NG
MHR06		+++	NG
MHR07		+++	NG
MHR08		+++	NG
MHR09		+++	NG
MHR10		+++	NG
MHR11		+++	NG
MHR12		+++	NG
P450		+++	NG

+++ large colonies; ++ medium colonies; + small colonies; NG = no colonies formed.

4. DISCUSSION

M. hyopneumoniae has been long recognised as difficult and unreliable to culture (Friis, 1971). Growth on solid medium has been considered particularly difficult with colonies growing to only 0.3–0.5 mm diameter in 7–10 days. This study was a systematic investigation to improve the medium for culture of this pathogen. The resulting improved medium was crucial in the successful development of transformation and transposon mutagenesis of *M. hyopneumoniae* as described by Maglennon et al. (2013a,b).

Establishment of the growth curve for *M. hyopneumoniae* in Friis medium was a pre-requisite for studies to allow efficient transformation. Using viable counts on the solid medium described here, rather than colour change units or the indirect methods of colour shift (Friis, 1975) or measuring ATP by luminometry (Stemke and Robertson, 1990; Calus et al., 2010), the growth curve could be accurately determined.

It has been considered that glucose is the primary carbon and energy source for *M. hyopneumoniae.* Supplementation of the medium with additional glucose did not improve the viability or growth rate. This was consistent with the results of Stemke and Robertson (1990) who found no increase in growth rate or peak viability. They used 27 mM glucose. Since physiological glucose concentrations are normally between 3 and 5 mM, we considered an increase to 4 and 8 mM to be appropriate for the medium. Quenching of the acid formed in the medium using NaOH had only a minor effect on the viability, which by 8 days incubation had reached the level of the control without added NaOH. Thus it is likely that depletion of key nutrients rather than acid inhibition is a factor limiting growth in culture.

Initial experiments indicated that the purity of the agar used to solidify modified Friis medium greatly affected the growth of *M. hyopneumoniae*. Bacteriological agar was found to prevent the growth of *M. hyopneumoniae* colonies. Gould et al. (1944) had reported a similar inhibitory effect on the growth of *Neisseria gonorrhoeae* following the solidification of liquid medium with agar. Ley and Mueller (1946) identified a component, similar to a fatty acid, present in agar following methanol extraction. This extracted component present within agar was found to inhibit the growth of *N. gonorrhoeae* in the absence of starch. Activated charcoal is known to adsorb several impurities such as fatty acids. Therefore, attempts were made to remove potential inhibitory factors from the agar such as free fatty acid-like inhibitors by treatment with activated charcoal. Since activated charcoal had no effect on the agar, we conclude that this was failure to remove inhibitory factor(s).

When the commercial ME supplement was solidified with agar other than the supplied agar component it was unable to support the growth of either *M. hyorhinis* or *M. hyopneumoniae*. This suggested that an essential constituent for growth was absent from the liquid supplement and only present within the agar component. Solidifying Friis medium with Purified agar treated with DEAE-dextran confers a substantial improvement on the commercially available medium for growth of *M. hyopneumoniae*. Since DEAE is a positively charged group able to bind negatively charged ions, its action may be explained by chelation of the agaropectin present in agar which may be inhibitory to growth of *M. hyopneumoniae*. Agaropectin is the mixture of smaller molecules, comprising alternating residues of L- and D-galactose modified with acidic side-groups (particularly sulfate). This is consistent with the beneficial use of purified agarose, which is depleted of agaropectin, over agar no. 1 in this study. No equivalent improvement was seen from DEAE in culture of *M. hyorhinis*, suggesting that it is less sensitive to the acidic components of agaropectin.

Friis (1971) described a selective medium to suppress *M. hyorhinis*. This requires rabbit antiserum which is expensive and not always available for regular culture. In testing the relative sensitivity of *M. hyopneumoniae* and *M. hyorhinis* to four antimicrobials, our results showed that the latter was consistently more sensitive to kanamycin. Williams (1978) had previously shown that both *M. hyorhinis* and *M. hyopneumoniae* were susceptible to kanamycin (0.1 and 0.5 μg/ml respectively). However, different isolates were used, and determination of the MIC by Williams was in liquid medium. Nevertheless, incorporation of 2 μg/ml of kanamycin into the agar medium appeared to prevent the growth of *M. hyorhinis* following 8 days incubation. When mixed in equal volumes with *M. hyopneumoniae* strain 325, the *M. hyorhinis* did not form colonies while controls, without the addition of kanamycin, yielded only *M. hyorhinis* colonies from the mixed culture. Screening of *M. hyopneumoniae* and *M. hyorhinis* strains showed that this difference in kanamycin susceptibility was present between all *M. hyopneumoniae* and *M. hyorhinis* strains tested.

The addition of kanamycin can be used in place of the hyperimmune rabbit serum and gives a clean selection for the culture of *M. hyopneumoniae*, effectively preventing the growth of *M. hyorhinis* while allowing colonies of *M. hyopneumoniae* to form. The relative kanamycin resistance of *M. hyopneumoniae* has greatly improved the primary isolation of *M. hyopneumoniae* from abattoir tissue samples originating from 15 different farms. False negatives, resulting from *M. hyorhinis* overgrowth have been overcome using this selective medium.

ACKNOWLEDGEMENTS

The authors acknowledge the BBSRC and Pfizer Animal Health (now Zoetis) for financial support of CASE studentships to BSC and LM-S. ANR, GAM and JGB were supported by BBSRC grant BB/G020744/1 of which 5% of the funds were contributed by Zoetis.

REFERENCES

1. Baumeister, A.K., Runge, M., Ganter, M., Feenstra, A.A., Delbeck, F., Kirchhoff, H., 1998. Detection of Mycoplasma hyopneumoniae in bronchoalveolar lavage fluids of pigs by PCR. J. Clin. Microbiol. 36, 1984–1988.
2. Calus, D., Maes, D., Vranckx, K., Villareal, I., Pasmans, F., Haesebrouck, F., 2010. Validation of ATP luminometry for rapid and accurate titration of Mycoplasma hyopneumoniae in Friis medium and a comparison with the color changing units assay. J. Microbiol. Methods 83, 335–340.
3. Djordjevic, S.P., Eamens, G.J., Romalis, L.F., Saunders, M.M., 1994. An improved enzyme linked immunosorbent assay (ELISA) for the detection of porcine serum antibodies against Mycoplasma hyopneumoniae. Vet. Microbiol. 39, 261–274.

4. Dubosson, C.R., Conzelmann, C., Miserez, R., Boerlin, P., Frey, J., Zimmermann, W., Häni, H., Kuhnert, P., 2004. Development of two real-time PCR assays for the detection of Mycoplasma hyopneumoniae in clinical samples. Vet. Microbiol.102, 55–65.
5. Feld, N.C., Qvist, P., Ahrens, P., Friis, N.F., Meyling, A., 1992. A monoclonal blocking ELISA detecting serum antibodies to Mycoplasma hyopneumoniae. Vet. Microbiol. 30, 35–46.
6. Friis, N.F., 1971. A selective medium for Mycoplasma suipneumoniae. Acta Vet. Scand. 12, 454–456.
7. Friis, N.F., 1975. Some recommendations concerning primary isolation of Mycoplasma suipneumoniae and Mycoplasma fiocculare a survey. Nord. Vet. Med. 27, 337–339.
8. Goodwin, R.F., Hurrell, J.M., 1970. Further observations on the problem of isolating Mycoplasma suipneumoniae from field cases of enzootic pneumonia in pigs. J. Hyg. (Lond.) 68, 313–325.
9. Gould, R.G., Kane, L.W., Mueller, J.H., 1944. On the growth requirements of Neisseria gonorrhoeae. J. Bacteriol. 47, 287–292.
10. Kobisch, M., Friis, N.F., 1996. Swine mycoplasmoses. Rev. Sci. Tech. OIE 15, 1569–1605.
11. Ley, H.L., Mueller, J.H., 1946. On the isolation from agar of an inhibitor for Neisseria gonorrhoeae. J. Bacteriol. 52, 453–460.
12. Lin, J.H., Chen, S.P., Yeh, K.S., Weng, C.N., 2006. Mycoplasma hyorhinis in Taiwan: diagnosis and isolation of swine pneumonia pathogen. Vet. Microbiol. 115, 111–116.
13. Maglennon, G.A., Cook, B.S., Matthews, D., Deeney, A.S., Bossé, J.T., Langford, P.R., Maskell, D.J., Tucker, A.W., Wren, B.W., Rycroft, A.N., 2013a. Development of a self-replicating plasmid system for Mycoplasma hyopneumoniae. Vet. Res. 44, 63.
14. Maglennon, G.A., Cook, B.S., Deeney, A.S., Bossé, J.T., Peters, S.E., Langford, P.R., Maskell, D.J., Tucker, A.W., Wren, B.W., Rycroft, A.N., 2013b. Transposon mutagenesis in Mycoplasma hyopneumoniae using a novel mariner-based system for generating random mutations. Vet. Res. 44, 124.
15. Marois, C., Dory, D., Fablet, C., Madec, F., Kobisch, M., 2010. Development of a quantitative Real-Time TaqMan PCR assay for determination of the minimal dose of Mycoplasma hyopneumoniae strain 116 required to induce pneumonia in SPF pigs. J. Appl. Microbiol. 108, 1523–1533.
16. Mattsson, J.G., Bergstrom, K., Wallgren, P., Johansson, K.E., 1995. Detection of Mycoplasma hyopneumoniae in nose swabs from pigs by in vitro amplification of the 16S rRNA gene. J. Clin. Microbiol. 33, 893–897.
17. Stemke, G.W., Robertson, J.A., 1990. The growth response of Mycoplasma hyopneumoniae and Mycoplasma flocculare based upon ATP-dependent luminometry. Vet. Microbiol. 24, 135–142.
18. Stemke, G.W., 1997. Gene amplification (PCR) to detect and differentiate mycoplasmas in porcine mycoplasmal pneumonia. Lett. Appl. Microbiol. 25, 327–330.
19. Tauraso, N.M., 1967. Effect of diethylaminoethyl dextran on the growth of Mycoplasma in agar. J. Bacteriol. 93, 1559–1564.
20. Thacker, E.L., Halbur, P.G., Ross, R.F., Thanawongnuwech, R., Thacker, B.J., 1999. Mycoplasma hyopneumoniae potentiation of porcine reproductive and respiratory syndrome virus-induced pneumonia. J. Clin. Microbiol. 37, 620–627.
21. Williams, P.P., 1978. In vitro susceptibility of Mycoplasma hyopneumoniae and Mycoplasma hyorhinis to fifty-one antimicrobial agents. Antimicrob. Agents Chemother. 14, 210–213.

CHAPTER 21

Comparison of the Serum Proteins and Immune Responses of Velogenic Newcastle Disease Virus Infected Chickens and Ducks

Chekwube Paul Eze[1], Vincent S. O. Shoyinka[1], John Osita Arinze Okoye[1], Wilfred Sunday Ezema[1], Innocent Okonkwo Ogbonna[2], Didacus Chukwuemeka Eze[1], Emmanuel Chukwudi Okwor[1], Ogbu Kenneth Ikejiofor[3]

[1]*Department of Veterinary Pathology and Microbiology, University of Nigeria, Nsukka, Nigeria*

[2]*Microbiology Unit, Department of Biological Sciences, University of Agriculture, Makurdi, Nigeria*

[3]*National Veterinary Research Institute, Vom, Nigeria*

ABSTRACT

The effect of velogenic Newcastle disease virus (vNDV) on the immune responses and serum proteins was investigated in six-week-old ducks and chickens. Results showed that weight loss was markedly significant ($p < 0.05$) from days 3 - 21 (PI) in chickens and mild ($p < 0.05$) on days 3 and 15 PI in ducks. The antibody response obtained showed significant ($p < 0.05$) increase in infected chickens (IC) than those of the infected ducks (ID). While the total serum protein and serum globulin increased significantly ($p < 0.05$) in IC on days 7 and 14 PI, they decreased significantly ($p < 0.05$) in ID only on day 21 PI. The immune responses and serum protein values in this experiment X-ray showed less susceptibility of ducks when compared with the chickens. This may be related to marked anorexia and severe dehydration observed in the latter consequent upon serum concentration. Ducks could be maintaining the endemicity of Newcastle disease (ND) as reservoir host.

Keywords: Velogenic Newcastle Disease, Experimental Infection, Immune Response, Serum Proteins, Ducks, Chickens

1. INTRODUCTION

Proteins are the most abundant compound in the serum comprising amino-acid building blocks and are in turn, building blocks for cells and tissues [1]. They are vital components of enzymes, hormones, antibodies and clotting agents [1]-[3]. Some proteins function as transporters, osmotic pressure and acid-alkaline balancers, and reserved energy sources. Due to these multiple functions, any condition that alters the serum proteins extends its tentacles on these roles played by proteins. Newcastle disease (ND) is a highly contagious and infectious disease that affects almost all avian species including poultry, cage and wild life bird species [4] -[6] . It is widespread affecting many continents of the world namely: Asia, Africa, Europe and America [7]. ND causes considerable economic losses, not only due to high flock mortality but also through the economic impact arising from trade restrictions and embargoes [8] [9] . It has been reported to affect other birds such as guinea fowls, quails, turkeys, pheasants and peacock [4]. Mortality due to ND ranges from negligible to as high as 100% depending on the form or pathogenicity of the virus [10]. Kudu 113, (Kuru duck 113) a strain of NDV isolated from apparently healthy ducks and characterized by Echeonwu et al. [11] in Nigeria, has been studied in chickens and guinea fowls [12]. Anseriformes are suspected to maintain the endemicity of ND around the world [7]. In view of these facts, a comparative study of the serum proteins and immune response of vNDV infected ducks and chickens was done.

2. MATERIALS AND METHODS

2.1. Flock History

One hundred and seventy day old birds hatched the same day were obtained comprising 70 ducklings from poultry section of National

Veterinary Research Institute NVRI Vom and 100 cockerel chicks from Zartech Hatchery. Brooding was done separately for the chicks and ducklings on deep litters under the same environmental conditions. The cockerels received IBD vaccine by intraocular route at days 10 and 24 post hatch (PH). Chicks' mash was given ad libitum to the birds from day old to 8 weeks PH. Growers' mash was given ad libitum also from 9 weeks PH until the end of the experiment. Water was allowed free choice.

2.2. NDV Inoculum

The VNDV strain, (Kudu-113) was acquired and used in the challenge experiment.

2.3. Experimental Design

At six weeks of age the chicks and ducklings were each randomly assigned into two groups of infected chicks (IC), uninfected chicks (UC) and infected ducks (ID), uninfected ducks (UD). The inoculum was reconstituted to give embryo lethal dose (ELD_{50}) titre of $10^{6.36}$ per ml. The chicks and ducks in groups IC and ID were inoculated intramuscularly (IM) with 0.2 ml of the inoculum (infected groups). The chicks and ducks in groups UC and UD received 0.2 ml of phosphate buffered solution IM (control groups). The infected and control groups were housed at different locations and maintained on deep litter system.

2.4. Live Body Weight

At day 3 PI, 10 birds from each group were randomly selected, marked, weighed and the weight recorded as live body weight (LBW) of each group. The marked birds were re-weighed at days 6, 9, 12, 15 and 21 PI.

2.5. Serology

Blood samples (3ml/bird) were collected from 10 birds in each group on days 0, 7, 14, and 21 PI through the wing vein, using sterile syringe. The sample bottles were stoppered, laid on near horizontal position, and allowed to clot. The serum samples were harvested into 2 ml vials. The duck sera were inactivated at 56°C in water bath for 30 minutes and with the chicken sera, use in the serological analysis.

2.5.1. Washing of Erythrocytes

Two ml of ND sero-negative chicken blood was collected from adult bird in a test tube containing EDTA and washed according to standard procedure. Using the formula CV = RxV/O, where CV = calculated volume (ml) of washed RBC, R = required % of RBC (0.5 %), V = volume (ml) of PBS intended to be used in the dilution and O = original PCV (%) of the RBC after washing, 0.5% of the washed chicken RBC was prepared [13] .

2.5.2. Haemagglutination (HA) Test

The standard HA technique as described by Beard, [13] was used and two hundred doses of La Sota ND vaccine was diluted in 10 ml of PBS (pH 7.0). The titre was then adjusted to 4 hemagglutinin units (HAU) for the HI test.

2.5.3. Haemagglutination Inhibition (HI) Test

The standard HI technique as described by Beard, [13] was used. The arithmetic method of geometric mean titre (GMT) was calculated mathematically for 2 fold dilution with the following formula: **GMT = antilog [(A − 1) × logB + logC**] where A= average end point well number, B = the factor (2.0) and C = the reciprocal of first dilution (2.0) [14] .

2.6. Serum Proteins

At days 0, 7, 14 and 21 PI, 10 birds from each group were randomly selected and at least 3 ml blood samples collected through the wing vein, into plain test tubes and serum samples harvested from the clotted blood into 2 ml vials and stored at −40°C for serum protein studies.

2.6.1. Determination of Total Serum Protein (TSP)

The TSP was determined using direct Biuret method. Biuret reagent containing NaOH, potassium iodide, copper (II) sulphate and sodium—potassium tartarate and Standard containing aqueous solution of protein, equivalent to 5 g/dL (50 g/L) were used [15] .

The total proteins were calculated as follows: **Total protein (g/dL) = absorbance of sample × 5/absorbance of standard.**

2.6.2. Determination of Serum Albumin Level (SAL)

The SAL was determined using bromocresol green method. Bromocresol green reagent containing bromocresol green, succinate buffer (pH 4.2), surfactants, preservatives and stabilizers, and standard containing aqueous solution equivalent to 5 g/dL (50 g/L) of albumin were used [15] .

The serum albumin was calculated as follows:

Serum albumin (g/dL) = absorbance of sample × 5/absorbance of standard.

2.6.3. Calculation of Serum Globulin Fractions (SGF)

The globulin fraction was calculated as the difference between total serum proteins and serum albumin level; **SGF (g/dL) = TSP – SAL.**

2.7. Data Analysis

The LBW, HI and serum protein values between and within groups were subjected to statistical analysis using independent sample t-test and the level of significance was determined and accepted at $p \leq 0.05$ for all the results using statistical product for service and solution (SPSS) version 16.0 computer software. The mean ± standard error of mean (SEM) of the results obtained in the experiment were calculated and presented in tables, graphs and charts.

3. RESULTS

3.1. Live Body Weight (LBW)

The result of live body weights of both species are shown on Table 1. The mean weights of IC were significantly lower ($p < 0.05$) than those of the control on days 3, 12, 15 and 21 PI, while the mean weights of ID were significantly lower ($p < 0.05$) than those of the control on days 3 and 15 PI.

3.2. Hemagglutination Inhibition (HI)

The PI HI values were determined and the results are shown in Table 2. On day 0 PI all the groups recorded zero HI titre. The HI titers of IC sera were significantly higher ($p < 0.05$) when compared with ID at days 7—(105.60 ± 20.27 and 22.40 ± 4.89), 14—(537.60 ± 59.73 and 32.00 ± 0.00), and 21—(119.20 ± 17.64 and 36.80 ± 6.33), PI respectively. Control chickens and UD had zero titre throughout the period of the experiment.

Table 1. Mean live body weight (g) ± SEM of chickens and ducks

Days PI	Infected chickens	Control chickens	Infected ducks	Control ducks
0	680.00 ± 23.81	680 ± 23.805	314.00 ± 9.91	314.00 ± 09.91
3	632.00 ± 18.61*	746.00 ± 17.59	283.00 ± 12.65*	349.50 ± 19.30
6	679.00 ± 28.61	745.00 ± 20.12	386.00 ± 24.13	356.00 ± 16.61
9	763.00 ± 51.45	835.00 ± 23.63	360.00 ± 20.98	384.50 ± 16.34
12	763.00 ± 46.60*	996.00 ± 31.70	363.00 ± 19.22	386.00 ± 19.56
15	795.00 ± 47.40*	1030.00 ± 41.95	358.00 ± 27.36*	414.50 ± 20.42
21	867.00 ± 38.73*	1044.00 ± 34.13	450.00 ± 19.72	448.50 ± 20.17

*Means values significantly different at $p < 0.05$ along the same row.

Table 2. Hemagglutination inhibition in chickens and ducks.

	Geometric mean titre ± SEM			
Days	IC	UC	ID	UD
D0 PI	000.00 ± 00.00	000.00 ± 00.00	00.00 ± 0.00	000.00 ± 00.00
D7 PI	105.60 ± 20.27	000.00 ± 00.00	22.40 ± 4.89*	000.00 ± 00.00
D14 PI	537.60 ± 59.73	000.00 ± 00.00	32.00 ± 0.00*	000.00 ± 00.00
D21 PI	119.20 ± 17.64	000.00 ± 00.00	36.80 ± 6.33*	000.00 ± 00.00

*Means values significantly different at $p < 0.05$.

3.3. Serum Proteins

The results of the serum proteins are shown in Table 3. The total serum proteins in infected chickens were significantly higher ($p < 0.05$) than the control on day 7 and 14 PI. By the day 21 PI, the infected chickens presented significantly lower TSP ($p < 0.05$) than the control group. In the infected ducks, the TSP were significantly lower ($p > 0.05$) than the control on days 7 and 21 PI.

The results of the SGF are shown in Table 4. The serum globulin protein fraction in infected chickens were significantly higher ($p < 0.05$) than the control on days 7 and 14 PI, but was significantly lower ($p > 0.05$) than the control on day 21 PI. In the infected ducks, the SGF was significantly lower ($p > 0.05$) than the control only on day 21 PI.

The results of the SAL are shown in Table 5. The serum albumin level in infected chickens was not significantly different ($p > 0.05$) when compared with the control throughout the experimental period. In the infected ducks, the SAL was significantly lower ($p > 0.05$) than the control only on day 7 PI.

4. DISCUSSION

Since maternal antibody could be detected in chickens up to 3 weeks of age [4] , the choice of 6 weeks as the age of inoculation in this experiment was to ensure that the antibodies did not interfere with the susceptibility of the birds in these two species [9] . The results of the HI tests showed that the virus elicited a progressive serological response and sero-conversion detected earlier from day 7 PI which is different from day 10 PI first response recorded by Piacenti et al. [16] and Igwe [9] . Significantly higher antibody titers were observed in the chickens when compared with the ducks, indicating that the immune system in chickens may be more responsive, but were more stable in ducks than that in the chickens since there was rapid antibody decay in the latter. However, both species presented protective levels of antibodies ($>2^3$) at day 21 PI as reported by Villagas and Purchase [14] . The results underscore variations between both species and may be due to dehydrations suffered by chickens. The response was in tandem with the results of the study of vNDV in chickens by Mishra et al. [17] , Okoye et al. [12] and Piacenti et al. [16] who recorded the highest antibody response on days 15 and 21 PI. Contrary to this result, Oladele et al. [18] detected the highest HI titers by day 4 PI using Kudu-113 strain. The variations may be due to differences in the immune status of the chickens [19] [20] and the laboratory procedures used in the HI test. One of the mechanisms by which antibodies fight against pathogens is neutralization, particularly viruses [21]. Neutralized viruses are unable to attach to surface receptors of target cells and are thus prevented from replication [22] [23].

Table 3. Total serum proteins ± SEM (g/dL × 10).

Days	IC	UC	ID	UD
0	4.09 ± 0.17	4.09 ± 0.17	5.25 ± 1.66	5.25 ± 1.66
7	6.00 ± 0.14*	5.34 ± 0.11	5.44 ± 0.12	5.76 ± 0.09
14	3.13 ± 0.12*	2.55 ± 0.40	3.81 ± 0.25	3.37 ± 0.13
21	2.55 ± 0.09*	3.61 ± 0.14	3.13 ± 0.05*	4.35 ± 0.13

*Means values significantly different at $p < 0.05$.

Table 4. Serum globulin proteins ± SEM (g/dL × 10).

Days	IC	UC	ID	UD
0	2.63 ± 0.19	2.63 ± 0.19	3.44 ± 0.11	3.44 ± 0.11
7	3.96 ± 0.27*	3.29 ± 0.21	4.10 ± 0.99	3.93 ± 0.17
14	1.63 ± 0.20*	1.13 ± 0.05	2.41 ± 0.24	1.80 ± 0.22
21	1.03 ± 0.12*	1.75 ± 0.21	1.42 ± 0.05*	2.61 ± 0.19

*Means values significantly different at $p < 0.05$.

The serum proteins elevation was more in chickens that presented higher antibody response and this may be associated with hyper-globulinemia due to sero-conversion to immunoglobulin [1] [3] [24] or hemo-concentration associated with the dehydration occasioned by the infection and reduced feed and water intake [25]. Serum proteins have been known to play important roles in infection following invasion of the body by pathogens such as virus, bacteria etc. [1] [2] [18] and as vital substrate for antibody formation [26] . Low level of albumin in serum in livestock may be due to heavy loss in urine or loss due to enteropathy (ulcers and diarrhea), or decreased production by the liver probably because of insufficient intake of protein in diet [1] [25] [27]. Serum albumin proteins which are the most abundant proteins in the clear fluid portion of the blood, typically decrease during inflammation [28] [29] and in a condition of anorexia, albumin is said to depress leading to osmotic disequilibrium and dehydration [29] [30]. There is a strong indication that ducks are far less susceptible to ND than chickens and may likely be maintaining the disease in poultry.

Table 5. Serum albumin proteins ± SEM (g/dL × 10).

Days	IC	UC	ID	UD
0	1.46 ± 0.10	1.46 ± 0.10	1.81 ± 0.10	1.81 ± 0.10
7	2.04 ± 0.16	2.03 ± 0.17	1.30 ± 0.07*	1.81 ± 0.10
14	1.50 ± 0.11	1.42 ± 0.56	1.40 ± 0.06	1.59 ± 0.08
21	1.49 ± 0.06	1.87 ± 0.19	1.68 ± 0.29	1.74 ± 0.12

*Means values significantly different at $p < 0.05$.

REFERENCES

1. Kaslow, E.J. (2011) Serum Proteins and Functions. Califonia (800), 633-2322. www.mbc.ca.gov
2. Eckersall, P.D. (2008) Proteins Proteomics and the Dysproteinemias. In: Kaneko, J.J., Harvey, J.W. and Bruss, M.L., Eds., Clinical Biochemistry of Domestic Animals, 6th Edition, Academic Press, San Diego, 117-155.
3. Schmidt, E.M.S., Paulillo, A.C., Locatelli-Dittrich, R., Beltrame, O. and Denadai, J. (2009) Serum Protein Profiles of Ring-Necked Pheasants Vaccinated or Not against Newcastle Disease. International Journal of Poultry Science, 8, 359-362. http://dx.doi.org/10.3923/ijps.2009.359.362
4. Alexander, D.J. and Gough, R.E. (2003) Newcastle Disease and Other Avian Paramyxovirus Infections. In: Saif, Y.M., Barnes, H.J., Glisson, J.R., Fadly, A.M., McDougal, L.R. and Swayne, D.E., Eds., Disease of Poultry, 11th Edition, Iowa State University Press, Ames, 63-87.
5. Terregino, C., Cattoli, G., Grossele, B., Bertoli, E., Tisato, E. and Capua, I. (2003) Characterization of Newcastle Disease Virus Isolates Obtained from Eurasian Collared Doves (Streptopelia decaocto) in Italy Short Communication. Avian Pathology, 32, 63-68. http://dx.doi.org/10.1080/0307 945021000070732
6. Vidanovic, D., Sekler, M., Asanin, R., Milic, N., Nisavic, J., Petrovic, T. and Savic, V. (2011) Characterization of Ve-logenic Newcastle Disease Viruses Isolated from Dead Wild Birds in Serbia during 2007. Journal of Wildlife Diseases, 47, 433-444. http://dx.doi.org/10.7589/0090-3558-47.2.433
7. Office International des Epizooties (OIE) (2009) Manual of Diagnostic Tests and Vaccines for Terrestial Animals. Newcastle Disease. OIE Standard Commission Publication, Office International Epizooties, 2009 Version, Part 2, Section 2.1, Chapter 2.3.14, 576-589.

8. Leslie, J. (2000) Newcastle Disease: Outbreak Losses and Control Policy Costs. Veterinary Record (Journal of British Veterinary Association), 146, 603-606. http://dx.doi.org/10.1136/vr.146.21.603
9. Igwe, A.O. (2009) Comparative Study of the Pathogenicity and Pathogenesis of a Local Nigerian Velogenic Newcastle Disease Virus in Guinea Fowls and Chickens. MSc Dissertation, University of Nigeria, Nsukka, 1-112.
10. Alexander, D.J., Bell, J.G. and Alders, R.G. (2004) Newcastle Disease A Technology Review: With Special Emphasis on Its Effects on Village Chickens FAO Animal Production and Health. Paper 161 Food and Agriculture Organization of the United Nations, Rome.
11. Echeonwu, G.O.N., Ireogbu, C.U. and Emeruwa, A.C. (1993) Recovery of Velogenic Newcastle Disease Virus from Dead and Healthy Free Roaming Birds in Nigeria. Avian Pathology, 22, 383-387. http://dx.doi.org/10.1080/03079459308418928
12. Okoye, J.O.A., Agu, A.O., Chineme, C.N. and Echeonwu, G.O.N. (2000) Pathological Characterization in Chicken of a Velogenic Newcastle Disease Virus Isolate from Guinea Fowl. Revue d'élevage et de Médecine Vétérinaire des pays Tropicaux, 53, 325-330.
13. Beard, C.W. (1980) Isolation and Identification of Avian Pathogens. 2nd Edition, American Association of Avian Pa-thologists, 129-135.
14. Villegas , P. and Purchase, H.G. (1989) Titration of Biological Suspension. In: A Laboratory Manual for the Isolation and Identification of Avian Pathogens, USA Kendal Hunt American Association of Avian Pathologists, Iowa, 186-190
15. Dumas, B.T. (1971) Diagnostic Reagent Kit for in Vitro Determination of Total Protein and Albumin in Serum (Code No25931). Clinical Chemistry Acta, 31, 87-96.
16. Piacenti, A.M., King, D.J., Seal, B.S., Zhang, J. and Brown, C.C. (2006) Pathogenesis of Newcastle Disease in Com-mercial and Specific Pathogen-Free Turkeys Experimentally Infected with Isolates of Different Virulence. Journal of Veterinary Pathology, 43, 168-178. http://dx.doi.org/10.1354/vp.43-2-168
17. Mishra, S., Kataria, J.M., Verma, K.C. and Sah, R.L. (2000) Response of Chickens to Infection with Newcastle Disease Virus Isolated from a Guinea Fowl. Tropical Animal Health and Production, 32, 276-284.
18. Oladele, S.B., Abdu, P., Nok, A.J., Esievo, K.A.N. and Useh, N.M. (2005) Haemagglutination Inhibition Antibodies rectal Temperature and Total Protein of Chicken Infected with a Local Nigerian Isolate of Velogenic Newcastle Disease Virus. Veterinary Research Communication, 29, 171-179. http://dx.doi.org/10.1023/B:VERC.0000047495.03341.2b
19. Scott, T.R. (2004) Our Current Understanding of Humoral Immunity of Poultry. Poultry Science, 83, 574-579. http://dx.doi.org/10.1093/ps/83.4.574
20. Dortmans, J.C.F.M., Koch, G., Rottier, P.J.M. and Peeters, B.P.H. (2011) Virulence of Newcastle Disease Virus: What Is Known So Far? Veterinary Research, 42, 122.
21. Grogan, K.B., Fernandez, R.J., Barranon, R.F.J. and Espinosa, H.G. (2008) Avian Immune System: A Brief Review. Merial Select, Gainesville, 1-12.
22. Nester, W.E., Anderson, G.D., Evans, C.R., Pearsall, N.N., Nester M.T. and Hurley, D. (2004) Microbiology: A Human Perspective. McGraw Hill, New York.
23. Talaro, K.P. (2005) Foundation in Microbiology. 5th Edition, The McGraw Hill Companies, New York.
24. Filipovic, N., Stojevic, Z., Milinkovic-Tur, S., Ljubic, B.B. and Zdelar-Tuk, M. (2007) Changes in Concentration and Fractions of Blood Serum Proteins of Chickens during Fattening. Veterinarski Archiv, 77, 319-326.
25. Ihedioha, J.I. and Chineme, C.N. (2005) Fundamental of Systemic Veterinary Pathology. Vol. 2, Great A P Express, Nsukka.
26. Nnadi, P.A., Eze, P.C. and Ezema, W.S. (2010) Influence of Delayed Feeding on the Performance Development and Response of Immune System to Newcastle Disease Vaccination in Chickens. International Journal of Poultry Science, 9, 669-674. http://dx.doi.org/10.3923/ijps.2010.669.674
27. Daramola, J., Adeboye, A., Fatoba, T. and Soladoye, A.O. (2003) Hematological and Biochemical Parameters of the West African Dwarf Goat. Department of Animal Production University of Ilorin Nigeria, Longman, Singapore City.

28. Lumeij, J.T. (2008) Avian Clinical Biochemistry. In: Kaneko, J.J., Harvey, J.W. and Bruss, M.L., Eds., Clinical Bio-chemistry of Domestic Animals, 6th Edition, Academic Press, San Diego, 839-872. http://dx.doi.org/10.1016/B978-0-12-370491-7.00030-1
29. Petersen, H.H., Nielsen, J.P. and Heegaard, P.M. (2004) Application of Acute Phase Protein Measurements in Veteri-nary Clinical Chemistry. Veterinary Research, 35, 163-187. http://dx.doi.org/10.1051/vetres:2004002
30. Thomson, R.G. (1984) General Veterinary Pathology.

CHAPTER 22

Genetic diversity of porcine group A rotavirus strains in the UK

Rebecca Chandler-Bostock[a], Laura R. Hancox[a], Sameena Nawaz[b], Oliver Watts[a], Miren Iturriza-Gomara[b], Kenneth M. Mellits[a,*]

[a]*University of Nottingham, School of Biosciences, Division of Food Science, Sutton Bonington Campus, Loughborough LE12 5RD, UK*

[b]*Virus Reference Department, Public Health England, London, NW9 5HT, UK*

ABSTRACT

Rotavirus is endemic in pig farms where it causes a loss in production. This study is the first to characterise porcine rotavirus circulating in UK pigs. Samples from diarrheic pigs with rotavirus enteritis obtained between 2010 and 2012 were genotyped in order to determine the diversity of group A rotavirus (GARV) in UK pigs. A wide range of rotavirus genotypes were identified in UK pigs: six G types (VP7); G2, G3, G4, G5, G9 and G11 and six P types (VP4); P[6], P[7], P[8], P[13], P[23], and P[32]. With the exception of a single P[8] isolate, there was less than 95% nucleotide identity between sequences from this study and any available rotavirus sequences.

The G9 and P[6] genotypes are capable of infecting both humans and pigs, but showed no species cross-over within the UK as they were shown to be genetically distinct, which suggested zoonotic transmission is rare within the UK. We identified the P[8] genotype in one isolate, this genotype is almost exclusively found in humans. The P[8] was linked to a human Irish rotavirus isolate in the same year. The discovery of human genotype P[8] rotavirus in a UK pig confirms this common human genotype can infect pigs and also highlights the necessity of surveillance of porcine rotavirus genotypes to safeguard human as well as porcine health.

Keywords: Rotavirus, Porcine, Phylogenetic, Zoonosis

1. INTRODUCTION

Rotaviruses have a broad host range that includes mammalian and avian species. In children, group A rotavirus (GARV) is the leading cause of severe gastroenteritis worldwide, and is associated with significant morbidity and mortality, with most children having been exposed by the time they are 5 years old (Tate et al., 2012). In pigs, rotavirus has a significant economic impact through loss in production and is most prevalent in neonatal pigs (<7 days) and piglets at the time of weaning (21–28 days) (Katsuda et al., 2006; Svensmark et al., 1989). Rotavirus can be transmitted zoonotically between pigs and humans. To date there are no reported studies of rotavirus genotypes in symptomatic UK pigs.

Rotaviruses belong to the *Reoviridae* family. They are non-enveloped, double stranded RNA viruses with a segmented genome. The 11 genome segments code for six structural proteins (VP1–4, 6–7) and six non-structural proteins (NSP1–6). There are eight different serogroups of rotavirus (Group A–H), all of which are found in animals or birds (Kindler et al., 2013; Molinari et al., 2014), but only A–C are found in humans (Estes and Cohen, 1989). Pigs are affected by rotavirus serogroups A, B, C, E and H (Molinari et al., 2014; Pedley et al., 1986). The outer capsid of the virus particle is constituted of VP7 (a glycoprotein) and VP4 (a protease sensitive protein), both elicit neutralising antibodies and form the basis of the dual classification of rotaviruses into G and P types, respectively (Estes and Cohen, 1989; Estes and Kapikian, 2007). To date, 27 G-types and 37 P-types of GARV have been identified (Matthijnssens et al., 2011; Trojnar et al., 2013).

Genotype diversity among rotavirus strains is generated by genetic drift, through the accumulation of point mutations, leading to genetic lineages within genotypes and monotypes within serotypes that possess altered epitopes and specific antibody recognition patterns (Coulson and Kirkwood, 1991). In addition, due to the segmented nature of the rotavirus genome, gene reassortment which can take place during co-infection with more than one strain can lead to further rotavirus strain diversity of co-circulating strains. The widespread presence of rotaviruses throughout the animal kingdom constitutes a large reservoir of rotavirus strains, and interspecies transmission combined with reassortment can lead to the emergence of novel or unusual strains that may spread globally. Numerous reports have described interspecies transmission leading to sporadic cases of human disease with rotaviruses from different animal species origin (Ben Hadj Fredj et al., 2013; Doan et al., 2013; Luchs et al., 2012; Mukherjee et al., 2013; Papp et al., 2013). The emergence of epidemiologically important strains such as G9P[8] globally, G10P[11] in India and G8P[4] in Africa, Europe and the USA, in the human population is postulated to have resulted from reassortment with animal strains leading to host adaptation and spread (De Donno et al., 2009; Iturriza-Gomara et al., 2000a; Jere et al., 2011; Leite et al., 2008; Nyaga et al., 2013; Pietsch et al., 2009; Ramani et al., 2009; Than et al., 2013; Weinberg et al., 2012). Worldwide, common porcine rotavirus genotypes are G3, G4, G5, G11 and P[6], P[7], P[13], P[19], P[23], P[26], P[27] (Martella et al., 2010). In Europe, genotypes G1–6, 9–12 and P[6]–P[10], P[13], P[22], P[23], P[27] and P[32] have been identified in pigs (Collins et al., 2010a,b; Midgley et al., 2012).

The aims of this study were to genotype rotavirus in symptomatic UK pigs, to determine the likelihood of zoonotic transmission between pigs and humans within the UK and to compare porcine rotavirus in the UK to genotypes prevalent in Europe and the rest of the world. The findings from the study will not, in themselves, improve biosecurity but will contribute to a better understanding of the potential threat of zoonosis.

2. METHODS

2.1. Sample collection

Porcine faecal and intestinal content samples were collected from UK pigs; 66% were obtained from the Animal Health Veterinary Laboratories Agency (AHVLA), 34% samples were referred directly to our lab from veterinarians. The samples obtained from the AHVLA had previously tested positive for rotavirus using gel electrophoresis. Other samples were suspected rotavirus infection and were confirmed using RT-PCR (described below). In total, there were 63 samples from 54 different locations between autumn 2010 and spring 2012. The distribution of these samples in the UK is shown in Fig. 1. All samples were obtained and analysed in accordance with the University of Nottingham ethical guidelines.

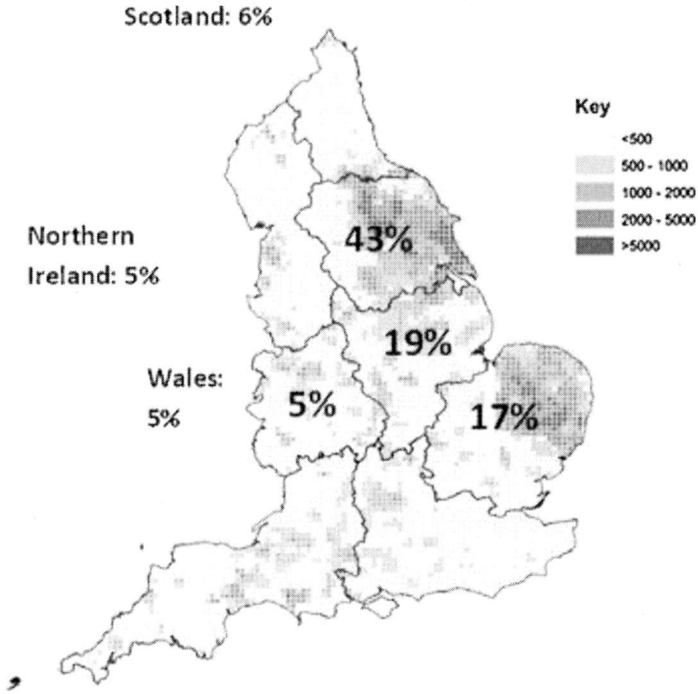

Figure. 1. Map of England showing the distribution of pigs per 5 km^2 in 2010 separated by region, adapted from DEFRA (2010). Percentages on map represent the percentage of samples from each region of England and the percentage taken from Scotland, Wales and Northern Ireland.

2.2. RNA preparation

Nucleic acid extraction was carried out with QiaXtractor platform (Qiagen) using the specified plastics and the VX reagent kit, as per manufacturers' instructions, from 10% faecal solutions in Dulbecco's modified Eagle's Medium (DMEM).

2.3. RT-PCR amplification of VP7 and VP4

VP7 and VP4 rotavirus genes were amplified from extracted nucleic acids by RT-PCR using methods and primers previously described by Gomara et al. (2001), Gray and Iturriza-Gomara (2011) and Gentsch et al. (1992). Samples producing a band for either VP7 or VP4 were considered positive. Samples that did not amplify in the VP7 and VP4 assays were considered negatives as they were also negative in a VP6-specific qPCR (Gomara et al., 2002). PCR products were purified using QIAQuick PCR Purification Kit (Qiagen) and the forward and reverse strands of VP7 and the VP8* portion of VP4 genes were sequenced using Sanger sequencing (MWG Eurofins) and the same primers as for amplification. Sequences have been added to Genbank database VP7 accession numbers KJ135124–KJ135172 and VP4 accession numbers KJ135173–KJ135220.

2.4. Sequence analysis

Genotypes were determined using the RotaC genotyping tool (Maes et al., 2009) and compared to similar sequences using NCBI BLASTn genbank database. Sequence alignments and phylogenetic trees were constructed using Mega6 and ClustalW.

3. RESULTS

3.1. Rotavirus genotypes in UK pigs

Porcine faecal samples with suspected rotavirus enteritis were obtained from pig producing regions within the UK (Fig. 1). The rotavirus genotypes determined for these samples are shown in Table 1. G4 and G5 were the most common VP7 genotypes, accounting for 25% (16/64) and 36% (23/64) of the strains, respectively. The most common VP4 genotypes were P[6] (33%, 21/64) and P[32] (27%, 17/64). Overall, the most common genotype combinations were G4P[6] and G5P[7].

Table 1. Rotavirus genotypes found amongst UK pigs.

	G2	**G3**	**G4**	**G5**	**G9**	**G11**	**Untyped**	**Total (%)**
P[6]			13	5	2	1		21 (33%)
P[7]			2	9				11 (17%)
P[8]				1				1 (2%)
P[13]		1		1				2 (3%)
P[23]				1				1 (2%)
P[32]		6		4	6		2	17 (27%)
Untyped	2	2	1	2	2	1		10 (16%)
Total (%)	2 (3%)	9 (14%)	16 (25%)	23 (36%)	10 (16%)	2 (3%)	2 (3%)	64

3.2. VP4 sequence and phylogenetic analysis

The single P[8] sequence from this study was most similar to a sequence isolated in Ireland in 2011 (Gunn et al., 2012) which shared 99.6% nucleotide identity. This sequence demonstrated high similarity to P[8] rotavirus strains of human origin from the UK and worldwide (93–99%) in lineage P[8]-3 (Fig. 2). The only porcine P[8] sequence available for comparison (Po|Brazil|1991|P[8]|AF052449.1) is Wa-like and found in another lineage (P[8]-1); (Iturriza-Gomara et al., 2000b).

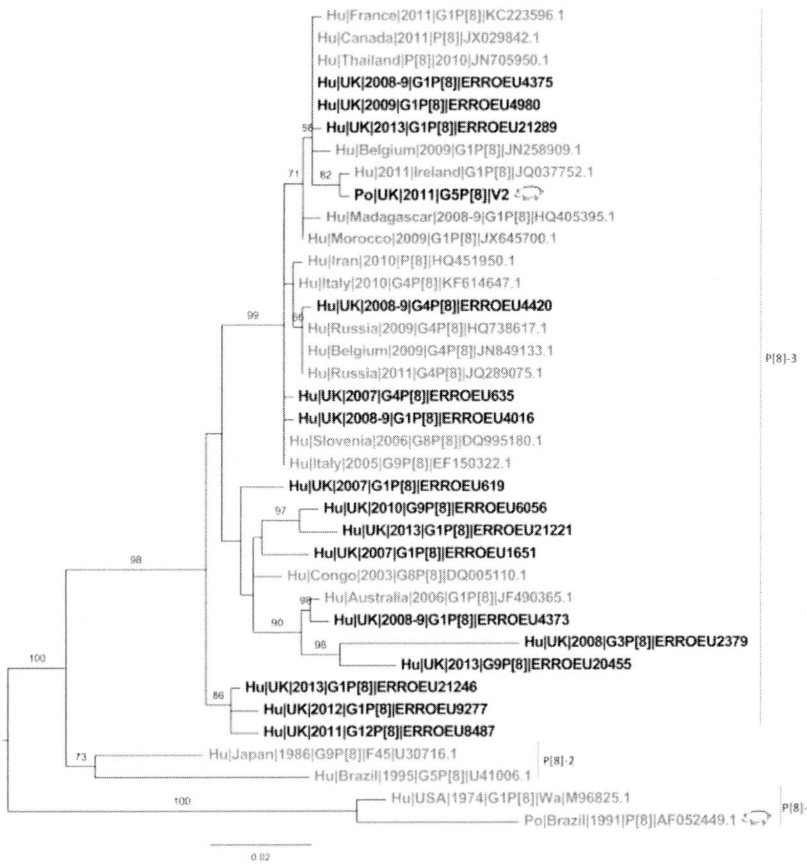

Figure. 2. Phylogenetic tree of P[8] sequences. The bars on the right indicate lineages within the P[8] genotype (Iturriza-Gomara et al., 2000a). UK sequences (black), non-UK (grey), porcine sequences (pig symbol), human sequences (no symbol). Bootstrap values more than 50% are shown (1000 pseudoreplicates).

P[6] was the most common (21/64) VP4 genotype (Table 1). Sequences from this study clustered with several P[6] lineages (Fig. 3) (Martella et al., 2006). Three sequences from this study (Po|UK|2011|G9P[6]|A1, Po|UK|2011|G9P[6]|B2 and Po|UK|2011|G5P[6]|E) were most similar to Gottfried strain but shared only 81.4–85.9% identity. Eight sequences from this study were most similar to P[6]-IV lineage but with only 87% identity they are a distinct sub-lineage. The majority of P[6] sequences from this study cluster with P[6]-I lineage sharing 87.1–94.2% identity to most similar reference sequences. There were also two sequences most similar to P[6]-III (84.5–91% identity).

P[32], the next most common genotype (17/54), was found in combination with G3, G5 and G9 (Table 1). The sequences from this study cluster separate to the reference sequences, except for Po|UK|2010|P[32]|E + 41. All sequences from this study except Po|UK|2010|P[32]|E + 41 shared <89% identity with a porcine isolate from the Republic of Ireland and <84% identity with three porcine isolates from Denmark.

3.3. VP7 sequence and phylogenetic analysis

The most common VP7 genotype was G5 (23/62) (Table 1). When analysed, the majority of G5 sequences from this study are clustered, together Fig. 4 sharing a maximum of 89.3% identity to any reference sequence but 92–99.8% identity to each other (Fig. 5). Two of the sequences from this study did not cluster with the rest; Po|UK|2010|G5P[32]|T3 and Po|UK|2011 |G5|Y1 were clustered with porcine sequences from Thailand (2005) and Italy (2005, 2006). Irish G5 sequences clustered separately to UK sequences from this study despite being the closest isolates geographically.

G9 rotavirus sequences (10/62) from this study formed two separate clusters (Fig. 6). The majority of G9 sequences from this study (8/10) clustered together, sharing <94% identity with any reference sequences. The remaining two G9 sequences from this study, Po|UK|2011|G9P[6]|A1 and Po|UK|2011|G9|A3, clustered separately to the other porcine UK sequences, instead with G9 sequences from the 1970s and 1980s isolated in India, China, Japan and USA (89–91% similarity).

Figure. 3. Phylogenetic trees of P[6] sequences. Roman numerals (I, II, III, IV) denote described lineages of P[6] (Martella et al., 2006). UK sequences (black), non-UK (grey), porcine sequences (pig symbol), human sequences (no symbol). UK porcine sequences are all from this study. Bootstrap values more than 50% are shown (1000 pseudoreplicates).

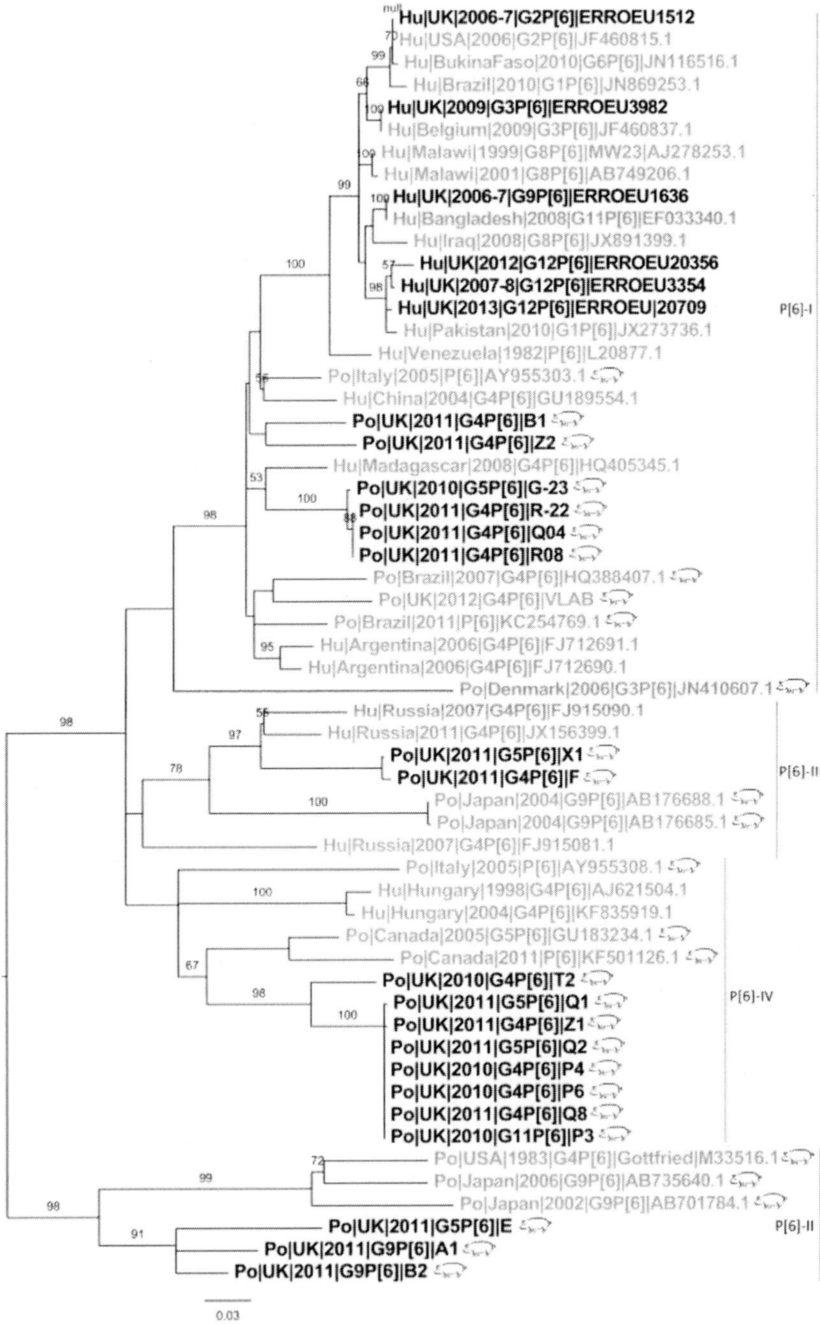
Hu|UK|2006-7|G2P[6]|ERROEU1512
Hu|USA|2006|G2P[6]|JF460815.1
Hu|BukinaFaso|2010|G6P[6]|JN116516.1
Hu|Brazil|2010|G1P[6]|JN869253.1
Hu|UK|2009|G3P[6]|ERROEU3982
Hu|Belgium|2009|G3P[6]|JF460837.1
Hu|Malawi|1999|G8P[6]|MW23|AJ278253.1
Hu|Malawi|2001|G8P[6]|AB749206.1
Hu|UK|2006-7|G9P[6]|ERROEU1636
Hu|Bangladesh|2008|G11P[6]|EF033340.1
Hu|Iraq|2008|G8P[6]|JX891399.1
Hu|UK|2012|G12P[6]|ERROEU20356
Hu|UK|2007-8|G12P[6]|ERROEU3354
Hu|UK|2013|G12P[6]|ERROEU|20709
Hu|Pakistan|2010|G1P[6]|JX273736.1
Hu|Venezuela|1982|P[6]|L20877.1
Po|Italy|2005|P[6]|AY955303.1
Hu|China|2004|G4P[6]|GU189554.1
Po|UK|2011|G4P[6]|B1
Po|UK|2011|G4P[6]|Z2
Hu|Madagascar|2008|G4P[6]|HQ405345.1
Po|UK|2010|G5P[6]|G-23
Po|UK|2011|G4P[6]|R-22
Po|UK|2011|G4P[6]|Q04
Po|UK|2011|G4P[6]|R08
Po|Brazil|2007|G4P[6]|HQ388407.1
Po|UK|2012|G4P[6]|VLAB
Po|Brazil|2011|P[6]|KC254769.1
Hu|Argentina|2006|G4P[6]|FJ712691.1
Hu|Argentina|2006|G4P[6]|FJ712690.1
Po|Denmark|2006|G3P[6]|JN410607.1
P[6]-I
Hu|Russia|2007|G4P[6]|FJ915090.1
Hu|Russia|2011|G4P[6]|JX156399.1
Po|UK|2011|G5P[6]|X1
Po|UK|2011|G4P[6]|F
Po|Japan|2004|G9P[6]|AB176688.1
Po|Japan|2004|G9P[6]|AB176685.1
Hu|Russia|2007|G4P[6]|FJ915081.1
P[6]-III
Po|Italy|2005|P[6]|AY955308.1
Hu|Hungary|1998|G4P[6]|AJ621504.1
Hu|Hungary|2004|G4P[6]|KF835919.1
Po|Canada|2005|G5P[6]|GU183234.1
Po|Canada|2011|P[6]|KF501126.1
Po|UK|2010|G4P[6]|T2
Po|UK|2011|G5P[6]|Q1
Po|UK|2011|G4P[6]|Z1
Po|UK|2011|G5P[6]|Q2
Po|UK|2010|G4P[6]|P4
Po|UK|2010|G4P[6]|P6
Po|UK|2011|G4P[6]|Q8
Po|UK|2010|G11P[6]|P3
P[6]-IV
Po|USA|1983|G4P[6]|Gottfried|M33516.1
Po|Japan|2006|G9P[6]|AB735640.1
Po|Japan|2002|G9P[6]|AB701784.1
Po|UK|2011|G5P[6]|E
Po|UK|2011|G9P[6]|A1
Po|UK|2011|G9P[6]|B2
P[6]-II
0.03

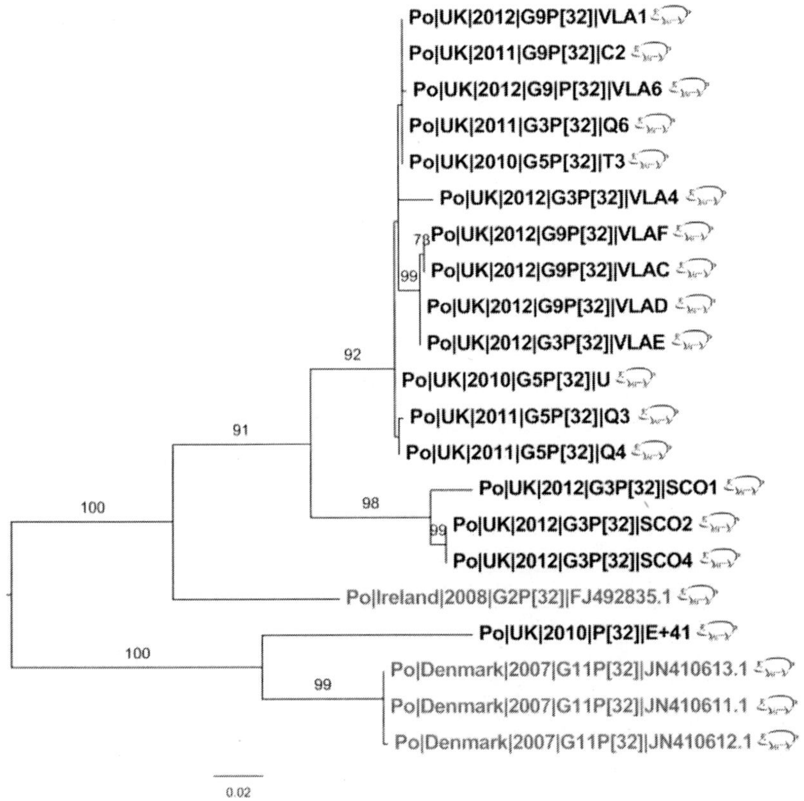

Figure. 4. Phylogenetic trees of P[32] sequences. UK sequences (black), non-UK (grey), all sequences are porcine (pig symbol), UK porcine sequences are all from this study. Bootstrap values more than 50% are shown (1000 pseudoreplicates).

4. DISCUSSION

4.1. High diversity of rotavirus strains in UK pigs

This study describes a wide range of rotavirus genotypes circulating within the UK pig population. Porcine sequences from this study and human rotavirus sequences from the UK were compared to determine the likelihood of zoonotic transmission of rotavirus between pigs and humans in the UK. Although most genotypes found in this study were similar to previous porcine studies, we also identified the P[8] genotype of VP4, which is almost exclusively found in humans, in one isolate (Martella et al., 2010).

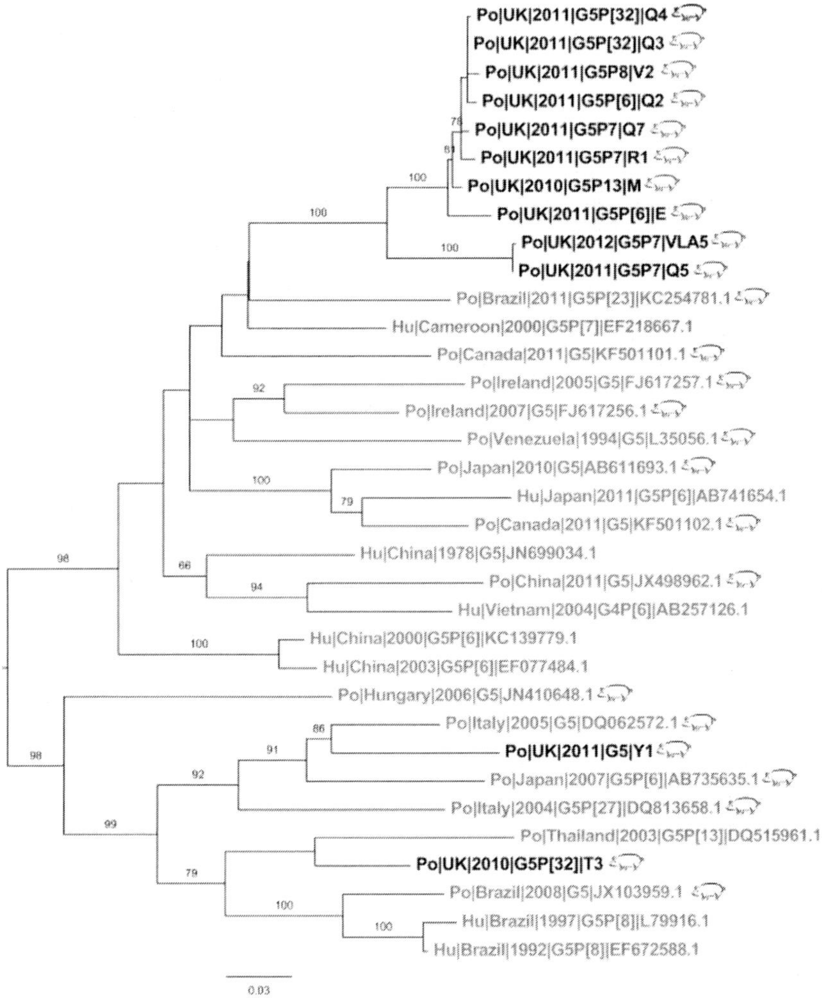

Figure. 5. Phylogenetic trees of G5 sequences. UK sequences (black), non-UK (grey), porcine sequences (pig symbol), human sequences (no symbol). UK porcine sequences are all from this study. Bootstrap values more than 50% are shown (1000 pseudoreplicates).

The samples from this study represent areas of the UK with the highest density of pig farms (Fig. 1) and hence this survey can be considered representative of rotavirus causing disease in pigs between 2010 and 2012 (DEFRA, 2010). No apparent link between geographical area and genotype was found in the UK. This suggests co-circulating genotypes are spread across the country and may indicate multiple introductions to farms rather than sustained outbreaks.

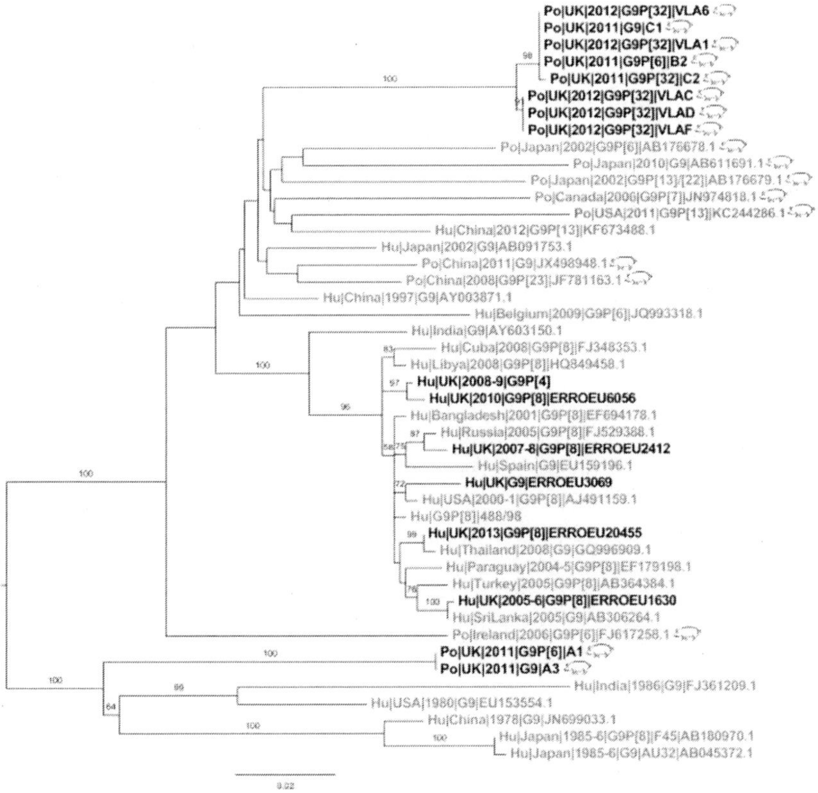

Figure. 6. Phylogenetic trees of G9 sequences. UK sequences (black), non-UK (grey), porcine sequences (pig symbol), human sequences (no symbol). UK porcine sequences are all from this study. Bootstrap values more than 50% are shown (1000 pseudoreplicates).

The vast majority sequences in this study have diverged by >5% from strains circulating in the rest of Europe and the world, for example G5 and G9 (Figs. 5 and 6), likely due to the relative isolation and limited import of pigs to the UK. Indeed the majority of "import" pigs come from the Republic of Ireland (Personal Communication, AHVLA); even so there was still >5% divergence between UK rotavirus and Irish rotavirus sequences at nucleotide level. This observation highlights effective biosecurity between UK pigs and those in neighbouring countries.

4.2. Common porcine genotypes in the UK are similar to those worldwide

G4P[6] and G5P[7] were the most common genotype combinations from this study, 20% and 14%, respectively, (Table 1) and they are frequently found in pigs (Martella et al., 2010). G3P[13] and G5P[13], also found in this study, are commonly found in pigs (Chan-It et al., 2008; Miyazaki et al., 2013; Saikruang et al., 2013; Steyer et al., 2008). Four samples from this study had the genotype combination G5P[23], this has only recently been identified in piglets in Brazil (Tonietti et al., 2013), even though both G5 and P[23] are common porcine genotypes (Collins et al., 2010b; Hong Anh et al., 2014; Martella et al., 2010).

4.3. Lack of interspecies reassortment of G9 and P[6] in the UK

The G9P[6] genotype has been associated with rotavirus outbreaks in children in the UK and more recently in Belgium (Iturriza-Gomara et al., 2000a; Zeller et al., 2012) and has previously been identified in pigs in the Republic of Ireland and Japan (Collins et al., 2010b; Teodoroff et al., 2005). In this study, porcine and human G9 sequences from the UK did not cluster together, neither did P[6] porcine and human UK sequences (Figs. 3 and 6). Therefore it is unlikely that the zoonotic transmission of G9 and P[6] rotavirus occurred within the UK.

Although the P[6] sequences from this study grouped with representative sequences of P[6] lineages I, II, III and IV (Fig. 3) (Martella et al., 2006), all sequences from this study shared <93% nucleotide identity to any reference sequence. Thus, UK porcine rotaviruses are divergent from European and global sequences. Moreover porcine P[6] sequences from this study did not cluster with the human sequences from the UK, demonstrating a lack of evidence for interspecies reassortment. The multiple lineages of P[6] circulating, suggests multiple introductions of this genotype into the UK pig herd. P[6] is an uncommon human rotavirus genotype in Europe (Cashman et al., 2012; Iturriza-Gomara et al., 2011; Lennon et al., 2008), which may also explain the lack of transmission between the general population and pigs, at least in the UK, who will typically have limited contact with live farm animals.

Worldwide, G9 emerged in humans in the mid 1990s, most likely from pigs as they are the only other species known to be infected by this genotype (Ramachandran et al., 2000). Porcine G9 sequences from the UK form two distinct clusters, both of which are in different lineages to the UK human G9 isolates, which clustered only with other human isolates (Fig. 6) This clear differential clustering of human and porcine rotavirus G9 strains from the UK suggests that interspecies transmission of this genotype was unlikely to have taken place within the UK, and that human and porcine G9 have arrived independently into the UK. The human UK sequences are recorded from 2006, it is impossible to say if there would have been interspecies transmission in this country before that date; however, it is believed that G9 strains only emerged as globally important human rotavirus in the mid 1990s. Prior to that G9 strains had been found in sporadic cases in several countries, and with relative frequency in India, often in association with asymptomatic neonatal infections and with strong evidence of zoonotic transmission (Bhan et al., 1993; Jain et al., 2001; Ramachandran et al., 1996).

4.4. The UK contains a rare P[32] genotype

The P[32] genotype was previously identified in Denmark and the Republic of Ireland (Collins et al., 2010a; Midgley et al., 2012). Although these countries are isolated geographically from the UK, pigs do circulate between these countries and the UK (AHVLA, Personal Communication, 2014), and thus is unsurprising that this genotype has been found in the UK as well. Despite having only been recently identified, P[32] is widespread in the UK (found in North Yorkshire, Derbyshire and Scotland). It occurs in combination with G3, G5 and G9, this range of genotypes were likely to be due to reassortment in the UK. The presence of P[32] suggests that UK strains form a distinct pattern different from other European and wider world strains, and therefore it is relevant to survey then as we have here to rationalize any worldwide vaccine.

4.5. UK porcine G5P[8] rotavirus is likely to be the result of a reassortment with human P8 and porcine G5 strains

G5P[8] is an uncommon genotype combination. There is only one incidence of G5P[8] isolated from pigs (Gouvea et al., 1999) and few examples of G5P[8] found in humans (Esona et al., 2004; Timenetsky et al., 1997). The main reservoir of the G5 genotype is pigs, but is found sporadically in horses, humans and cattle. P[8] however is the most prevalent human genotype (da Silva et al., 2011; Esona et al., 2004; Martella et al., 2010) it has sporadically been found in pigs and sheep but it was the only genotype found in this study not commonly associated with porcine (Fitzgerald et al., 1995; Gouvea et al., 1999; Halaihel et al., 2010).

The porcine P[8] sequence from this study was most similar to the Irish sequence Hu|2011|Ireland|G1P[8]||JQ037752.1 from an elderly patient in Ireland as part of a study of a rotavirus outbreak in a care home, isolated as

G1P[8] (Gunn et al., 2012). This is notable as it was isolated the same year as the P[8] sequence from this study (G5P[8]) was isolated from a pig farm in North Yorkshire. These two sequences share 99.6% nucleotide identity and 99.4% amino acid identity, with only one amino acid difference between the two sequences. This strongly suggests that the two sequences are derived from the same source and that interspecies transmission occurred directly or indirectly to give G5P[8] rotavirus in a neonatal pig. The VP6 genotype of the P[8] isolate from this study was I1, also a human genotype, suggesting the pig was infected with a human rotavirus containing a porcine G5 segment.

All the P[8] sequences from the UK were found in lineage P[8]-3, which is notable considering P[8] is found commonly in the UK (Iturriza-Gomara et al., 2001). The other porcine P[8] (Po|Brazil|1991|P[8]AF052449.1) clustered with P[8]-1 lineage (Iturriza-Gomara et al., 2000b) distinct from the UK P[8] sequence from this study (Fig. 2). P[8] is almost exclusively found in human infections, suggesting adaption of P[8] to the human host,. As with norovirus (NoV) histo-blood group antigens (HBGA's) may determine susceptibility to infection by rotaviruses of different P-types (Huang et al., 2005, 2012). The same HBGA receptors are used for viral attachment of NoV in pigs and humans such as H type 1 receptor (Tian et al., 2007). Huang et al. (2012) postulated that P[8] rotavirus share HBGA receptors with norovirus in the human gut. P[8] and P[4] rotavirus have shown specificity to both Lewis-b and H-type 1 HBGA's. Therefore the lack of P[8] in pigs may be related to pigs having HBGA other than Lewis b. As we have found evidence for zoonotic transmission, a detailed study of the binding capacity of different genotypes to human and animal blood receptors may allow us to better understand constraints to interspecies transmission and predict which strains are more likely to pass between species.

5. CONCLUSION

This study has highlighted a gap in the knowledge regarding UK porcine rotavirus strains, and has found a lack of transmission of porcine rotavirus between the UK and the rest of the world. This information will be useful in the rationalization of genotypes for vaccines to protect UK pigs. An effective vaccine would add significant value to the farming industry (Svensmark et al., 1989), increasing yield of pork, and also may potentially reduce zoonotic transmission. Also, the important discovery of human genotype P[8] rotavirus in a UK pig not only confirms this common human genotype can infect pigs but also highlights the necessity of surveillance of porcine rotavirus genotypes to safeguard human health as well as porcine health.

ACKNOWLEDGEMENTS

We would like to thanks Ben Strugnell (AHVLA, Thirsk) and Susanna Williamson (AHVLA, Bury St. Edmunds) for providing rotavirus positive faecal samples used in this study. We would also like to thank Jonathan Ball for help with phylogenetic analysis and Janet Daly for critically reading the manuscript. This work was supported by the University of Nottingham, BBSRC Doctoral Training Grant.

REFERENCES

1. Ben Hadj Fredj, M., Heylen, E., Zeller, M., Fodha, I., Benhamida-Rebai, M., Van Ranst, M., Matthijnssens, J., Trabelsi, A., 2013. Feline origin of rotavirus strain, Tunisia, 2008. Emerg. Infect. Dis. 19, 630–634.
2. Bhan, M.K., Lew, J.F., Sazawal, S., Das, B.K., Gentsch, J.R., Glass, R.I., 1993. Protection conferred by neonatal rotavirus infection against subse-quent rotavirus diarrhea. J. Infect. Dis. 168, 282–287.
3. Cashman, O., Collins, P.J., Lennon, G., Cryan, B., Martella, V., Fanning, S., Staines, A., O'Shea, H., 2012. Molecular characterization of group A rotaviruses detected in children with gastroenteritis in Ireland in 2006–2009. Epidemiol. Infect. 140, 247–259.
4. Chan-It, W., Khamrin, P., Saekhow, P., Pantip, C., Thongprachum, A., Peerakorne, S., Ushijima, H., Maneekarn, N., 2008. Multiple combina-tions of P 13-like genotype with G3, G4, and G5 in porcine rotaviruses. J. Clin. Microbiol. 46, 1169–1173.
5. Collins, P.J., Martella, V., Buonavoglia, C., O'Shea, H., 2010a. Identification of a G2-like porcine rotavirus bearing a novel VP4 type, P[32]. Vet. Res. 41, 73.
6. Collins, P.J., Martella, V., Sleator, R.D., Fanning, S., O'Shea, H., 2010b. Detection and characterisation of group A rotavirus in asymptomatic piglets in southern Ireland. Arch. Virol. 155, 1247–1259.
7. Coulson, B.S., Kirkwood, C., 1991. Relation of VP7 amino acid sequence to monoclonal antibody neutralization of rotavirus and rotavirus mono-type. J. Virol. 65, 5968–5974.
8. da Silva, M.F.M., Tort, L.F.L., Gomez, M.M., Assis, R.M.S., Volotao, E.D., de Mendonca, M.C.L., Bello, G., Leite, J.P.G., 2011. VP7 gene of human rotavirus A genotype G5: phylogenetic analysis reveals the existence of three different lineages worldwide. J. Med. Virol. 83, 357–366.
9. De Donno, A., Grassi, T., Bagordo, F., Idolo, A., Cavallaro, A., Gabutti, G., 2009. Emergence of unusual human rotavirus strains in Salento, Italy, during 2006–2007. BMC Infect. Dis. 9, 43.
10. DEFRA, 2010. Maps of Livestock Populations in 2000 and 2010 across England. DEFRA, London, UK.
11. Doan, Y.H., Nakagomi, T., Aboudy, Y., Silberstein, I., Behar-Novat, E., Nakagomi, O., Shulman, L.M., 2013. Identification by full-genome analysis of a bovine rotavirus transmitted directly to and causing diarrhea in a human child. J. Clin. Microbiol. 51, 182–189.
12. Esona, M.D., Armah, G.E., Geyer, A., Steele, A.D., 2004. Detection of an unusual human rotavirus strain with G5P[8] specificity in a Camer-oonian child with diarrhea. J. Clin. Microbiol. 42, 441–444.
13. Estes, M.K., Cohen, J., 1989. Rotavirus gene structure and function. Micro-biol. Rev. 53, 410–449.
14. Estes, M.K., Kapikian, A.Z., 2007. Rotaviruses: Fields Virology, Fifth ed., 2. Lippincott Williams & Wilkins, Philadelphia, PA, USA, , pp. 1917–1974.

15. Fitzgerald, T.A., Munoz, M., Wood, A.R., Snodgrass, D.R., 1995. Serological and genomic characterisation of group A rotaviruses from lambs. Arch. Virol. 140, 1541–1548.
16. Gentsch, J.R., Glass, R.I., Woods, P., Gouvea, V., Gorziglia, M., Flores, J., Das, B.K., Bhan, M.K., 1992. Identification of group-A rotavirus gene-4 types by polymerase chain-reaction. J. Clin. Microbiol. 30, 1365–1373.
17. Gomara, M.I., Cubitt, D., Desselberger, U., Gray, J., 2001. Amino acid substitution within the VP7 protein of G2 rotavirus strains associated with failure to serotype. J. Clin. Microbiol. 39, 3796–3798.
18. Gomara, M.I., Wong, C., Blome, S., Desselberger, U., Gray, J., 2002. Molec-ular characterization of VP6 genes of human rotavirus isolates: correlation of genogroups with subgroups and evidence of indepen-dent segregation. J. Virol. 76, 6596–6601.
19. Gouvea, V., Lima, R.C., Linhares, R.E., Clark, H.F., Nosawa, C.M., Santos, N., 1999. Identification of two lineages (WA-like and F45-like) within the major rotavirus genotype P[8]. Virus Res. 59, 141–147.
20. Gray, J., Iturriza-Gomara, M., 2011. Rotaviruses. Methods Mol. Biol. (Clif-ton, NJ) 665, 325–355.
21. Gunn, L., Feeney, S.A., Cashman, O., Collins, P.J., Coyle, P.V., O'Shea, H., 2012. Molecular characterization of group A rotavirus found in elderly patients in Ireland; predominance of G1P[8], continued presence of G9P[8], and emergence of G2P[4]. J. Med. Virol. 84, 2008–2017.
22. Halaihel, N., Masia, R.M., Fernandez-Jimenez, M., Ribes, J.M., Montava, R., De Blas, I., Girones, O., Alonso, J.L., Buesa, J., 2010. Enteric calicivirus and rotavirus infections in domestic pigs. Epidemiol. Infect. 138, 542– 548.
23. Hong Anh, P., Carrique-Mas, J.J., Van Cuong, N., Hoa, N.T., Lam Anh, N., Duy, D.T., Hien, V.B., Vu Tra My, P., Rabaa, M.A., Farrar, J., Baker, S., Bryant, J.E., 2014. The prevalence and genetic diversity of group A rotaviruses on pig farms in the Mekong Delta region of Vietnam. Vet. Microbiol. 170, 258–265.
24. Huang, P., Farkas, T., Zhong, W., Tan, M., Thornton, S., Morrow, A.L., Jiang, X., 2005. Norovirus and histo-blood group antigens: demonstration of a wide spectrum of strain specificities and classification of two major binding groups among multiple binding patterns. J. Virol. 79, 6714– 6722.
25. Huang, P., Xia, M., Tan, M., Zhong, W., Wei, C., Wang, L., Morrow, A., Jiang, X., 2012. Spike protein VP8* of human rotavirus recognizes histo-blood group antigens in a type-specific manner. J. Virol. 86, 4833– 4843.
26. Iturriza-Gomara, M., Cubitt, D., Steele, D., Green, J., Brown, D., Kang, G., Desselberger, U., Gray, J., 2000a. Characterisation of rotavirus G9 strains isolated in the UK between 1995 and 1998. J. Med. Virol. 61, 510–517.
27. Iturriza-Gomara, M., Dallman, T., Banyai, K., Bottiger, B., Buesa, J., Die-drich, S., Fiore, L., Johansen, K., Koopmans, M., Korsun, N., Koukou, D., Kroneman, A., Laszlo, B., Lappalainen, M., Maunula, L., Marques, A.M., Matthijnssens, J., Midgley, S., Mladenova, Z., Nawaz, S., Poljsak-Pri-jatelj, M., Pothier, P., Ruggeri, F.M., Sanchez-Fauquier, A., Steyer, A., Sidaraviciute-Ivaskeviciene, I., Syriopoulou, V., Tran, A.N., Usonis, V., V.A.N.R, M., A.D.E.R, Gray J., 2011. Rotavirus genotypes co-circulating in Europe between 2006 and 2009 as determined by EuroRotaNet, a pan-European collaborative strain surveillance network. Epidemiol. Infect. 139, 895–909.
28. Iturriza-Gomara, M., Green, J., Brown, D.W., Desselberger, U., Gray, J.J., 2000b. Diversity within the VP4 gene of rotavirus P[8] strains: implications for reverse transcription-PCR genotyping. J. Clin. Micro-biol. 38, 898–901.
29. Iturriza-Gomara, M., Isherwood, B., Desselberger, U., Gray, J., 2001. Reas-sortment in vivo: driving force for diversity of human rotavirus strains isolated in the United Kingdom between 1995 and 1999. J. Virol. 75, 3696–3705.
30. Jain, V., Das, B.K., Bhan, M.K., Glass, R.I., Gentsch, J.R., 2001. Great diversity of group A rotavirus strains and high prevalence of mixed rotavirus infections in India. J. Clin. Microbiol. 39, 3524–3529.
31. Jere, K.C., Mlera, L., O'Neill, H.G., Potgieter, A.C., Page, N.A., Seheri, M.L., van Dijk, A.A., 2011. Whole genome analyses of African G2, G8, G9, and G12 rotavirus strains using

sequence-independent amplification and 454(R) pyrosequencing. J. Med. Virol. 83, 2018–2042.
32. Katsuda, K., Kohmoto, M., Kawashima, K., Tsunemitsu, H., 2006. Frequen-cy of enteropathogen detection in suckling and weaned pigs with diarrhea in Japan. J. Vet. Diagn. Invest. 18, 350–354.
33. Kindler, E., Trojnar, E., Heckel, G., Otto, P.H., Johne, R., 2013. Analysis of rotavirus species diversity and evolution including the newly deter-mined full-length genome sequences of rotavirus F and G. Infect. Genet. Evol. 14, 58–67.
34. Leite, J.P., Carvalho-Costa, F.A., Linhares, A.C., 2008. Group A rotavirus genotypes and the ongoing Brazilian experience: a review. Mem. Inst. Oswaldo Cruz 103, 745–753.
35. Lennon, G., Reidy, N., Cryan, B., Fanning, S., O'Shea, H., 2008. Changing profile of rotavirus in Ireland: predominance of P 8 and emergence of P 6 and P 9 in mixed infections. J. Med. Virol. 80, 524–530.
36. Luchs, A., Cilli, A., Morillo, S.G., Carmona Rde, C., Timenetsky Mdo, C., 2012. Rare G3P[3] rotavirus strain detected in Brazil: possible hu-man-canine interspecies transmission. J. Clin. Virol. 54, 89–92 (the official publication of the Pan American Society for Clinical Virology).
37. Maes, P., Matthijnssens, J., Rahman, M., Van Ranst, M., 2009. RotaC: a web-based tool for the complete genome classification of group A rota-viruses. BMC Microbiol. 9.
38. Martella, V., Banyai, K., Ciarlet, M., Iturriza-Gomara, M., Lorusso, E., De Grazia, S., Arista, S., Decaro, N., Elia, G., Cavalli, A., Corrente, M., Lavazza, A., Baselga, R., Buonavoglia, C., 2006. Relationships among porcine and human P[6] rotaviruses: evidence that the different human P[6] lineages have originated from multiple interspecies transmission events. Virology 344, 509–519.
39. Martella, V., Banyai, K., Matthijnssens, J., Buonavoglia, C., Ciarlet, M., 2010. Zoonotic aspects of rotaviruses. Vet. Microbiol. 140, 246–255. Matthijnssens, J., Ciarlet, M., McDonald, S.M., Attoui, H., Banyai, K., Brister, J.R., Buesa, J., Esona, M.D., Estes, M.K., Gentsch, J.R., Iturriza-Gomara, M., Johne, R., Kirkwood, C.D., Martella, V., Mertens, P.P.C., Nakagomi, O., Parreno, V., Rahman, M., Ruggeri, F.M., Saif, L.J., Santos, N., Steyer, A., Taniguchi, K., Patton, J.T., Desselberger, U., Van Ranst, M., 2011. Unifor- mity of rotavirus strain nomenclature proposed by the Rotavirus Classification Working Group (RCWG). Arch. Virol. 156, 1397–1413.
40. Midgley, S.E., Banyai, K., Buesa, J., Halaihel, N., Hjulsager, C.K., Jakab, F., Kaplon, J., Larsen, L.E., Monini, M., Poljsak-Prijatelj, M., Pothier, P., Ruggeri, F.M., Steyer, A., Koopmans, M., Bottiger, B., 2012. Diversity and zoonotic potential of rotaviruses in swine and cattle across Europe. Vet. Microbiol. 156, 238–245.
41. Miyazaki, A., Kuga, K., Suzuki, T., Kohmoto, M., Katsuda, K., Tsunemitsu, H., 2013. Annual changes in predominant genotypes of rotavirus A detected in the faeces of pigs in various developmental stages raised on a conventional farm. Vet. Microbiol. 163, 162–166.
42. Molinari, B.L., Lorenzetti, E., Otonel, R.A., Alfieri, A.F., Alfieri, A.A., 2014. Species H rotavirus detected in piglets with diarrhea, Brazil, 2012. Emerg. Infect. Dis. 20, 1019–1022.
43. Mukherjee, A., Mullick, S., Deb, A.K., Panda, S., Chawla-Sarkar, M., 2013. First report of human rotavirus G8P[4] gastroenteritis in India: evi-dence of ruminants-to-human zoonotic transmission. J. Med. Virol. 85, 537–545.
44. Nyaga, M.M., Jere, K.C., Peenze, I., Mlera, L., van Dijk, A.A., Seheri, M.L., Mphahlele, M.J., 2013. Sequence analysis of the whole genomes of five African human G9 rotavirus strains. Infect. Genet. Evol. 16, 62–77 (Journal of Molecular Epidemiology and Evolutionary Genetics in Infectious Diseases).
45. Papp, H., Borzak, R., Farkas, S., Kisfali, P., Lengyel, G., Molnar, P., Melegh, B., Matthijnssens, J., Jakab, F., Martella, V., Banyai, K., 2013. Zoonotic transmission of reassortant porcine G4P[6] rotaviruses in Hungarian pediatric patients identified sporadically over a 15year period. Infect. Genet. Evol. 19C, 71–80 (Journal of Molecular Epidemiology and Evolutionary Genetics in Infectious Diseases).
46. Pedley, S., Bridger, J.C., Chasey, D., McCrae, M.A., 1986. Definition of two new groups of atypical rotavirus. J. Gen. Virol. 67, 131–137.

47. Pietsch, C., Petersen, L., Patzer, L., Liebert, U.G., 2009. Molecular char-acteristics of German G8P[4] rotavirus strain GER1H-09 suggest that a genotyping and subclassification update is required for G8. J. Clin. Microbiol. 47, 3569–3576.
48. Ramachandran, M., Das, B.K., Vij, A., Kumar, R., Bhambal, S.S., Kesari, N., Rawat, H., Bahl, L., Thakur, S., Woods, P.A., Glass, R.I., Bhan, M.K., Gentsch, J.R., 1996. Unusual diversity of human rotavirus G and P genotypes in India. J. Clin. Microbiol. 34, 436–439.
49. Ramachandran, M., Kirkwood, C.D., Unicomb, L., Cunliffe, N.A., Ward, R.L., Bhan, M.K., Clark, H.F., Glass, R.I., Gentsch, J.R., 2000. Molecular characterization of serotype G9 rotavirus strains from a global col-lection. Virology 278, 436–444.
50. Ramani, S., Iturriza-Gomara, M., Jana, A.K., Kuruvilla, K.A., Gray, J.J., Brown, D.W., Kang, G., 2009. Whole genome characterization of reassortant G10P[11] strain (N155) from a neonate with symptomatic rotavirus infection: identification of genes of human and animal rotavirus origin. J. Clin. Virol. 45, 237–244 (the official publication of the Pan American Society for Clinical Virology).
51. Saikruang, W., Khamrin, P., Chaimongkol, N., Suantai, B., Kongkaew, A., Kongkaew, S., Ushijima, H., Maneekarn, N., 2013. Genetic diversity and novel combinations of G4P 19 and G9P 19 porcine rotavirus strains in Thailand. Vet. Microbiol. 161, 255–262.
52. Steyer, A., Poljsak-Prijatelj, M., Barlic-Maganja, D., Marin, J., 2008. Human, porcine and bovine rotaviruses in Slovenia: evidence of interspecies transmission and genome reassortment. J. Gen. Virol. 89, 1690–1698.
53. Svensmark, B., Nielsen, K., Dalsgaard, K., Willeberg, P., 1989. Epidemio-logical-studies of piglet diarrhea in intensively managed Danish Sow Herds .4. Acta Vet. Scand. 30, 63–76.
54. Tate, J.E., Burton, A.H., Boschi-Pinto, C., Steele, A.D., Duque, J., Parashar, U.D., 2012. 2008 estimate of worldwide rotavirus-associated mortal-ity in children younger than 5 years before the introduction of universal rotavirus vaccination programmes: a systematic review and meta-analysis. Lancet Infect. Dis. 12, 136–141.

INDEX